NITROGEN FIXATION BY LEGUMES IN MEDITERRANEAN AGRICULTURE

Developments in Plant and Soil Sciences

1. J. Monteith and C. Webb, eds.,
Soil Water and Nitrogen in Mediterranean-type Environments. 1981. ISBN 90-247-2406-6
2. J.C. Brogan, ed.,
Nitrogen Losses and Surface Run-off from Landspreading of Manures. 1981. ISBN 90-247-2471-6
3. J.D. Bewley, ed.,
Nitrogen and Carbon Metabolism. 1981. ISBN 90-247-2472-4
4. R. Brouwer, I. Gašparíková, J. Kolek and B.C. Loughman, eds.,
Structure and Function of Plant Roots. 1981. ISBN 90-247-2510-0
5. Y.R. Dommergues and H.G. Diem, eds.,
Microbiology of Tropical Soils and Plant Productivity. 1982. ISBN 90-247-2624-7
6. G.P. Robertson, R. Herrera and T. Rosswall, eds.,
Nitrogen Cycling in Ecosystems of Latin America and the Caribbean. 1982. ISBN 90-247-2719-7
7. D. Atkinson et al., eds.,
Tree Root Systems and their Mycorrhizas. 1983. ISBN 90-247-2821-5
8. M.R. Sarić and B.C. Loughman, eds.,
Genetic Aspects of Plant Nutrition. 1983. ISBN 90-247-2822-3
9. J.R. Freney and J.R. Simpson, eds.,
Gaseous Loss of Nitrogen from Plant-Soil Systems. 1983. ISBN 90-247-2820-7
10. United Nations Economic Commission for Europe.
Efficient Use of Fertilizers in Agriculture. 1983. ISBN 90-247-2866-5
11. J. Tinsley and J.F. Darbyshire, eds.,
Biological Processes and Soil Fertility. 1984. ISBN 90-247-2902-5
12. A.D.L. Akkermans, D. Baker, K. Huss-Danell and J.D. Tjepkema, eds.,
Frankia Symbioses. 1984. ISBN 90-247-2967-X
13. W.S. Silver and E.C. Schröder, eds.,
Practical Application of Azolla for Rice Production. 1984. ISBN 90-247-3068-6
14. P.G.L. Vlek, ed.,
Micronutrients in Tropical Food Crop Production. 1985. ISBN 90-247-3085-6
15. T.P. Hignett, ed.,
Fertilizer Manual. 1985. ISBN 90-247-3122-4
16. D. Vaughan and R.E. Malcolm, eds.,
Soil Organic Matter and Biological Activity. 1985. ISBN 90-247-3154-2
17. D. Pasternak and A. San Pietro, eds.,
Biosalinity in Action: Bioproduction with Saline Water. 1985. ISBN 90-247-3159-3
18. M. Lalonde, C. Camiré and J.O. Dawson, eds.,
Frankia and Actinorhizal Plants. 1985. ISBN 90-247-3214-X
19. H. Lambers, J.J. Neeteson and I. Stulen, eds.,
Fundamental, Ecological and Agricultural Aspects of Nitrogen Metabolism in Higher Plants. 1986. ISBN 90-247-3258-1
20. M.B. Jackson, ed.,
New Root Formation in Plants and Cuttings. 1986. ISBN 90-247-3260-3
21. F.A. Skinner and P. Uomala, eds.,
Nitrogen Fixation with Non-Legumes. 1986. ISBN 90-247-3283-2
22. A. Alexander, ed.,
Foliar Fertilization. 1986. ISBN 90-247-3288-3
23. H.G. v.d. Meer, J.C. Ryden and G.C. Ennik, eds.,
Nitrogen Fluxes in Intensive Grassland Systems. 1986. ISBN 90-247-3309-X
24. A.U. Mokwunye and P.L.G. Vlek, eds.,
Management of Nitrogen and Phosporus Fertilizers in Sub-Saharan Africa. 1986. ISBN 90-247-3312-X
25. Y. Chen and Y. Avnimelech, eds.,
The Role of Organic Matter in Modern Agriculture. 1986. ISBN 90-247-3360-X
26. S.K. De Datta and W.H. Patrick Jr., eds.,
Nitrogen Economy of Flooded Rice Soils. 1986. ISBN 90-247-3361-8
27. W.H. Gabelman and B.C. Loughman, eds.,
Genetic Aspects of Plant Mineral Nutrition. 1987. ISBN 90-247-3494-0
28. A. van Diest, ed.,
Plant and Soil: Interfaces and Interactions. 1987. ISBN 90-247-3535-1
29. United Nations, ed.,
The Utilization of Secondary and Trace Elements in Agriculture. 1987. ISBN 90-247-3546-7
30. H.G. v.d. Meer, R.J. Unwin, G.C. Ennik and T.A. van Dijk, eds.,
Animal Manure on Grassland and Fodder Crops. Fertilizer or Waste? 1987. ISBN 90-247-3568-8
31. N.J. Barrow,
Reactions with Variable-Charge Soils. 1987. ISBN 90-247-3589-0
32. D.P. Beck and L.A. Materon, eds.,
Nitrogen Fixation by Legumes in Mediterranean Agriculture. 1988. ISBN 90-247-3624-2

Nitrogen Fixation by Legumes in Mediterranean Agriculture

Proceedings of a workshop on Biological Nitrogen Fixation on Mediterranean-type Agriculture, ICARDA, Syria, April 14–17, 1986

Edited by

D.P. BECK and L.A. MATERON
The International Center for Agricultural Research in the Dry Areas, (ICARDA), Aleppo, Syria

1988 **MARTINUS NIJHOFF PUBLISHERS**
a member of the KLUWER ACADEMIC PUBLISHERS GROUP
DORDRECHT / BOSTON / LANCASTER

Distributors

for the United States and Canada: Kluwer Academic Publishers, P.O. Box 358, Accord Station, Hingham, MA 02018-0358, USA
for the UK and Ireland: Kluwer Academic Publishers, MTP Press Limited, Falcon House, Queen Square, Lancaster LA1 1RN, UK
for all other countries: Kluwer Academic Publishers Group, Distribution Center, P.O. Box 322, 3300 AH Dordrecht, The Netherlands

Library of Congress Cataloging in Publication Data

Nitrogen fixation by legumes in Mediterranean agriculture.

(Developments in plant and soil sciences ; v. 32)
1. Legumes--Congresses. 2. Legumes--Mediterranean Region--Congresses. 3. Nitrogen--Fixation--Congresses. 4. Rhizobium--Congresses. 5. Symbiosis--Congresses. I. Beck, D. P. (Douglas P.) II. Materon, L. A. (Luis A.) III. International Center for Agricultural Research in the Dry Areas. IV. Series.
SB317.L43N565 1987 635'.65 87-24830

ISBN 90-247-3624-2
ISBN 90-247-2405-8 (series)

Copyright

PRINTED IN THE NETHERLANDS

D.P. Beck and L.A. Materon (editors), Nitrogen Fixation by Legumes in Mediterranean Agriculture
© 1988 ICARDA – Printed in the Netherlands. ISBN 90 247 3624 2

Foreword

In the traditional cereal/fallow rotations of west Asia and north Africa one of the most promising options for increasing production is to replace the fallow with a food, pasture, or forage legume. Such crops are able to fix enough atmospheric nitrogen to supply their own needs and to contribute to the cereal, at the same time providing protein-rich food in the form of seeds or animal products. Biological nitrogen fixation is central to this process.

The food legumes-faba beans, lentils, and chickpeas – are important crops in the region, especially for poor families. Not only are the seeds important for human food but also the straw is used for feeding animals. The great importance of livestock in farming systems is an indicator of the potential significance of forage and pasture legumes in replacing fallows, especially in areas too dry for food legumes.

On April 14-17, 1986, an interdisciplinary workshop was organized at ICARDA headquarters in Aleppo, to examine the scope for regional research and development of biological nitrogen fixation in food, forage, and pasture legumes in semi-arid Mediterranean agriculture. The workshop was successful in bringing together regional researchers in microbiology, plant physiology, agronomy and breeding with counterpart scientists from other institutions worldwide. Objectives of discussions included appraisal of current technologies and regional research, planning an integrated strategy toward new technology development, potential collaboration with national programs and effective mechanisms for the transfer of technology. ICARDA's future research and training priorities for nitrogen fixation in legumes were also discussed.

Although it was not possible in a limited time to completely examine all relevant factors, the contributions and recommendations presented in this volume provide ample treatment of the workshop objectives.

ICARDA greatly appreciates the assistance given by UNDP in sponsoring the workshop and financing this publication. These proceedings are a valuable contribution to knowledge about nitrogen fixation by legumes and represent regional cooperation in a dynamic way.

Mohamed A. Nour
Director General

D.P. Beck and L.A. Materon (editors), Nitrogen Fixation by Legumes in Mediterranean Agriculture
 ISBN 90 247 3624 2

Acknowledgements

The workshop on which these proceedings are based was conceived of jointly by ICARDA and the United Nations Development Programme (UNDP). We express our sincere thanks to UNDP for their financial support of the workshop and for publication of these proceedings.

We are especially grateful to Ms. Fiona Thomson for her considerable editorial assistance and figure preparation; without her perseverance this publication might have been long delayed. Also, we thank all those who assisted in the organization of this workshop, as well as all participants, whose contributions made it a success.

The Editors

D.P. Beck and L.A. Materon (editors), Nitrogen Fixation by Legumes in Mediterranean Agriculture
© 1988 ICARDA – Printed in the Netherlands. ISBN 90 247 3624 2

Table of Contents

II. Genetic and Physiological Aspects of Legumes and Rhizobia

III. Ecology of Rhizobium and Inoculation

D.P. Beck and L.A. Materon (editors), Nitrogen Fixation by Legumes in Mediterranean Agriculture
© 1988 ICARDA – Printed in the Netherlands. ISBN 90 247 3624 2

List of Participants

ABDEL-GHAFFAR, AHMED	Soil and Water Science Dpt. Faculty of Agriculture University of Alexandria Chatby, Alexandria EGYPT
ACIKGOZ, ESVET	Faculty of Agriculture Uludag University P.K. 343 Bursa TURKEY
AMARGER, NOELLE	INRA 7, Rue Sully 21034 Dijon, Cedex FRANCE
BAREA, JOSE M.	Estacion Experimental del Zaidin, CSIC Profesor Albareda 1, Apdo 419 Granada 18008 SPAIN
BECK, DOUGLAS P.	ICARDA P.O. Box 5466 Aleppo SYRIA
BEN-ALI, NOUREDINNE	OEP Cite Sidi Layouni Kairouan TUNISIA 3100
BEYENE, DESTA	Holetta Agricultural Research Center Institute of Agricultural Research Holetta ETHIOPIA
BRHADA, FATIHA	Faculty of Sciences Mohammed V University B.P. 1014 Rabat MOROCCO
BROCKWELL, JOHN	CSIRO G.P.O. Box 1600 Canberra, ACT 2601 AUSTRALIA
BROUGHTON, WILLIAM J.	LBMPS University of Geneve 1, ch. de l'imperatrice CH-1292 Chambesy-Geneve SWITZERLAND
CLEYET-MAREL, JEAN C.	INRA 9, Place Viala 34060 Montpellier, Cedex FRANCE

COCKS, PHILIP S.	ICARDA P.O. Box 5466 Aleppo SYRIA
DANSO, SETH K.A.	Soil Fertility, Irrigation and Crop Production Section FAO/IAEA Wagramerstrasse 5 P.O. Box 200 Vienna AUSTRIA
DAOUD, DEIFALLAH M.	Soils and Irrigation Dpt. Faculty of Agriculture University of Jordan Amman JORDAN
EAGLESHAM, ALLAN R.	Boyce Thompson Institute Tower Road Ithaca, N.Y. 14853 USA
ELBERN-NGUYEN, B.	ATSAF Hans-Bockler Strasse 5 D-5300 Bonn WEST GERMANY
ENGIN, MUJGAN	Faculty of Agriculture Cukurova University Adana TURKEY
HUBBELL, DAVE H.	Soil Science Department University of Florida 2169 McCarty Hall Gainesville, FL 3261 USA
KOOPMAN, GERRIT, J.	ICARDA P.O. Box 5466 Aleppo SYRIA
LIE, TEK AN	Department of Microbiology Agricultural University Hasselink van Suchtelenweg 4 6703 CT Wageningen THE NETHERLANDS
MATERON, LUIS A.	ICARDA P.O. Box 5466 Aleppo SYRIA
MOAWAD, HASSAN	National Research Center Dokki, Cairo EGYPT
MOHAMMED, ISMAILI	Faculty of Sciences Sidi Mohammed Ben Abdellah University B.P. 4050 Meknes MOROCCO

MUKHTAR, NURI O	Soil Science Section Gezira Research Station Wad Medani SUDAN
MYTTON, LANCE	Welsh Plant Breeding Station Plas Gogerddan Aberystwyth Dyfed SY23 3EB Wales UNITED KINGDOM
OLIVARES, JOSE	Estacion Experimental del Zaidin, CSIC Profesor Albareda 1, Apdo 419 Granada 18008 SPAIN
PAPASTYLIANOU, I.	Agricultural Research Institute Nicosia CYPRUS
PRZEKOP, LAWRENCE	ICARDA P.O. Box 5466 Aleppo SYRIA
ROUGHLEY, RODNEY J.	Gosford Agricultural Station Pacific Highway West Gosford, NSW 2250 AUSTRALIA
RUIZ-ARGUESO, TOMAS	ETSIA-Microbiologia Universidad Politecnica 28040 Madrid-3 SPAIN
SAAD, ADNAN	Agricultural Research Center Douma SYRIA
SAXENA, MOHAN	ICARDA P.O. Box 5466 Aleppo SYRIA
SHAYYA, MONA	American University of Beirut P.O. Box 11-0236/574 Beirut LEBANON
SOLH, MAHMOUD B.	ICARDA Office B.P. 6299 Rabat-Instituts Rabat MOROCCO
SPRENT, JANET I.	Biological Sciences Dpt. University of Dundee Dundee DD1 4MN Scotland UNITED KINGDOM
SYLVESTER-BRADLEY, R.	CIAT Apdo, Aereo 6713 Cali COLOMBIA

THOMPSON, JOHN A.	Agricultural Research Center RMB 944 Tamworth, N.S.W. 2340 AUSTRALIA
VINCENT, JAMES M.	102 Northwood Rd. Northwood, N.S.W. 2066 AUSTRALIA
WEAVER, RICHARD W.	Soil and Crop Sciences Dpt. Texas A & M University College Station, TX 77843 USA
WERY, JACQUES	INRA 9, Place Viala 34060 Montpellier, Cedex FRANCE
WITTY, JOHN	Welsh Plant Breeding Station Plas Gogerddan Aberystwyth Dyfed SY23 3EB Wales UNITED KINGDOM
ZAROUG, MAHGOUB G.	Range Management and Fodder Production Section FAO Via delle Terme di Caracalla 00100 Rome ITALY

Part I

Biological Nitrogen Fixation in the Mediterranean Region

D.P. Beck and L.A. Materon (editors), Nitrogen Fixation by Legumes in Mediterranean Agriculture
© 1988 ICARDA – Printed in the Netherlands. ISBN 90 247 3624 2

The Role of Pasture and Forage Legumes in Livestock Based Farming Systems

P.S. COCKS
ICARDA, Aleppo, Syria

Summary. This paper discusses recent research on forage legumes in rotation with cereals and the use of pasture legumes in the ley farming system. The potential benefits of introducing pasture and forage legumes into cereal rotations are also discussed.

Introduction

Legumes grown for animal feed form two distinct groups: forage legumes which are sown and harvested in the same year, and pasture legumes which become part of the permanent flora of the farming system. In West Asia and North Africa, forage legumes comprise vetches (*Vicia sativa*, *V. narbonensis*, *V. villosa* subsp. *dasycarpa*), chickling (*Lathyrus sativus*, *L. ochrus*), and peas (*Pisum sativum*) while pasture legumes comprise several medic (*Medicago* spp.) and clover (*Trifolium* spp.) species. In this paper, I will consider both groups as part of a farming system in which cereals (either wheat or barley) are grown in rotation with forage or pasture legumes.

Role of Forage Legumes

It has been customary to think of forage crops as hay crops. Indeed the yield of forage legumes frequently reaches 5t/ha (Fig. 1). Most authorities are of the opinion that forage legumes should be mixed with cereals (Hadjichristodoulou 1973) for two reasons. Firstly, the mixture is more upright which facilitates harvesting, and secondly the yield is higher. However, recent results from rotation experiments indicate that mixtures grown in rotation with cereals may depress cereal yields in direct proportion to the amount of cereal in the mixture (Table 1). Furthermore, yields tend to be higher only in soils where available nitrogen is high or small amounts of nitrogen fertilizer are added.

The economic benefits of including forages in rotations is indicated from data obtained from simulated farms at Tel Hadya, ICARDA's research station south of Aleppo. The 'farms' were managed by scientists and, as far as possible, local farming methods were compared with simple improvement. The

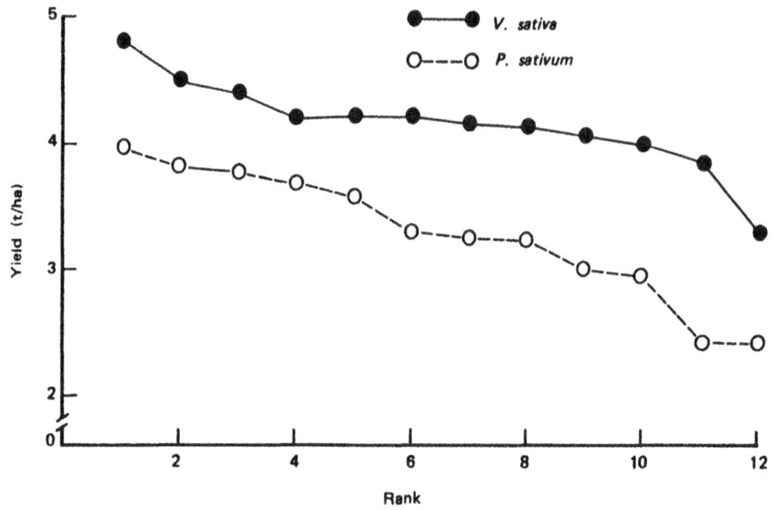

Fig. 1. Herbage yields of the 12 best species of *Vicia sativa* and *Pisum sativum* at Tel Hadya, 1982/83 (ICARDIA 1983).

results enable us to consider theoretically four types of farms; a traditional barley/fallow farm, a farm where the fallow was replaced by a forage crop (*Vicia sativa*), with each farm having access to 10 ha of either unimproved or improved marginal land, the latter improved with superphosphate. On each of the four farms, the sheep were managed at low, medium, and high levels of nutrition, calculated on the basis of known requirements for protein and metabolizable energy.

The results (Fig. 2) indicate that using vetch to replace fallow and superphosphate on marginal land increased farm profitability through increasing sheep numbers. Precisely the reverse was true of feeding at high levels of nutrition. Here, the effect was to increase individual animal productivity allowing for highest profitability without increasing flock size. It is worth noting that cereal production was not reduced by replacing fallow with vetch, a result which is confirmed by other experiments in areas drier than Tel Hadya. Indeed, it seems likely that the effect of fallow on cereal yields has

Table 1. Effect of proportion of legume in cereal/legume forage mixtures on subsequent grain yields of barley (kg/ha): yields are means of 2 years.

Legume/cereal ratio	Vetch/cereal	Pea/cereal
0:100	986	1069
33:66	1209	1196
50:50	1317	1299
66:33	1340	1370
100:0	1536	1806

Data of A.E. Osman (personal communication).

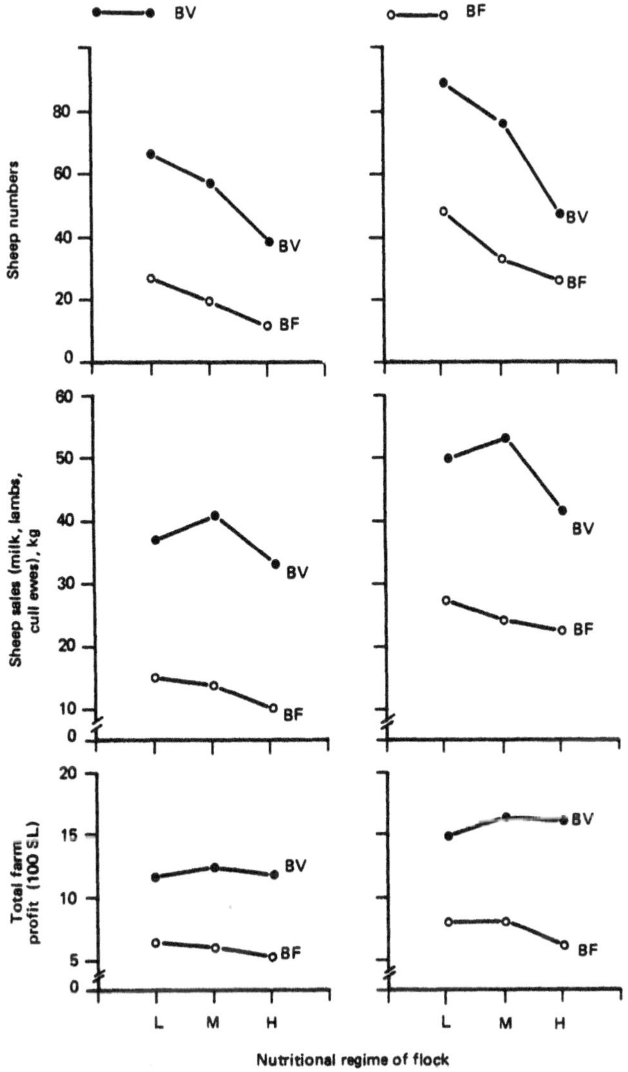

Fig. 2. Sheep numbers, sheep sales (kg of product), and total farm profitability (100 SL/ha) of farms using barley/vetch (BV) and barley/fallow (BF) rotations and unimproved (left) and improved (right) marginal lands. L, M, and H represent flocks of low, medium, and high nutritional regimes, respectively. (Data of T.L. Nordblom and E.F. Thomson.)

more to do with increased nitrogen availability than with water storage (Cooper 1983).

Of great interest is the recent use of forages for grazing. At Breda, a village near Aleppo where rainfall is 280 mm, fields of vetch and chickling have been successfully used to fatten lambs through direct grazing. The results indicate that this practice can be used profitably and, since it does not depend on specialized hay-making equipment, is a practical alternative to fallows for poor farmers. It is also interesting to note that in these fields inoculating the forage crops with rhizobia at double and treble the normal rates has resulted in big increases in herbage yield.

Role of Pasture Legumes

An alternative to using forage crops is to use pastures of self-regenerating annual legumes, especially medics (*Medicago* spp.). Based on ley farming in southern Australia, the concept is to reduce the costs of annual resowing and harvesting of forage crops by using medics able to regenerate naturally and which are harvested directly by grazing (Puckridge and French 1983). Although often seen as a simple transfer of technology, there are many problems with ley farming, including lack of adapted cultivars, inability of medics to nodulate with wild rhizobia, inappropriate grazing management, and the need for large tillage machinery. At least in north Syria, some of these problems have been resolved by using local medic (Abd El Moneim and Cocks 1986) and rhizobia, grazing to conserve seed, and using farmers' own tillage equipment.

In Australia, the two most widely sown species are *M. truncatula* (barrel medic) and *M. littoralis* (strand medic) both of which, in their native habitats, are commonly found near the sea. Indeed *M. littoralis*, as its name implies, is restricted to the coasts. That they are successful in southern Australia is not surprising since its climate is similar to that of the littoral zone of the Mediterranean Sea, especially Morocco, Tunisia, and Libya. In general, when re-introduced they succeed where they are native and fail elsewhere, especially in Iraq (Radwan *et al.* 1978), Syria, Jordan, and the high plateau of Algeria (Saunders 1976).

It was with this in mind that ICARDA developed the concept of using native medics. Based on the assumption that if a particular species existed on natural rangelands it was likely to be adapted to adjacent arable land, the distribution of medics (and other annual legumes) native to Syria was studied with special reference to the incidence of frosts and soil fertility. It was soon realized that in the cold dry areas the most abundant species are *M. rigidula* and *M. rotata*, the former on soils developed from limestone and the latter on soils developed from basalt. *M. polymorpha* is also abundant although its relationship to soil type is less clear.

The next step was to find out if these species are as productive as the Australian cultivars, whether they are adapted to cereal/pasture rotations, and their response to grazing. Herbage production of annual medics has been measured at Tel Hadya for several years, the results in 1981 being typical. In that year, the total herbage production of 124 accessions of nine different species was compared. The yield of *M. rigidula* was 3.1–9.7 t/ha compared with that of *M. truncatula* which was 0.8–2.8 t/ha and a single cultivar of *M. scutellata* (cv Robinson) which yielded only 1.3 t/ha (ICARDA 1982). Subse-

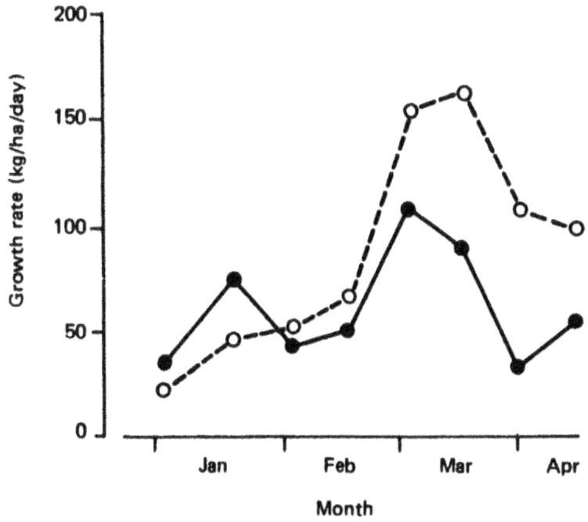

Fig. 3. Growth rate (kg/ha/day) during winter and spring of swards of *M. rigidula*, a local species (broken line) and *M. truncatula*, an imported species (solid line), at Tel Hadya, 1984.

quently (in 1984), the seasonal growth rate of a promising ecotype of *M. rigidula* was compared with *M. truncatula* cv Jemalong. Although Jemalong's growth in the mild winter of 1984 was slightly superior to that of *M. rigidula*, its spring growth was much less and its total yield was 25% less (Fig. 3). The growth rate of *M. rigidula* was equal to that of subterranean clover in southern Australia (Cocks 1974).

The final step in selecting medics is to choose ecotypes adapted to the farming systems in which they will be used i.e., able to survive in cereal rotations where the pasture phase is grazed. In this context, it is important to understand that pasture production depends on plant numbers (especially in

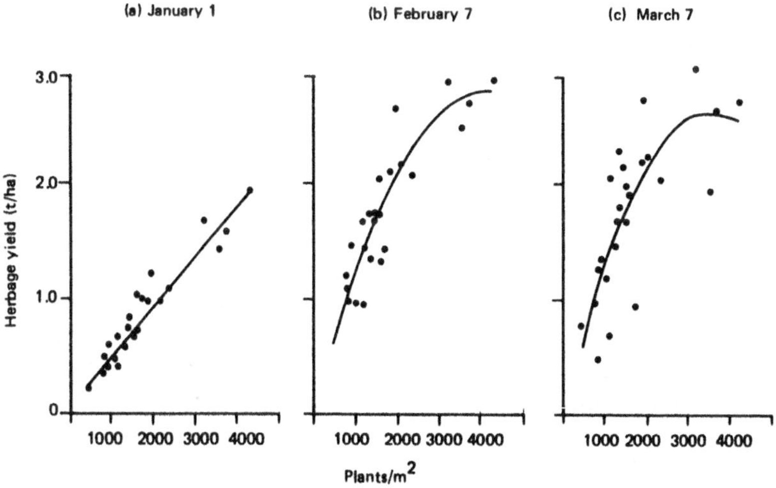

Fig. 4. The relationship between herbage yield of regenerating pastures and plant density (a) early winter, (b) late winter, (c) early spring (Abd El Moneim and Cocks 1986).

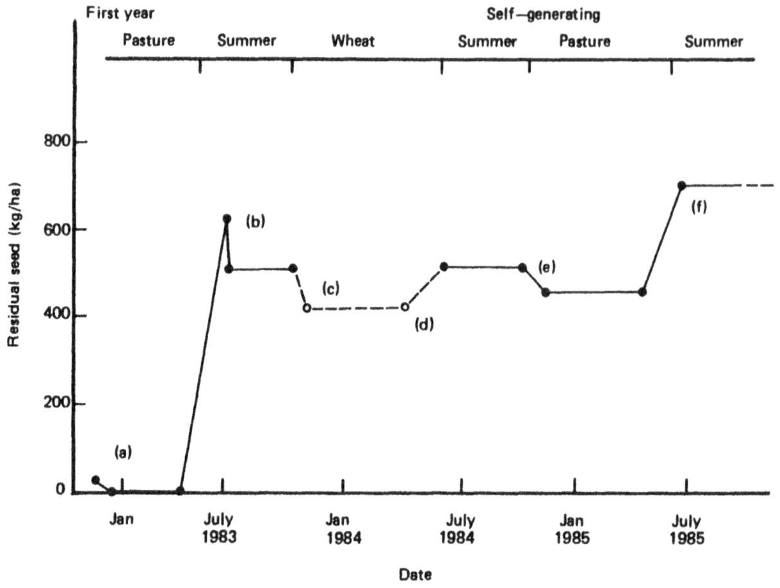

Fig. 5. Changes in seed population over 3 yars in a medic/cereal rotation where *M. rigidula* was the most important medic. The point (a) is the weight of seed originally sown, (b) is the yield of seed at the end of the first year (the fall is caused by the sample harvested), the fall at (c) is the amount germinated in the year of crop, the increase at (d) is the amount set under the crop, the fall at (e) is the amount which germinated to form pasture in the third year, and the increase at (f) is the amount of seed set in the third year. The dashed lines and open circles in the second year are estimates based on the level of seed dormancy at point (c) Abd El Moneim and Cocks 1986).

winter) and that selection for adapted genotypes is selection for high densities of self-regenerating plants. This concept is illustrated in Fig. 4, where winter production is closely related to plant population irrespective of which of the 25 genotypes was examined. Plant population itself is dependent on seed production, survival, and dormancy.

That *M. rigidula* is able to survive in rotation with cereals is shown in Fig. 5. In this experiment, seed production and residual seed were monitored in the first year pasture, the second year wheat crop, and the third year regenerating pasture (Abd El Moneim and Cocks 1986). Note that 10–15% of the seed germinated each year, that even in the wheat, germinated seed was replaced by seed from weedy medics, and that after the third year there was a residual population of more than 700 kg/ha. Even assuming only 10% germination in the fifth year of the rotation, some 70 kg/ha of seed is likely to germinate. In the third year of this experiment, germination of the most productive genotypes was 150 kg/ha, so it seems that *M. rigidula* is extremely well adapted to survive in a pasture/crop rotation.

In an experiment to compare profits from medics with those of other crops (Table 2), milk yield from ewes on medic pasture was recorded at 2-week intervals from the beginning of March until the end of lactation in July. During this period, the pasture supported 20 sheep/ha and produced nearly

Table 2. Costs, returns, and profit of various crops grown in the Tah region during 1985 (Syrian pounds/ha).

	Costs		Returns	Profit
	Establishment	Harvest *		
Wheat	654	267	2791	1870
Barley	801	388	3037	1848
Lentils	542	1154	2200	504
Sesame	400	209	1323	714
Water melons	527	368	1869	974

* Includes mid-season (sprays, etc.) costs.

1500 kg milk/ha. Valued at 3 SL/kg, the total value of milk was 4400 SL or nearly 40% more than the gross return from wheat or barley. Costs have not been taken into account, and the comparison of one farmer's returns with the average return of 48 farmers is clearly not valid. Nevertheless, the potential for increasing farm profitability using medics is similar to that in southern Australia where the system also increased soil fertility and ultimately cereal yields.

Conclusions

In conclusion, I would like to briefly discuss the potential benefits of introducing pasture and forage legumes into cereal rotations.

Carter (1978) estimated that approximately 30 million ha of fallow land in North Africa and West Asia are suitable for pasture and forage legumes. If 70% of this land were sown, he estimated that approximately 80 million tonnes of herbage would become available, enough to feed 100 million ewes. He went on to calculate that there would be a total increment of 1.4 million tonnes of nitrogen to the soil per year, which is 65% greater than the amount currently applied as fertilizer in Algeria, Tunisia, Libya, Jordan, Syria, Turkey, Iran, and Afghanistan combined.

Some of Carter's assumptions have been tested at Tel Hadya. For example, we have recorded herbage yields of up to 10 t/ha where Carter assumed only 4 t/ha. Even on first year pasture or forage, sheep can be grazed at an average of 5/ha and regenerating pasture is sure to carry more: Carter's assumption was only 3/ha. There is little doubt that this is a very potent method of integrating cereals and livestock with enormous potential benefit to farmers in the Mediterranean region.

No discussion of the role of pasture and forage legumes can ignore their nitrogen input. *M. noeana* and *M. rigidula* fixed 140 and 150 kg/ha, respectively in recent experiments (ICARDA 1984). However, because of the long-term nature of changes in soil fertility, detailed analysis of their effects on cereal yields is not yet possible. In Australia, where there is more experience of legume farming, the rate of nitrogen accretion under subterranean clover is about 50 kg/ha/year for periods of up to 40 years (Russell 1960; Donald and Williams 1954). In rotation with cereals, Carter (1978) states that 70 kg/ha/year is a reasonable estimate, and Puckridge and French (1983) quote several authors who measured rates of increase varying from 30 to 150 kg/ha/year.

ICARDA now has sufficient confidence to proceed with developing farming systems based on pastures and forages. Livestock productivity is being measured under closely controlled conditions at Tel Hadya and the new systems are being introduced to farmers at five villages in Syria. Results are promising, and in one village the fallows on whole farms are being planted to forage legumes. The coming few years should see the successful introduction of pasture and forage legumes to large numbers of farms in West Asia.

References

Abd El-Moneim, A.M., and Cocks, P.S. 1986. Adaptation of *Medicago rigidula* to a cereal-pasture rotation in north west Syria. Journal of Agricultural Science, Cambridge 107: 179–186.

Carter, E.D. 1978. A review of the existing and potential role of legumes in farming systems of the Near East and North African region. A Report to ICARDA, Aleppo, Syria.

Cocks, P.S. 1974. Potential production of grass and clover monocultures in a Mediterranean-type environment – an experimental approach. Australian Journal of Agricultural Research 25: 835–846.

Cooper, P.M. 1983. Crop management in rainfed agriculture with special reference to water use efficiency. Proceedings 17th Conference International Potash Institute, Bremen, Germany.

Donald, C.M. and Williams, C.H. 1954. Fertility and productivity of a podzolic soil as influenced by subterranean clover (*Trifolium subterraneum* L.) and superphosphate. Australian Journal of Agricultural Research 5: 664–687.

Hadjichristodoulou, A. 1973. Production of forage from cereals, legumes and their mixtures under rainfed conditions in Cyprus. Technical Bulletin 14, Agricultural Institute, Nicosia, Cyprus.

ICARDA (International Center for Agricultural Research in the Dry Areas). 1982. Annual report 1982. ICARDA, Aleppo, Syria. 215pp.

ICARDA (International Center for Agricultural Research in the Dry Areas). 1984. Annual report 1984. ICARDA, Aleppo, Syria. 344pp.

Puckridge, D.W. and French, R.J. 1983. The annual legume pasture in cereal – ley farming systems of southern Australia: a review. Agriculture, Ecosystems and Environment 9: 229–267.

Radwan, M.S., Al-Fakhry, A.K. and Al-Hasan, A.M. 1978. Some observations on the performance of annual medics in northern Iraq. Mesopotamia Journal of Agriculture 13: 55–67.

Russell, J.S. 1960. Changes in the long-term experimental plots at Hybybolite, South Australia. 1. Changes in pH, total nitrogen, organic carbon, and bulk density. Australian Journal of Agricultural Research 11: 902–926.

Saunders, D.A. 1976. Early management issues in establishing wheat-forage legume rotations. Proceedings 2nd Scientific Conference, Baghdad, Iraq.

D.P. Beck and L.A. Materon (editors), Nitrogen Fixation by Legumes in Mediterranean Agriculture
© 1988 ICARDA – Printed in the Netherlands. ISBN 90 247 3624 2

Food Legumes in the Mediterranean Type of Environment and ICARDA's Efforts in Improving their Productivity

M.C. SAXENA
ICARDA, P.O. Box 5466, Aleppo, SYRIA

Summary. Although a Mediterranean type climate, characterized by hot dry summers and cool wet winters, occurs in some parts of all the continents in the world, the largest single contiguous area with this type of climate is found in the region of West Asia and North Africa. The International Center for Agricultural Research in the Dry Areas has the responsibility for improving agricultural production in the dry areas of this region through crop improvement and farming system research and training. Faba bean, kabuli-chickpea, and lentil are the three major food legumes in the region. Their importance stems from the fact that they are an important source of high quality protein in the diets of the masses in the region, their byproducts are a valuable source of nutritious fodder for livestock, they reduce the dependence of the cropping system on nitrogenous fertilizers because of their ability to fix atmospheric nitrogen in association with rhizobia, and they have a general beneficial effect on the productivity of the land when introduced as a break crop in cereal dominated crop sequences. ICARDA has world-wide responsibility for research in improvement of faba bean and lentil and a joint responsibility, with the International Crop Research Institute for Semi-Arid Tropics (ICRISAT), for the improvement of kabuli-type chickpea. Research efforts in the past have led to the development of genotypes and production techniques for all three crops that ensure increased and more stable productivity in the dry areas of the West Asia and North Africa region and in other countries having a Mediterranean type climate.

Introduction

The Mediterranean type of climate is characterized by hot, dry summers alternating with cool and wet winters. Such a climate is found at middle latitudes (between 30-40°) in all continents (Fig. 1). However, the largest single contiguous area experiencing this climate is in the region of West Asia and North Africa. It stretches from Morocco at the Atlantic coast in the west around the Mediterranean to Pakistan in the east where the tropical monsoon region starts. It is bound by a cool temperate region in the north and the arid Sahara in the south. A total of 23 countries fall within this region which is inhabited by more than 400 million people. Because of variations in topogra-

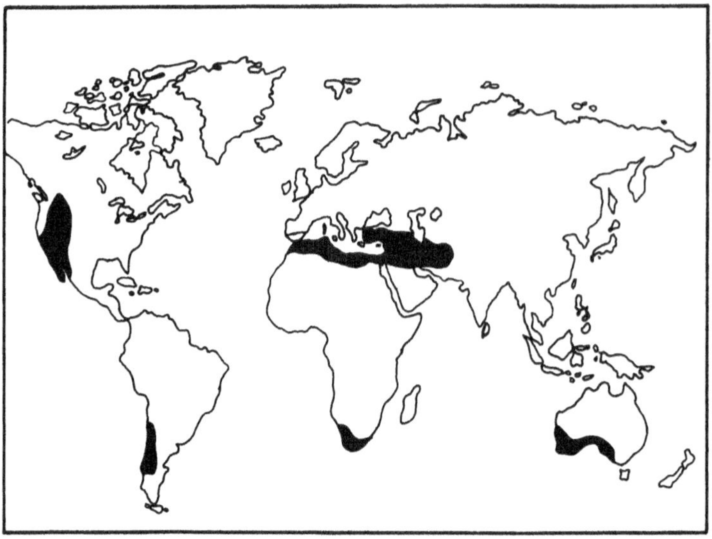

Fig. 1. Areas in the world with a Mediterranean-type climate (shaded area).

phy and the influence of several major geographical features, the region exhibits great differences in vegetation and agroecological conditions owing to variations in precipitation and thermal regimes. A large proportion of the total area experiences annual rainfall ranging from less than 200 mm to about 600 mm. The reliability of rainfall is low and there is also considerable spatial variability. All these features make agriculture a risky proposition in a major part of this region.

The International Center for Agricultural Research in the Dry Areas (ICARDA) has been assigned by the Consultative Group on International Agricultural Research (CGIAR) the responsibility for spearheading a research thrust to improve agricultural productivity in this region. Since its establishment in 1977, the Center has been working closely with the national agricultural research systems (NARS) in the region to achieve this goal..pg;

Role of Food Legumes in the Mediterranean Region

The rainfed farming system in the dry areas of the region is dominated by the cultivation of cereals and the use of annual pastures to raise livestock, mainly sheep. Depending on precipitation, the intensity of cropping may range from 33% to more than 100%. Food legumes are also an important component of the cropping system in the region. Although a variety of food legumes are grown, those that dominate are chickpea (*Cicer arietinum* L.), faba bean (*Vicia faba* L.), and lentil (*Lens culinaris* L.). The area and production statistics for the major food legumes are shown in Table 1. Whereas chickpea is the most

Table 1. Annual area and production of different pulses (food legumes) in various counties of the ICARDA region, some European countries having Mediterranean climate and the world as a whole, average of 1982 and 1984.

Country	Area (thousand ha)						Production (thousand t)					
	Total pulses	Dry bean	Dry pea	Chick-pea	Faba bean	Len-til	Total pulses	Dry bean	Dry pea	Chick-pea	Faba bean	Len-til
North Africa	1758	63	198	296	770	137	1751	58	155	235	1002	80
Algeria	120	2	10	39	47	17	75	1	4	44	23	4
Egypt	160	6	3	10	118	6	341	14	6	16	266	8
Ethiopia	886	41	147	129	362	60	909	31	123	95	525	42
Libya	8	–	1	–	7	–	10	–	2	–	8	–
Morocco	359	11	37	55	155	52	230	8	20	40	111	25
Sudan	77	3	–	3	16	–	89	4	–	3	22	–
Tunisia	148	–	–	60	65	2	97	–	–	37	47	1
West Asia	3295	435	51	1379	56	850	2439	393	45	827	98	684
Afganistan	21	–	–	–	–	–	34	–	–	–	–	–
Cyprus	6	1	–	1	2	–	6	1	–	1	3	–
Iran	364	152	48	71	–	68	272	128	40	56	–	34
Iraq	36	7	–	16	6	5	31	5	–	12	9	5
Jordan	16	–	–	3	1	10	12	–	–	2	1	8
Lebanon	10	1	–	2	–	4	10	1	–	3	–	3
Pakistan	1387	159	–	933	–	77	647	81	–	410	–	30
Syria	190	6	1	58	8	65	149	12	1	36	14	44
Turkey	1265	109	2	295	39	621	1278	165	4	307	71	560
Icarda Region	5053	498	249	1675	826	987	4190	451	200	1062	1100	764
Europe												
France	189	16	112	–	52	8	725	32	517	–	160	15
Greece	65	27	1	11	5	6	83	36	1	12	7	6
Italy	222	46	3	12	149	1	287	73	4	15	180	1
Portugal	306	238	–	29	32	–	73	38	–	11	20	–
Spain	454	116	5	83	57	65	330	74	5	54	59	37
World	66804	25820	7948	9996	3296	2352	46571	15271	9515	6314	4208	1565
All developed	10248	2160	5484	140	363	220	12540	2216	7179	95	527	188
All developing	56556	23659	2463	9857	2932	2133	34031	13054	2836	6219	3681	1377

Source: 1984 FAO Production Year Book Vol. 38, 1985: 134-141, FAO of UN, Rome.

dominant crop in the West Asia region, faba bean dominates in North Africa. Lentil also occupies an important place in West Asia and all these crops are important in those European countries with Mediterranean climate (Table 1).

Ecologically, lentil is better adapted to the drier side of the rainfall spectrum in the region. It is generally grown as a rainfed crop in areas where seasonal precipitation is around 300 mm. Chickpea is grown rainfed where precipitation is around 450 mm per annum. Rainfed faba bean is raised where annual rainfall is more than 500 mm.

The importance of these crops lies in the fact that, as there are hardly any economic alternatives in the areas of their adaptation, they are a major source of good quality protein in the common diet. Also their byproducts provide nutritious fodder for livestock, they reduce the dependence of the cropping system on nitrogenous fertilizers because of their ability to fix atmospheric nitrogen in association with rhizobia, and they serve as a good break crop in the continuous cereal rotations to prevent yield decline and improve sustainability of productivity of the agricultural system. The importance of these food legumes in the human diet in the West Asia and North Africa region has been well documented (Aykroyd and Doughty 1964; Doughty and Walker 1982; Tannous *et al.* 1979). So also is the case with the value of legume straw, particularly lentil straw, as an alternative source of sheep feed in the dry areas of this region (Nordblom and Halimeh 1982). Therefore only those aspects of the food legumes that relate to symbiotic nitrogen fixation and their role in affecting the productivity of the succeeding cereal crops in the rotation will be highlighted here.

Symbiotic Nitrogen Fixation

Biologically fixed nitrogen was estimated for different food legumes under rainfed conditions of Tel Hadya, northern Syria in the 1980/81 and 1981/82 seasons using ^{15}N isotopic dilution technique (Fried and Middleboe, 1977). The total seasonal precipitation was 345 mm in 1980/81 and 338 mm in 1981/82 and 1981/82 was colder than 1980/81. In 1980/81 biologically fixed nitrogen ranged from 75.0 kg/ha in winter chickpea to 107.1 kg/ha in lentil treated with carbofuron to protect nodules from damage by the larvae of *Sitona* weevil (Table 2). In 1981/82 the values were much reduced because of severe frost that adversely affected crop growth. The values of biologically fixed nitrogen ranged from 27.9 kg/ha (in winter chickpea) to 80.7 kg/ha (in pea) in that season. More than 80% of total recoverable plant nitrogen in all the crops in 1980/81 and more than 70% in 1981/82 (except in winter chickpea) was derived from biologically fixed nitrogen. It is clear from these results that the adapted food legumes can fix a considerable amount of nitrogen and can thus serve to ensure influx of nitrogen into the production systems in dry areas where farmers are averse to using high cost inputs such as nitrogen fertilizer because of risks associated with uncertainty of rainfall.

Table 2. Yield of biologically fixed nitrogen (BNF) through recoverable dry matter of several food legumes at maturity, Tel Hadya 1980/81 and 1981/82.

Crop	1980/81 BNF (kg/ha)	%N from BNF	1981/82 BNF (kg/ha)	%N from BNF
Faba bean	90.5	89.9	53.8	76.8
Lentil	83.9	86.2	49.6	73.9
Lentil + Carbofuran	107.1	90.2	62.7	80.0
Winter chickpea	75.0	86.2	27.9	50.8
Pea	95.9	87.5	80.7	79.0

These results have also highlighted the importance of controlling *Sitona* weevil in improving BNF in lentil. Field observations, before the start of the above studies, had shown that depending upon the season, the root nodules of lentil could be damaged by the larvae of *Sitona* spp. to varied degrees and application of 1.5 kg a.i./ha of carbofuron at the time of sowing could completely eliminate this damage. Studies on BNF in the 1982/83 season further confirmed the observations of the previous two seasons as in this season also the BNF with carbofuron treated lentil was 99.3 kg N/ha in contrast to 88.0 kg N/ha with untreated check. In order to confirm that the improvement in BNF in lentil occurred because of control of nodule damage by *Sitona* weevil, studies were carried out during the 1982/83 and 1983/84 seasons on the seasonal profile of acetylene reduction activity (ARA) of a carbofuron treated and untreated lentil crop. Results in Fig. 2 showed that

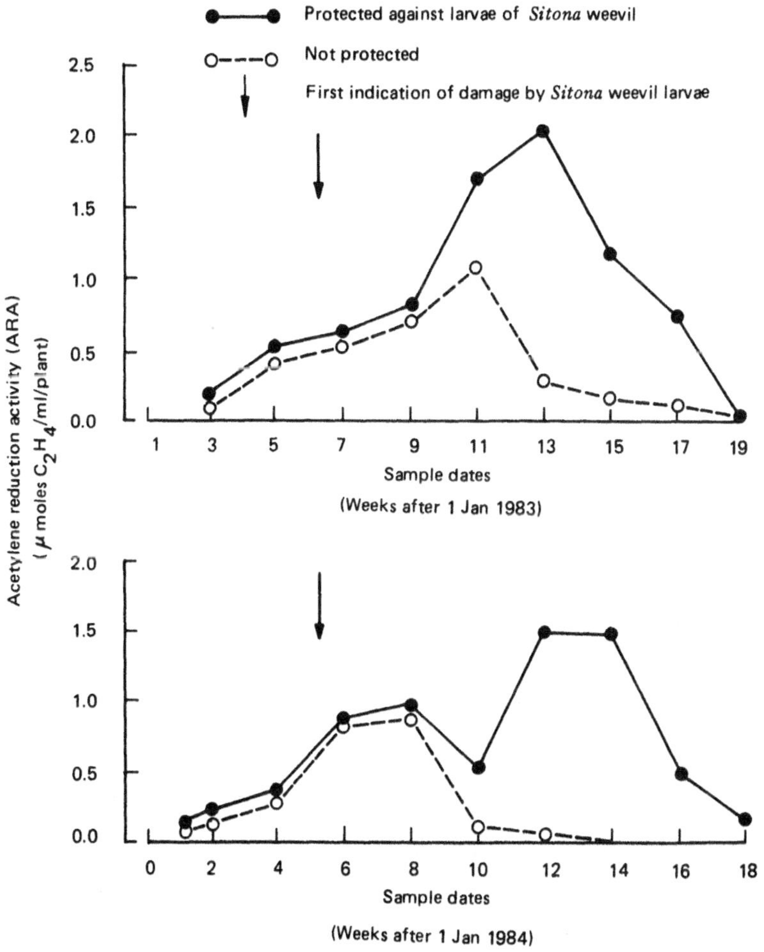

Fig. 2. Effect of control of *Sitona* weevil larvae on acetylene reduction (C_2H_4 production) values in lentils in 1983 and 1984.

ARA values fell early on in untreated crop following the onset of damage to the nodules by *Sitona* spp. larvae (Saxena *et al.* 1985). Also, comparison of carbofuron with another insecticide (heptachlor) revealed that the effect was nonspecific to carbofuron (ICARDA 1985).

Residual Effect of Legumes on Subsequent Cereal Grop

Continuous cropping of cereals with 33 to 100% cropping intensity, depending upon seasonal precipitation, is quite common in the drier areas of the ICARDA region. Yields are low because of depletion in soil fertility and development of unfavourable soil physical and biological conditions. That food and forage legumes, as a break crop in this system, could improve productivity is evident from the data in Fig. 3 from a study on the residual effect of the 1980/81 season cropping treatments on the grain yield of rainfed wheat raised at different rates of directly applied nitrogen in the 1981/82 season at Tel Hadya, northern Syria. The grain yield of wheat following lentil, faba bean and dry pea was significantly higher than that following wheat or even fallow at all levels of directly applied nitrogen, although the differences were higher when no nitrogen was applied. The nitrogen status of the soil was extremely low and the wheat crop following wheat showed severe nitrogen deficiency at 0 and 30 kg N/ha treatments. The data in Fig. 3 clearly show that under rainfall of about 340 mm, a two course rotation of faba bean or pea

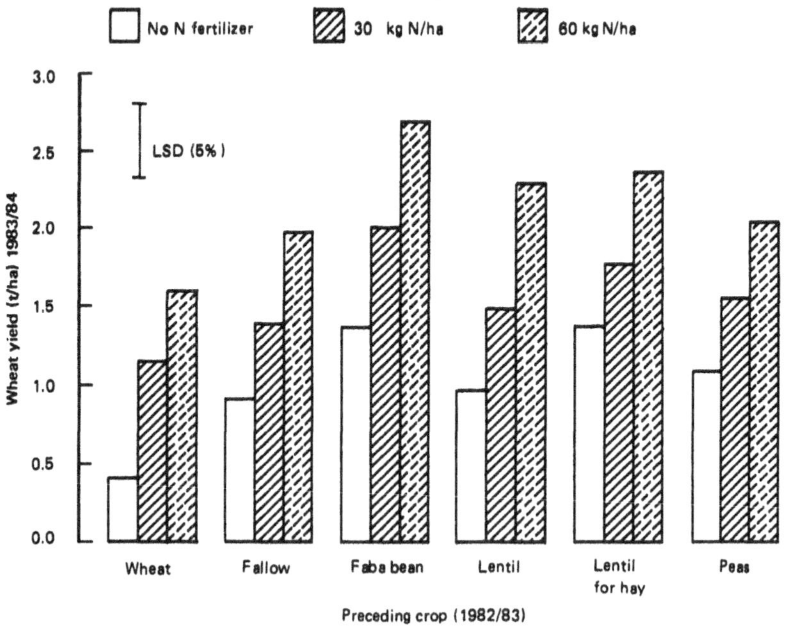

Fig. 3. Grain yield of wheat receiving three levels of nitrogen fertilizer (0, 30, and 60 kg/ha) following wheat, fallow, peas, faba bean, and lentil, Tel Hadya, 1983/1984.

or lentil with cereal can be very productive. This productivity benefit is in addition to the yield of symbiotically fixed nitrogen in the whole system through the direct yield of the legume component in the rotation. The increase in productivity of cereal following the legumes cannot entirely be attributed to reduced depletion of soil nitrogen by the legume crop in contrast to the cereal, because the shape of nitrogen response curve of wheat (Fig. 3) is more or less the same irrespective of the preceeding crop and there seem to be some other beneficial effects of legumes that need to be studied. Notwithstanding that, it is clear that dry pea, faba bean, lentil, and winter chickpea can play a very significant role in increasing the productivity of the cropping system in the dry areas of the region.

Problems of Food Legume Production in the ICARDA Region and ICARDA's Efforts in Developing Effective Solutions

General production constraints

Some of the major constraints to the production of food legumes in the region, and outside, are the inherently low yield potential of the existing landraces, high instability in yield from year to year because of susceptibility to pests and diseases, poor production techniques, high cost of hand harvesting, and lack of suitable methods for mechanical harvesting. The national programs need support to overcome these constraints since the development of research facilities and trained manpower for food legume research has started receiving due attention only recently. The best support that ICARDA's Food Legume Improvement Program (FLIP) can give to national programs is by developing improved genetic material and suitable production techniques which can then be modified or, if needed, improved by national programs through adaptive research, before making them available to farmers.

Objectives of ICARDA's Food Legume Improvement Program

The crop improvement work at ICARDA has the following objectives:

1. Collection, maintenance, and evaluation of a diverse set of germplasm from which further genetic improvement can be made.
2. Development and distribution of advanced lines, segregating populations, and other genetic material for the development of cultivars having increased seed as well as total biological yield potential and yield stability; appropriate phenology to suit the growing conditions; resistance to common diseases, pests, *Orobanche*, and temperature and moisture stresses; growth characters that permit mechanized harvest; and acceptable or improved nutritive value and cooking quality of seed.

3. Conducting relevant research on genetics, cytogenetics, physiology, and seed quality to support the work on cultivar development.
4. Development of appropriate cultural practices for different genotypes and agroecological conditions by conducting research on production agronomy and physiology, pathology, entomology, microbiology, weed management, mechanization, etc.
5. Training scientists from national programs in food legume improvement, developing an international network of food legume scientists, and disseminating technical information through workshops, conferences, information services, research reports, and other publications.

Faba bean improvement

Instability in yield and low productivity are the two major constraints in faba bean production. Besides various environmental stresses (frost, heat, drought) the following major diseases contribute a great deal to the instability and reduction in the productivity of the crop:

Chocolate spot (*Botrytis fabae*, *B. cinerea*)
Ascochyta blight (*Ascochyta fabae*)
Rust (*Uromyces fabae*)
Leaf spot (*Alternaria* spp. and *Cercospora* spp.)
Powdery mildew (*Erysiphe polygoni*, *Leveilulla taurica*)
Downy mildew (*Peronospora* sp.)
Root rot/wilt (*Fusarium* spp., *Rhizoctonia* spp., *Scelerotinia* spp., *Helminthosporium* spp., etc.)
Bacterial diseases (e.g. *Xanthomonas* spp.)
Viruses (BYMV, CMV, BBMV, PLRV, etc.)

To date, research has concentrated mostly on chocolate spot and ascochyta blight. Promising sources of resistance to these diseases have already been identified and are now being tested widely and incorporated into well-adapted genetic backgrounds. The program will continue its efforts to identify further resistant sources and use them in breeding work besides studying the mechanism of resistance, pathogen biotype, and inheritance of resistance to these two diseases. Work will be intensified on the next most important diseases: rust, root rot/wilt, and virus diseases.

Broomrape, a parasitic angiosperm (*Orobanche crenata*, *O. aegyptiaca*) is probably the single most important biotic factor limiting faba bean yields in the region. Research on identifying sources of resistance to broomrape and development of appropriate alternative control measures, including chemical control, has been very rewarding and has been conducted in collaboration with the Egyptian national program. Giza 402 has been released by the Egyptian program as the first *Orobanche* tolerant faba bean cultivar.

Stem nematode (*Ditylenchus dipsaci*), a damaging seed-borne nematode, is widespread in West Asia and North Africa. Following the development of suitable screening techniques, the development of resistance to this pest is expected to yield good results. In the meanwhile effective chemical control measures have been developed.

Among the wide range of insect pests of faba beans, special attention is devoted to aphids and stem borers. Resistance to both these pests is already being sought. There is still a great need, however, for detailed pest surveys in the region to firmly establish the importance and distribution of these and other insect pests.

The vegetative growth of cultivars of faba bean currently grown under favourable environments is excessive, which often results in lodging and reduced harvest index. The program is developing varieties with terminal inflorescence so that faba bean may become biologically a more efficient and agronomically better manageable crop. Already, determinate lines developed so far have shown yield potential comparable with the best indeterminate types, although the original determinate mutant from Sweden used for this work was very low yielding and non-adapted to the Mediterranean environment. Improvement in disease resistance and seed size has also been effected. Incorporating an independent vascular supply trait in this and indeterminate backgrounds, thereby reducing flower and young pod drop, is another strategy being adopted by the program for improving the productivity and yield stability of faba bean varieties. Development of highly autofertile lines of faba bean is also receiving due attention to reduce the dependence of the crop on insect pollinators.

With respect to quality, protein and cooking quality levels are monitored to ensure that acceptable standards are maintained. In addition, following the results of current surveys and other research, the objectives of breeding cultivars with reduced vicine and convicine levels is expected to be included in the program. These two compounds are strongly implicated as the causal agents of favism syndrome in susceptible humans.

A special project on wide-crossing in *Vicia* spp. is being conducted in collaboration with the University of Reading, UK. If successful techniques are developed, the results of interspecific hybridization are expected to make a very major impact on the breeding program a few years hence, as it will open avenues for using enormous variability that had never been used in the past. Also, attempts are being made to identify male gametocides to explore the possibility of exploiting heterosis in the crop, now that high yielding autofertile lines have been identified.

Agronomic studies, in collaboration with national program scientists have led to the development of improved production techniques that are being field tested in such applied projects as the Nile Valley Project in Egypt, Sudan, and Ethiopia.

Lentil Improvement

The ICARDA region accounts for nearly half of the total lentil area in the world. The largest producers in the region are Turkey and Syria, and thus ICARDA in Aleppo is well situated for research on this important crop. Of the three food legume species addressed by the program, lentil is grown in the driest areas, and can be found in rainfall zones down to about 250 mm rainfall per year. Both of the two major subgroups of the species are important in the region: the large-seeded (var. *macrosperma*) and small-seeded (var. *microsperma*) types. Approximately equal emphasis is placed on the two types.

Improving yield and yield stability are the main objectives of the program. There are comparatively few pest and disease problems. One very important objective is to devise means of reducing the labor requirement for harvesting the crop. Three main approaches to the problem are being followed. First, through breeding, it is aimed to develop cultivars which have substantially reduced pod-drop and dehiscence characteristics, which are taller than conventional types, have pods borne higher off the ground, and which do not lodge. These types would enable harvesting to take place over a longer period and would be more amenable to mechanical cutting, including direct combine-harvesting. The second approach is through agronomy: good seed-bed preparation, rolling, seeding practices, fertilizer use, etc. can all contribute to alleviating the problem. The third approach is the development of appropriate machinery. Some work has been carried out in the past on developing a lentil puller, and on testing a reciprocating cutter bar and a standard bean cutter. These have shown considerable promise and will be further developed and tested in the future. Work on all three approaches to the harvesting problem has made considerable progress and the developed technology is increasingly moving to an on-farm trial/demonstration phase.

Another aspect of the research aims at developing cold-tolerant lentil cultivars for high elevations. Some success has already been achieved in this regard in collaborative trials with the Turkish National Program. In view of the substantial yield advances which should be possible by switching from spring to autumn planting, this research will continue to receive a high priority for the higher elevations. Over the next few years screening techniques will be developed and standardized and collaborative projects will be established with organizations working in appropriate environments.

Of the important pests and diseases of lentils, research emphasis has been on *Orobanche* sp., rust (primarily for the Indian subcontinent, Ethiopia, and Latin America), and root rot/wilt. Sources of resistance are being sought and work in this area will increase over the next few years. Ethiopia has released ILL 358 as a lentil variety with improved yield and resistance to rust. Work on *Sitona* control has resulted in valuable information to improve the yield and biological nitrogen fixation as already described earlier.

Lentil has a narrow adaptation to environment. Studies on the nature of adaptation have already been initiated both at ICARDA and in collaboration with the University of Reading, UK. As more information becomes available in the next few years, new screening techniques, zoning of the region, etc. will become possible. In view of the narrow adaptation, research in Syria is of comparatively little relevance to the needs of the crop in more southerly locations, particularly in Egypt, Sudan, Ethiopia, and the Indo-Pakistan subcontinent. Hence a collaborative research program is already underway in Egypt, Sudan, Ethiopia, and Pakistan to develop breeding material adapted to these conditions.

Research on crop quality places emphasis on maintaining adequate nutritional and cooking quality levels. Additionally, research has been initiated on other quality parameters, especially hulling percent. Straw is a very important by-product of the crop, and can even exceed the value of the grain, as already mentioned earlier. Studies on straw quality have shown that environmental effects far exceed those due to genotypes.

Agronomy has been an important focus of the program, but resources will increasingly be switched, over the next few years, to more detailed physiological studies on aspects such as drought resistance, cold tolerance, growth habit, and adaptation.

Kabuli Chickpeas

The chickpea program at ICARDA is a joint program with ICRISAT. Kabuli types account for about 15% of the total world chickpea production and are widely preferred in southern Europe and Latin America in addition to most countries of the ICARDA region. The ICARDA chickpea program also takes a lead in research on the disease ascochyta blight, caused by *Ascochyta rabiei*. In the future the chickpea program will continue to place major emphasis on kabuli types, and will devote an increasing proportion of its effort to the large-seeded types.

A notable success of the program has been in demonstrating the feasibility of switching chickpea in West Asia and North Africa from the normal spring-planted to an autumn-planted crop. Highly promising materials with reasonable cold tolerance and resistance to ascochyta blight have been developed, which are essential to the success of the new system. With this change in sowing time large increases in yield, by nearly 100%, can be obtained. The potential of increasing the productivity of kabuli chickpea in several Mediterranean countries is shown by the yield performamce of some of the ICARDA developed chickpea lines in one of International Yield Trials for winter sowing conducted during 1984/85 (Table 3). Several countries in the region have now released varieties adapted for winter sowing.

Studies on identifying different biotypes of the ascochyta blight pathogen in the region strongly suggest the presence of different physiological races. Hence breeders are attempting to develop broad based resistance by combining resistance genes from different parents.

Table 3. Yield performance in 1984/85 International Yield Trial on Winter sown Chickpeas for Mediterranean Areas (CIYT-W-MR) in some Southern European and North-African Countries in contrast to the national average yields.

Country (Location)	Yield (kg/ha)		
	Average of the country	Mean of trial	Yield of best entry
Greece (Larisa)	1125	2536	3325
Italy (Capalbio)	1150	4441	5562
Portugal (Elvas)	303	2972	4245
Spain (Cordoba)	559	3816	5577
Morocco (Kara)	600	2313	2750
Tunisia (Beja)	733	2231	3667

As research on ascochyta blight becomes well established, greater attention will be paid in the future to other diseases such as root rot/wilt, virus diseases, and possibly certain others such as botrytis grey mold. *Orobanche* sp. and nonparasitic weeds are a major problem on the winter-sown crop compared to the normal spring-sown one. More research efforts are being devoted to this. Root-knot and cyst nematodes have also been found to damage the crop in the region and host resistance is being sought in collaboration with Italian scientists.

The research on the two major insect pests, pod borers (*Heliothis* spp.) and leaf miners has been underway. Breeding for resistance has been given highest priority.

There is a strong need to increase cold tolerance in the breeding materials and considerable efforts have been made in the past few years. Good sources of cold tolerance have already been identified from a screening of the germplasm at Ankara, Turkey, and at ICARDA sites at Tel Hadya and Terbol.

Winter chickpea can be grown economically in much drier areas than is possible for the spring crop. Thus, new areas will be opened up for chickpea cultivation. In these areas the soil will mostly be lacking the necessary chickpea rhizobia populations and the crop will greatly benefit from inoculation. Attention is being paid to the identification of suitable strains of *Rhizobium* and development of commercial inoculant production facilities in the region to provide the necessary quantities of inoculum.

Agronomy work on chickpea has concentrated largely on the winter crop. Large-scale research/demonstration activities on the components of production of winter chickpea in West Asia and North Africa are underway to develop and evaluate the practice at the farm level.

The high cost and nonavailability of labor for harvesting chickpea, although less serious than for lentil, still poses a constraint to production in some areas. The development of tall cultivars suitable for mechanized harvesting is receiving attention by the program. Also, adaptation of conventional cereal combines for harvesting chickpea has been feasible.

References

Aykroyd, W.R. and Doughty, J. 1964. Legumes in Human Nutrition. FAO Nutritional Studies No. 19, FAO, Rome, pp. 138.

Doughty, J. and Walker, A. 1982. Legumes in Human Nutrition. Food and Nutrition Paper 20, FAO, Rome, pp 152.

Fried, M. and Middleboe, V. 1977. Measurement of amount of nitrogen fixed by a legume crop. Plant and Soil 47:713.

Nordblom, T. and Halimeh, H. 1982. Lentil crop residue make a difference. LENS 9:8-10.

Saxena, M.C., Stephens, J.H. and Cardona, C. 1985. Some studies on biological nitrogen fixation by food legumes in dry areas of northern Syria. Paper presented at 25th Science Week, 2-7 November 1985, Damascus, Syria. Available from ICARDA, Aleppo, P.O. Box 5466, Ayria., pp. 13.

Tannous, R., Abu-Shakra, S. and Hallab, A.H. 1979. Nutritional quality and importance of food legumes in the Middle Eastern diet. Pages 29-32 *in* G.C. Hawtin and G.J. Chancellor (eds.), Proceedings of a workshop held at the University of Aleppo, Aleppo, Syria 2-7 May 1978, IDRC-126e, Ottawa, Canada.

D.P. Beck and L.A. Materon (editors), Nitrogen Fixation by Legumes in Mediterranean Agriculture
© 1988 ICARDA – Printed in the Netherlands. ISBN 90 247 3624 2

Development of a Regional Program on Biological Nitrogen Fixation to Enhance Agricultural Production in the Near East

M.G. ZAROUG
FAO, Rome, Italy

Summary. The Near East region comprises a large total land area in which legumes are important in improving forage quality and hence livestock production, as well as for human food.

Programs initiated by countries in the region to increase forage production and improve feed quality are described. The importance of biological nitrogen fixation (BNF) is highlighted and FAO activities in the field of BNF are outlined. Future action in the form of training, research, and cooperation with various institutions, organizations, and regional programmes is discussed.

Introduction

The Near East region comprises twenty-six countries (Table 1) covering a total land area of about 1736 million ha (Table 2) with permanent pastureland forming 22% of this total. Raising livestock is an important traditional activity of nomadic, transhumant, and settled rural communities. According to the FAO Production Yearbook (1984), the region is populated by 68.2, 12.9, 6.4, 20.2, and 25.7% of the world's camels, buffalos, cattle, sheep, and goats, respectively.

The legumes in the arid and semi-arid tropical pastureland of the region include those belonging to the genera *Tephrosia*, *Crotalaria*, *Indigofera*, and

Table 1. FAO member countries in the Near East region.

Afghanistan	Morocco
Algeria	Oman
Bahrain	Pakistan
Cyprus	Qatar
Djibouti	Saudi Arabia
Egypt	Somalia
Iran	Sudan
Iraq	Syria
Jordan	Tunisia
Kuwait	Turkey
Lebanon	UA Emirates
Libya	Yemen AR
Mauritania	PDR Yemen

Table 2. Land use in the Near East region*.

	Land area (×1000 ha)	Percentage of total area
Total	1736492	
Permanent pasture	385346	22.19
Arable	115597	6.65
Cereals	56984	3.28
Pulses	4528	0.26

* FAO (1984b).

Acacia. Sub-tropical and temperate pastures are rich in *Medicago*, *Vicia*, *Onobrychis*, and *Hedysarum*. The role of these legumes in improving forage quality and thus livestock performance is of great importance. The standing hay with its seeds and pods provides livestock with the crude protein badly needed during the dry season.

Faced with a general increase in livestock populations and feed shortages, several member countries have initiated programmes to increase forage production and improve dry season feed quality. Such programmes include:

(i) rangeland improvement by seeding with selected grasses and legumes, planting adaptable fodder trees and shrubs, and applying soil moisture conservation techniques;

(ii) utilizing marginal fallow land to establish self-regenerating pasture; and.

(iii) introducing fodder legumes in crop rotations. A range of indigenous and exotic legumes have been used, such as alfalfa, annual medics, clovers, sainfoin, vetches, lablab, clitoria, stylos, leucaena, and acacias.

Arable land in the region is estimated to be about 115 million ha (Table 2) with cereals and pulses occupying about 57 and 4.5 million ha, respectively. Additional areas have been commissioned for agricultural production, mainly for growing cereals and other food crops to meet the increasing demands of an expanding human population. These areas comprise marginal lands with serious constraints such as low and fluctuating precipitation, salinity and/or alkalinity, and erosion. In view of the low organic matter and nitrogen contents of the arid and semi-arid lands of the Near East, there is an increasing dependence on synthetic nitrogenous fertilizers to improve production.

It is estimated that crops in the region consumed about 3.8 million tonnes of these fertilizers during 1983/84 (FAO 1984). Several countries in the region produce part or all of their nitrogen fertilizer requirements and, in a few of the oil-producing countries, there is a surplus which is exported. However, purchase of nitrogenous fertilizers is a burden for the non-oil producing countries which are interested in improving yields of the different non-legume crops.

Food legumes such as lentils, chickpeas, cowpeas, pigeonpeas, faba beans, peas, beans, and mung beans, are produced in traditional farming areas for domestic use. However, when there is a surplus, they are normally sent to local markets. Crops such as soybean and groundnuts are increasing in importance for oil production and, in the case of the former crop, for the poultry feed industry. Food legumes contribute significantly to the diet of rural populations by providing vegetable protein, oil, and calories and, with localized shortages and increasing meat prices in the region, their nutritional importance is becoming more appreciated even by urban dwellers.

Food legumes are also important for livestock which can utilize straws of crops such as groundnut, cowpea, lablab, pigeon pea, and mung bean. Such straws are valuable as dry season feed in areas where no supplements are available. Food legumes may also contribute to improved soil fertility, particularly when straw is consumed *in situ*.

Biological nitrogen fixation (BNF) should be looked upon as a low-cost and effective method to increase food production, particularly in traditional farming areas where synthetic nitrogen fertilizer may be too expensive for farmers. The importance of BNF in producing good quality forage from legume and grass/legume mixtures should not be overlooked as an efficient tool for improving livestock production.

FAO Activities in the Field of Biological Nitrogen Fixation

FAO has, through its Regular Programme and Field Project activities, initiated or assisted in the promotion of food legume production and the expansion of areas under such crops in developing countries, including those of the Near East. Also there are continuous efforts to encourage the use of adaptable forage legumes for establishing improved pastures on marginal lands and in rotation with cereals and other crops. Realizing the importance of BNF to both food and forage legumes and in enhancing soil fertility, FAO provides assistance to interested member countries to acquire such technology and use it optimally.

A FAO-UNEP meeting on BNF held in Rome 1980 stressed the fact that full utilization of legumes depends on developing effective rhizobial inoculants and establishing a practical system of distribution to farmers, with the required information on practical use. Through its Technical Cooperation Programme (TCP) and subcontracts from its Regular Programme funds FAO, in collaboration with national institutions and research centres, initiated activities to promote on-farm demonstrations. The aim was to illustrate to small farmers the value of inoculants, inoculation techniques, and BNF in improving food and forage legume production.

FAO played a leading role in organizing regional working groups and associations which comprise researchers and interested technicians from member countries, to facilitate the exchange of information, experience, and materials. Also, FAO has produced, alone or in collaboration with other organizations, practical manuals on BNF and has trained national staff from developing countries in the skills of *Rhizobium* strain selection and evaluation, inoculation, and inoculant production. Such training is done by FAO experts and consultants in the field and through short-term courses organized in collaboration with other UN organizations and research centres.

Through assistance from FAO, BNF investigations have been linked to a development objective to improve arid rangelands. Efforts were made to select early flowering cultivars of *Medicago littoralis* and to extend its distribution beyond the 150 mm isohyte. An integral part of this programme were BNF studies to identify strains highly effective on selected medic ecotypes that can be used later to produce the inoculant needed for large-scale seeding operations.

Current Activities of the Regional Programme

The activities on BNF recently commenced by the Regional Programme aim to support national programmes in conducting adaptive research and demonstrations directed towards the transfer of BNF technology to farmers. The purpose is to increase the availability of nitrogen for legume and non-legume crops at low cost. Ultimately, the goal is to achieve a satisfactory level of livestock integration in crop production systems and increased production.

Trials comparing the effectiveness of locally-isolated and introduced rhizobial strains are reported from several countries of the Near East. The benefits of inoculation have been proven and in some cases results and findings have been extended to farmer groups. These trials are mentioned as an indication that activities in member countries will always be taken into consideration in orientating regional activity on BNF.

Member countries in the Near East can be roughly divided into three categories according to their present activities on BNF and inoculant production:

(i) Countries with established institutions dealing with BNF research, production and testing of inoculants, and inoculation techniques. Inoculants are produced for farmers' use, at least for major crops.

(ii) Countries with identifiable institutions carrying out BNF research. Inoculants are produced for demonstration purpose and plans for supplying farmers with inoculants are underway.

(iii) Countries in which limited and inconsistent trials on BNF are carried out by individual researchers or projects. Inoculants are imported when needed.

FAO assistance could be provided to all interested member countries. The regional activity should look towards possible collaboration with countries in the first category for future BNF short-term training courses. Countries in the second category should be supported in implementing on-farm demonstrations and small-scale inoculant production. The third category of countries should be assisted in identifying priorities for applied research and demonstration trials.

Funds have already been allocated to one national institution, to strengthen its current efforts to produce rhizobial inoculants and demonstrate their value through on-farm trials using both forage and food legumes.

Support for the activities of national institutions in BNF and inoculant production and use is also provided by the Regional Officer and by FAO consultants during visits to member countries. On-going activities, constraints, and training requirements are discussed, and ways of overcoming problems are outlined. Whenever applicable, FAO consultants who are soil microbiologists demonstrate to national staff techniques related to rhizobial selection and testing and nodulation assessment. Efforts are also made to assist some national institutions to determine priorities for inoculant production and demonstrate inoculant responses based on the importance of crops and the areas they cover.

Legumes are examined in the field, with the national staff, to observe nodulation patterns and growth of food and forage legumes. Nodule effectiveness is also assessed by examining nodule colour and observing top growth and foliage colour. In one country, such examination revealed extensive nodule damage by larvae of insects, such as *Sitona* and *Rivellia*, on roots of lentils, peas, and particularly on faba beans. The attention of national staff was drawn to the fact that control of such larvae would likely result in a better contribution of those nodules to legume growth and production.

The importance of BNF for range seeding and pasture establishment activities was the theme of an independant presentation during the Expert Consultation on Rangeland Rehabilitation and Development in the Near East, organized by the Regional Programme during October 1985. In view of the importance of BNF, the Consultation recommended regional training on BNF, technical meetings, and exchange of strains between member countries.

Future Action

A shortage of national technicians with reasonable knowledge of BNF, inoculants, and inoculation techniques is slowing down the transfer of this technology to farmers. Short-term training should be carried out in collaboration with research centres and well-equipped national institutions. Such courses should be designed to familiarize participants with the methodology of *Rhizobium* isolation, tests for effectiveness, assessment of nitrogen fixation efficiency, proper inoculant handling, inoculation techniques, and on-farm demonstrations using both forage and food legumes.

Funds should be provided by concerned organizations and research centres to support national programmes aiming to acquire this technology for the benefit of farmers' groups and increasing agricultural production.

A regional working group on BNF should be established with the assistance of all interested organizations and research centres to facilitate the exchange of technical information and materials. This group could meet periodically in the different countries concerned to exchange views and share experiences.

Efforts will be made to identify, on a country basis, the need for assistance in BNF research and inoculant production with collaboration between interested organizations and research centres operating in the region. Eventually, countries could be grouped according to similarities e.g. environments, food and forage crops, language etc., to facilitate assistance, particularly in training.

Efforts made by some member countries in the Near East to produce inoculants should be encouraged so that they can supply neighbouring countries which, because of lack of experience, facilities, and funds, cannot establish their own inoculant production units. Subregional inoculant production centres should be considered.

National institutions should emphasize the multidisciplinary approach to optimize the utilization of BNF. Also, more attention needs to be given to the vital research-extension-farmer links, so that adoption of BNF technology will meet farmers' needs and expectations.

Researchers and extension officers should identify practical means for effectively utilizing BNF technology to increase production of intercropping systems in traditional farming areas.

The region is witnessing increased interest in trees and shrubs for the rehabilitation of degraded arid and semi-arid rangelands, sand dune fixation, agroforestry, soil fertility improvement, and production of fuel and construction wood. Several of these are legumes, such as *Prosopis*, *Acacia*, *Indigofera*, and *Leucaena*. Efforts should be made to identify how BNF technology can assist in the establishment and healthy growth of some of these species.

References

FAO (Food and Agriculture Organization). 1984a. Fertilizer Yearbook. FAO, Rome, Italy.
FAO (Food and Agriculture Organization). 1984b. Production Yearbook. FAO, Rome, Italy.

Discussion

Smith

It is important to know the current status of BNF in the Near East and North African region. In addition to the valuable statistics which you presented, do you have an estimate of the percentage of total plant production which can currently be directly attributed to BNF?

Zaroug

I have no estimate to offer at this stage.

Mohamed

What is the specific program of FAO in the development of inoculant production?

Zaroug

I have mentioned in my presentation the assistance provided by FAO to interested member countries with programmes for inoculant production. This is done to increase the capabilities of national programmes in demonstrating the usefulness of inoculants and inoculation for legumes grown by local farmers and to facilitate local inoculant production.

Abdel-Ghaffer

FAO activities are always carried out in cooperation with Ministries of Agriculture. Could these activities include faculties of agriculture of the universities in the region?

Zaroug

I do not agree, as FAO's interest is to develop agricultural production in developing countries and assistance is provided upon request without discriminating between national institutions. Several examples of FAO's assistance to member countries and collaboration with universities in those countries could be mentioned.

Moawad

Do you have any programmes for inoculation of cereals using nitrogen fixers other than *Rhizobium*?

Zaroug

I did not mention anything on inoculation of cereals. I suggested that nitrogen availability to cereals could be improved through nitrogen fixation by legumes so reducing dependence on costly nitrogenous fertilizers.

D.P. Beck and L.A. Materon (editors), Nitrogen Fixation by Legumes in Mediterranean Agriculture
 ISBN 90 247 3624 2

Maximizing Biological Nitrogen Fixation by Forage and Pasture Legumes in Semi-Arid Areas

L.A. MATERON

International Center for Agricultural Research in the Dry Areas (ICARDA), P.O. Box 5466, Aleppo, Syria

Summary. While properly nodulated pasture legumes are among the most efficient nitrogen fixers they are generally neither productive nor persistent in the absence of appropriate rhizobia. Selection and evaluation of effective rhizobia to maximize nitrogen fixation in pasture legumes involve 1) regional surveys of the production area, 2) strain collection, 3) strain selection, 4) strain identification, 5) field evaluation of strains, and 6) inoculant production. In this paper, a set of procedures is presented for each of the stages of the selection and evaluation process, with particular reference to annual *Medicago* spp. (medics).

Introduction

In terms of biological nitrogen fixation (BNF), pasture legumes are among the more efficient members of the Leguminosae. Several pasture species can fix nitrogen in excess of their own needs and BNF rates can be high (Gibson 1977). This is the result of an effective symbiosis between the legume and specific strains of *Rhizobium*, and pasture legumes are neither productive nor persistent without effective nodulation by appropriate rhizobia. Cultivars of agronomic interest are the result of selections made mainly for improved production, disease resistance, and cold tolerance (hardiness). BNF has appeared as a breeding and selection concern only recently and awaits exploitation (Smith and Knight 1984; Mytton *et al.* this workshop).

Apart from their economic value, pasture legumes are important because they improve soil fertility. In Australia, it has been estimated that pasture legumes add 30–160 kg N/ha to the soil each year, and may fix a total of 150 kg N/ha (Puckridge and French 1983). Clearly, if fixation occurs at this rate, its impact on the fertility of soils in which the organic matter is less than 1% will result in greatly improved yields and stability (Materon and Cocks 1987). To obtain the necessary strains of *Rhizobium* for inoculations, an evaluation and characterization process concurrent with host plant selections becomes mandatory. No standardized procedures have been developed to accomplish this but there are sufficient examples of such efforts from several researchers to provide what can be regarded as the most useful approach for evaluation and characterization (Vincent 1970; Date 1975; Halliday 1984). This approach can be divided into six stages: 1) regional survey of the production area, 2) strain collection, 3) strain selection, 4) strain identification, 5) field evaluation of strains, and 6) inoculant production.

This paper will briefly address the essential methods, guidelines, and approaches for each stage and the author's experiences with annual species of *Medicago* and annual clovers will be used as examples.

Regional Survey of the Area

A vigorous legume plant at a particular site is frequently the result of a very effective symbiotic association between the host plant and a strain of *Rhizobium*. Seed collection alone will not provide the full potential of a genotype in subsequent evaluations unless some attempt is made to reconstitute an equally effective symbiosis. Hence, a survey should be conducted in the planning stages of selection and evaluation programs and must include the entire area where the legume host occurs naturally or in regions where the legume has been introduced. The objective is to ensure that all of the soil types representing the major production areas are included and these may be grouped into geographical zones (Roughley *et al.* 1976) or micro-climate regions (Hagedorn 1978). It is necessary to obtain sufficient soil and plant samples, select potential field sites, and to become familiar with the history of the host plant in the area.

Strain Collection

Specific sampling sites should be chosen within each major subregion and characterized on the basis of agronomic history, current use, soil type, drainage, and chemical composition of the soil. Sampling ought to include those sites at which the host species 1) grows at present, 2) has grown in the past, and 3) has never grown (not applicable for clovers and medics in most Mediterranean countries where many of these species occur naturally). Soil analyses should be as complete as possible for both physical and chemical properties. Many of these are standardized (Page *et al.* 1982) and emphasis should be placed on soil pH, P, K, and exchangeable bases.

Rhizobial isolates should be taken from nodules on plants in the field, from soil around the roots, and from soil cores taken from sites where no plants have grown. When collecting from roadsides and other disturbed sites, note that the soil conditions, plant performance, and nodulation may not be typical of the region. Seek out legumes that appear to be successful in an ecosystem pertinent to the collector's interest, such as native pastures and/or stable natural ecosystems. Having located a plant from which to collect nodules, subsequent procedures depend very much on the duration of the collection trip. For trips of 1-14 days, collected nodules must be prevented from decomposing and protected from invasion by soil microorganisms which interfere with subsequent isolation procedures. The nodules may be conserved in vials containing a desiccant and these must have caps that provide an air-tight seal. The desiccant (anhydrous calcium chloride or silica gel) should occupy one-quarter to one-third of the volume of the container and be held in place by a cotton wool plug. The desiccant must not come in contact with the nodule samples. The volume of nodules and/or roots should not exceed the volume of desiccant (Date and Halliday 1979). Special precautions should be taken when recovering rhizobia from these nodules. At ICARDA we have found that the optimum disinfection time for small, desiccated nodules (e.g. medics and clovers) is 25-30 seconds: longer exposures are usually lethal to nodule rhizobia.

Soil must be examined for indigenous rhizobia which can nodulate the host species. Plant infection assays using the most probable number technique in either plastic pouches or glass tubes will provide an adequate estimate of rhizobial population size (Vincent 1970). The more prevalent strains can be recovered from the tubes containing the most dilute sample in the plant infection assay tests. At least ten isolations per sampling site should be included, all from nodules of the healthiest plants that can be found. It is often valuable to analyze soil data against the rhizobia and host plant measurements since this can yield information on the levels of fertility needed to maintain productive stands and the types of climate-related stress which occur most frequently (Materon and Hagedorn 1984).

Strain Selection

To be a successful inoculum, a strain must be able to form effective nitrogen fixing nodules on the hosts for which it is recommended under the conditions of soil and climate in which the legume crop is to be grown (Halliday 1984). In the past, this has been the most important criterion for strain selection because only effective, nitrogen fixing strains should be used when assessing other characters such as competitiveness in nodule formation (Marques-Pinto *et al.* 1974; Materon and Hagedorn 1982) and survival and multiplication in the soil, particularly in the absence of the host (Chatel *et al.* 1968). Additional criteria include prompt, effective nodulation over a range of root temperatures and the ability to grow well in culture, grow and survive in peat (Brockwell *et al.*1986), survive on the seed (Weaver *et al.* 1985), and multiply rapidly in the host rhizosphere (Brockwell, personal communication). Other criteria are pH and pesticide tolerance, and a requirement for nodulation in the presence of combined nitrogen (Date 1975) and the presence of the hydrogenase system (Maier 1986). Use of the last criterion in the screening of medic and clover rhizobia has proved to be impractical because these microorganisms apparently do not posses hydrogen uptake systems. Examination of the effectiveness of the rhizobial isolates on those host plant genotypes which are grown in the region is of primary importance. However, other rhizobial strains, from whatever source, should be included in the evaluations. These diverse rhizobia will usually produce wide variation in effectiveness tests when examined on different species of the same genus, as in the case of annual medics (Materon and Cocks 1987). In general, rhizobia isolated originally from the same genus, and sometimes species, as the legume for which a superior strain is being sought, emerge from selection programs as the best strains for use in legume inoculants.

At ICARDA we have found that the acetylene reduction method is not suitable for screening isolates for nitrogenase activity in various species of annual medics. These plants produce nodules which are usually small and easily detached from the roots so samples are incomplete and not reliable for analysis. The problem is more critical if the plants develop in the heavy-textured soils of semi-dry areas. Witty and Minchin (this workshop) present a full assessment of the sources of errors associated with the acetylene reduction method.

Effectiveness ratings of isolates can be placed on arbitrary scales of dry matter yields and should be examined on both a sampling site and host plant basis. The soil analyses are frequently useful in determining which type of stress situations to simulate in order to select strains which can persist under field conditions. A rating should be developed for each strain based on effectiveness, specificity, and stress survival. This should greatly reduce the number of strains which will be suitable for consideration in field trials. However, the management of field trials to select rhizobia is difficult and costly, even when the number of strains under test is small. When not justified, the use of sophisticated, time-consuming, and costly methods such as the acetylene reduction assay, ^{15}N dilution technique, and ureide determinations should be avoided. By adapting a screening procedure, undesireable strains can be progessively eliminated from an initially high number of entries to give a relatively small number of promising strains for testing in the field. This is one way to reconcile the requirements that selection be from a diverse genetic base, and that strains also be assessed under field conditions.

Strain Identification

In order to adequately measure strain performance in the field it is necessary to determine what proportion of the nodules of the host plant is occupied by the inoculum strain. These strains must somehow be distinguished from each other and from indigenous rhizobia. Nodule characteristics may be used under some circumstances, such as in the recognition of an effective strain in competition with ineffective strains or where a particular strain/host combination gives a distinctive nodule color (Vincent 1970). The degree of pigmentation of the shoots of anthocyanin – rich legumes, such as some varieties of *Trifolium subterraneum* and *Medicago tribuloides*, can be used to give a very early indication of strain effectiveness or ineffectiveness (Brockwell 1958).

In dealing with effective rhizobia, serology (Vincent 1970) and antibiotic resistance (Obaton 1971) have been the most widely used methods, although with pasture legumes the antibiotic method has received more attention as there are cross-reaction problems with serology. It appears that a wide range of antibiotics are suitable for markers in the rhizobia (Brockwell *et al.* 1977; Schwinghamer and Dudman 1973), although each strain must be examined for marker retention (after mutant selection) through plant passage and persistence with time and stress. If mutant selection is avoided, resistance to intrinsic levels of antibiotic may be useful (Josey *et al.* 1979). There are serological procedures, for example gel immunodiffusion (Vincent 1970), fluorescent antibodies (Trinick 1969), and enzyme-linked immunosorbent assay (ELISA) (Tchan 1982), which can distinguish between closely related and non identical strains. One must be aware that all serological methods are vulnerable to single mutational events which alter the cell surface, and alternatives to these identifying properties can be selected for inadvertently in nature (e.g., phage resistance). In practice, this might mean that an apparently insignificant alteration in serological properties, possibly selected for by experimental conditions, would render a strain unrecognizable by these methods. The selected method for the identification of strains should be adaptable, reliable and convenient enough for rapid and reproducible analyses.

Regardless of the precautions taken to avoid cross-contamination of field plots, the strains will eventually become mixed in volunteer stands and identification of each strain is the only way the evaluation can proceed. Strain identification is also essential to follow the behavior of strains under competition for nodule formation and monitor their seasonal persistence in the soil. It is also a valuable tool in the evaluation of inoculants.

Field Evaluation of Strains

The objective of field evaluation is to determine the effect of inoculation on a particular plant parameter, such as the percentage of nodules formed by an effective strain, nodule location and mass, total nitrogen, herbage mass, and seed yield. Thus, the need for inoculation must be determined as a prerequisite to *Rhizobium* strain selection programs and simple need-for-inoculation trials must be carried out in any new situation. Such trials provide an indication of the criteria to be used for strain selection trials. For most purposes rhizobia should be grown in yeast mannitol broth, added preferably to sterilized, neutral-pH peat and then inoculated onto the seed with an adhesive. At ICARDA we have observed that coating inoculated medic seed with fine calcium carbonate (lime) may afford some protection to rhizobia, regardless of the alkalinity of the soil (ICARDA 1986). Other coating agents are currently being evaluated.

Need-for-inoculation trials should consist of small plots with at least four replicates per treatment. Randomized block and split plot designs are frequently used although in some cases it may be necessary to group each treatment in a non-random fashion to minimize cross-contamination between strains. This is feasible if the soil at the field site does not vary greatly across the plot area (Vincent 1970). Uninoculated control plots should generally be placed at least several meters away and ought to consist of 1) no fertilizer, 2) PK fertilizer, and 3) PK fertilizer plus added nitrogen to ensure that the host will grow satisfactorily when nitrogen is supplied.

Another aspect of field inoculation is the prevention of cross-contamination of rhizobia between plots. If soil is not moved mechanically by some means between plots, cross-contamination is not likely to occur. Border rows may be used as in fertility and plant breeding trials but soil movement by water or wind is often difficult to control. Hence, a non-legume should be planted along the borders between plots.

In most complex field trials, herbage samples should be collected regularly during the growing season so that a growth curve can be constructed for the host plant. Similarly, periodic observations of nodulation patterns, herbage nitrogen content, and in some cases, measurements of nitrogenase activity should be recorded. Nodule occupancy tests need to be performed as a function of time to determine the competitive behavior of the strains. Frequently, large differences will be found between strains and these should correlate well on both an individual plant and a plot herbage yield basis. In addition, the fertility status of soils used for these strain evaluation trials should be characterized using soil tests and the population size of indigenous rhizobia determined.

Inoculant Production

Strains selected on the basis of excellent performance with a particular host and environment are then considered for inclusion in single-strain inoculants. Some of these strains may be highly host specific or possess a wide spectrum of infectivity with a wide range of hosts.

In general, the production of peat inoculants can be divided into the following stages: selecting suitable strains and their genetic maintenance; harvesting and preparing the peat; growing the broth inoculum; mixing the broth with peat; and packaging the culture and providing suitable storage conditions (Vincent 1970; Roughley and Pulsford 1982). Like all biological products legume inoculants are prone to loss of quality due to variation in the organism concerned and unforeseen factors affecting some aspect of growth or survival. It is therefore essential that a quality control system be established. The standards of quality adopted need to reflect both the number of rhizobia required for successful nodulation and the ability of the manufacturers to maintain, produce, and supply such cultures (Roughley and Pulsford 1982).

In some instances, it may be necessary to develop an inoculum with more than one strain e.g., when inoculating several species of medics. These multi-strain inoculants may be successful providing the strains used are equally effective on a particular host. However, there may be problems such as microbial antagonism and competition for nodule formation by the less effective and less compatible strains. Commercial firms and/or research institutes may participate or assist in the production and distribution of inoculants to farmers.

References

Brockwell, J.A. 1958. The use of an anthocyanin-rich variety of barrel medic, *Medicago tribuloides* Desr. for prompt assessment of strain effectiveness in *Rhizobium meliloti*. Journal Australian Institute of Agricultural Science 24: 342–346.

Brockwell, J., Dudman, W.F., Gibson, A.H., Hely, F.W., and Robinson, A.C. 1968. An integrated programme for the improvement of legume inoculant strains. Pages 103–114 *in* Transactions of the Ninth International Congress of Soil Science, Vol. 2. Adelaide, Australia.

Brockwell, J.E., Schwinghamer, A. and Gault, R.R. 1977. Ecological studies of root-nodule bacteria introduced into field environments. V. A critical examination of the stability of antigentic and streptomycin-resistant markers for identification of strains of *Rhizobium trifolii*. Soil Biology and Biochemistry 9: 19–24.

Chatel, D.L., Greenwood, R.M., and Parker, C.A. 1968. Saprophytic competence as an important character in the selection of *Rhizobium* for inoculation. Pages 65–73 *in* Transactions of the Ninth International Congress of Soil Science, Vol. 2. Adelaide, Australia.

Date, R.A. 1975. Principles of *Rhizobium* strain selection. Pages 137–150 *in* Symbiotic Nitrogen Fixation in Plants. IBP No. 7. (Nutman, P.S., ed.). Cambridge University Press, UK.

Date, R.A. and Halliday, J. 1979. Collection of strains of *Rhizobium*. Pages 21–26 *in* Handbook for the Collection, Preservation, and Characterization of Tropical Forage Germplasm Resources. CIAT, Cali, Colombia.

Gibson, A.H. 1977. The influence of the environment and managerial practices on the legume-*Rhizobium* symbiosis. Pages 1087-1104 *in* A Treatise on Dinitrogen Fixation, Section IV. (Hardy, W.F., ed.). John Wiley and Sons, New York, USA.

Hagedorn, C. 1978. Effectiveness of *Rhizobium trifolii* populations associated with *Trifolium subterraneum* L. in southwest Oregon soils. Soil Science Society of America Journal 42: 447–451.

Halliday, J. 1984. Principles of *Rhizobium* strain selection. Pages 155–171 *in* Biological Nitrogen Fixation (Alexander, M., ed.). Plenum Publishing Co., New York, USA.

ICARDA (International Center for Agricultural Research in the Dry Areas) 1986. Annual Report 1985. ICARDA, Aleppo, Syria.

Josey, D.P., Beynon, J.L., Johnston, A.W.B. and Beringter, J.E. 1979. Strain identification in *Rhizobium* using intrinsic antibiotic resistance. Journal of Applied Bacteriology 46: 313–350.

Maier, R.J. 1986. Biochemistry, regulation, and genetics of hydrogen oxidation in *Rhizobium*. Pages 17-38 *in* Critical Reviews in Biotechnology. Vol. 3. CRC Press, London, UK.

Marques-Pinto, C., Yao, P.Y. and Vincent, J.M. 1974. Nodulating competitiveness amongst strains of *Rhizobium meliloti* and *R. trifolii*. Australian Journal of Agricultural Research 25: 317–329.

Materon, L.A., and Hagedorn, C. 1982. Competitiveness of *Rhizobium trifolii* strains associated with red clover (*Trifolium pratense* L.) in Mississippi soils. Applied and Environmental Microbiology 44: 1096–1101.

Materon, L.A., and Hagedorn, C. 1984. Responses to inoculation with genetically-marked strains of *Rhizobium trifolii*. Communications in Soil Science and Plant Analysis 15: 33–47.

Materon, L.A. and Cocks, P. 1987. Constraints to biological nitrogen fixation in ley farming systems designed for west Asia. *in* Microbiology in Action (Murell, W.C. and Kennedy, I.R. eds.). Research Studies Press, John Wiley and Sons, New York (in press.)

Obaton, M. 1971. Utilisation de mutants spontanes resistants aux antibiotiques pour l'etude ecologique des *Rhizobium*. Comptes Rendus de l'Academie des Sciences (Paris) 2720: 2630–2633.

Page, A.L., Miller, R.H. and Keeney, D.R. (eds.). 1982. Methods in Soil Analyses. Part 2. Chemical and Microbiological Properties. Agronomy No. 9. American Society of Agronomy, Madison, Wisconsin, USA.

Puckridge, D.W. and French, R.J. 1983. The annual legume pasture in cereal-ley farming systems in southern Australia: a review. Agriculture, Ecosystems and Environment 9: 229–267.

Roughley, R.J., Blowes, W.M. and Herridge, D.F. 1976. Nodulation of *Trifolium subterraneum* by introduced rhizobia in competition with naturalized strains. Soil Biology and Biochemistry 8: 403–407.

Roughley, R.J. and Pulsford, D.J. 1982. Production and control of legume inoculants. Pages 193–207 *in* Nitrogen Fixation in Legumes (Vincent, J.M., ed.). Academic Press, Sydney, Australia.

Schwinghamer, E.A. and Dudman, W.F. 1973. Evaluation of spectinomycin-resistance as a marker for ecological studies with *Rhizobium* spp. Journal of Applied Bacteriology 36: 263–272.

Smith, G.R. and Knight, W.E. 1984. Breeding legumes for improved N_2-fixation. Pages 1–23 *in* Applications of Genetic Engineering to Crop Improvement (Collins, G.B. and Petolino, J.G., eds.). Martinus Nijhoff/Dr. W. Junk Publishers, The Hague, The Netherlands.

Tchan, Y.T. 1982. Application of some newer serological techniques. Pages 27-32 *in* Nitrogen Fixation in Legumes (Vincent, J.M., ed.). Academic Press, Sydney, Australia.

Trinick, M.J. 1969. Identification of legume nodule bacteria by the fluorescent antibody reaction. Journal of Applied Bacteriology 32: 181–186.

Vincent, J.M. 1970. A Manual for the Practical Study of Root-Nodule Bacteria. IBP Handbook No. 15. Blackwell Scientific Publications, Oxford, UK.

Weaver, R.W., Materon, L.A., Krautmann, M.E. and Rouquette, F.M. 1985. Survival of *Rhizobium trifolii* in soil following inoculation of arrowleaf clover. MIRCEN Journal 1: 311–318.

D.P. Beck and L.A. Materon (editors), Nitrogen Fixation by Legumes in Mediterranean Agriculture
 ISBN 90 247 3624 2

The Use of Biological Nitrogen Fixation to Develop Livestock and Cereal Production in Morocco

M. ISMAILI and A. BENTASSIL

Universite Sidi Mohammed Ben Abdellah Meknes, Morocco

Summary. The ley farming system was introduced in Morocco in 1985 when legume seed was distributed to a number of farmers who agreed to cooperate in the project. Results are encouraging and more than 20000 ha will be seeded in 1986.Studies on the nitrogen nutrition of grass-legume mixtures showed that the grass took up more soil nitrogen than the legume which was also less competitive. The effect of water stress was more severe in the legume than in the grass. In drought stress studies, *Medicago truncatula* was only affected by severe drought, resisting drought at lower levels to ensure regeneration. The number of nodules was reduced by drought stress but increased after watering.

A number of rhizobial isolates from different regions of Morocco will be screened for drought resistance and competitive ability.

Introduction

The development of rangelands and pastures is one of the priorities of agriculture in Morocco. Nitrogen input to arid and semi-arid regions is one of the main factors which controls production in these areas. However the use of fertilizer nitrogen was found not to be feasible or effective because of the pattern of precipitation and the cost of combined nitrogen.

Since the 1960's alfalfa pastures have been grown in all irrigated areas as part of a rotation system, and in the rainfed areas, legumes and biological nitrogen fixation have gained importance since 1980. The legume pasture produces high quality forage for milking cows and could produce 2000–2500 forage units/ha compared with 500 forage units/ha from fallow land. It also improves cereal and livestock production, as well as soil nitrogen.

Much research has been conducted in different regions of Morocco (Settat, Beni Mellal, Safi, and Karia Tissa) to investigate the adaptability of legume species to different soil types and arid conditions, and to study the need for inoculation. A number of annual legumes, such as *Medicago truncatula*, *M. polymorpha*, *M. littoralis*, *M. scutellata*, and *Trifolium subterraneum*, have been selected for use in a program of fallow land replacement with legume pasture in regions of 300 mm or more annual rainfall.

The comparison of Australian and Moroccan conditions, which was completed by experts from Australia and the Ministry of Agriculture and Agrarian

Reform (Chatterton and Chatterton 1986), showed that the only way to develop forage and cereal production in more than 3 million ha of arid and semi-arid lands is to use the ley farming system developed in Western Australia.

The Ley Farming Program

During 1985, the extension of the ley farming system started. In the first year, the aim was to seed 35000 ha with annual legume species. The rotation system used is described below.

First season

In the autumn the legume pasture is seeded, and during the winter it provides green forage for livestock. In the spring it is grazed but not severely, and in summer it provides dry forage.

Second season

During the autumn a cereal is cultivated which is harvested in spring. In the summer, the cereal stubbles are grazed. Only shallow cultivation is used.

Nine thousand farmers agreed to take part in the program and about 760 tonnes of annual pasture legume seeds were imported and distributed to farmers. During 1985, 20000 ha were successfully seeded with annual *Medicago* species. Seed production was high, and was estimated to be about 400 kg/ha. Farmers were pleased with the results and a larger area will be seeded with annual *Medicago* during 1986.

Research Programs on Biological Nitrogen Fixation (BNF)

The Department of Biology of the School of Sciences and the Agronomy Department of the National School of Agriculture in Meknes are conducting research on three aspects of BNF: nitrogen nutrition of grass-legume mixtures, selection of rhizobial strains for drought resistance, and the effect of water stress on biological nitrogen fixation.

The use of grass-legume mixtures is studied in greenhouse and field experiments using *Lolium rigidum* and *Trifolium alexandrium*. The replacement series method is used with different proportions of legume and grass e.g., 100:0, 75:25, 50:50, 25:75, and 0:100, and different levels of soil nitrogen and water stress treatments are used. Results show that the number of legume plants decreases with time, suggesting that the legume is less competitive than the grass under all treatments. The grass took up more soil nitrogen than the legume when in mixture and after three clippings the mixture was not stable. In the mixture the effect of water stress was more severe on the legume than the grass.

Screening for rhizobial strains started with the collection of isolates taken from different annual *Medicago* species in different regions of Morocco. About 300 isolates were collected and will be tested under greenhouse and field conditions. These isolates will be screened for resistance to drought and competitive ability in the soil. The inoculum production phase is supported by FAO.

In this paper experiments concerning the effect of water stress on nitrogen fixation will be presented in more detail.

Materials and Methods

Seeds of *Medicago truncatula* var. Borung were planted in 10 liter pots filled with soil from the experimental farm of the National School of Agriculture in Meknes. Plants were grown in the greenhouse and naturally inoculated with soil. Four water-stress treatments (T_1–T_4) were imposed by withholding water for periods of 6, 12, 18, and 24 days. The control was continuously irrigated near field capacity, by weight. Treatments were started 69 days after planting. The number of flowers, pod and dry weights, and total nitrogen of shoots and roots were determined before, during, and after stress. Nitrogen fixation was estimated from the number of nodules and total plant nitrogen.

Results

Water stress treatments affected the flowering period and the number of flowers per plant (Fig. 1). The effect on the number of flowers showed that water stress could affect the regeneration and seed bank potentials of *M. truncatula*. The number of pods was only reduced in the more severe water stress treatments, T_3 and T_4 (Table 1). Pod weight was also reduced only in treatments T_3 and T_4 (Table 2). However, at the end of the experiment the pod weight of T_3 was not different from the control, showing the high resistance to drought of *M. truncatula*. Apparently only severe drought affects seed production in *M. truncatula*, as it resists stress by producing more seeds to ensure regeneration.

The dry weight of shoots and roots decreased significantly and in a similar manner during stress treatment (Fig. 2). Similar results were found in our previous work on siratro (*Macroptilium atropurpureum* D.C. Urb.) (Ismaili *et al.* 1983). Total shoot and root nitrogen also decreased in a similar manner, as did dry weight (Fig. 3). Periods of drought reduced forage production without significantly affecting seed production, showing that under water stress *M.*

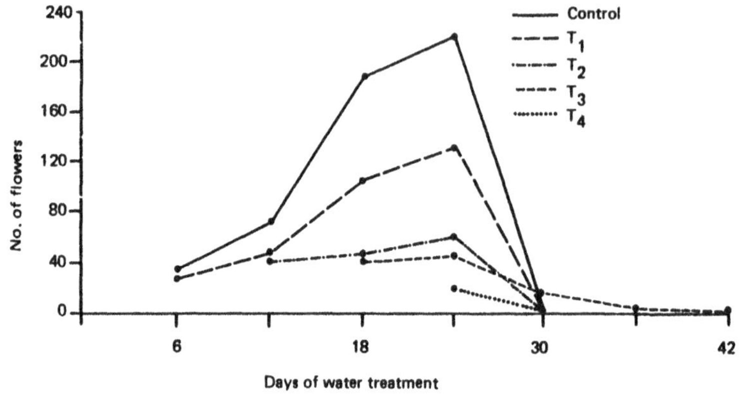

Fig. 1. Number of flowers at each clipping of *Medicago truncatula* grown under different water stress treatments.

truncatula directed all its potential towards pod formation and filling.

The number of nodules was reduced by water stress treatments from 42 nodules/plant in the control to 12 nodules/plant in T_4 indicating a decrease in nitrogen fixation by the plant. After rewatering, the number of nodules increased showing a recovery of the nitrogen fixing system through the formation of new nodules. Water stress reduced nitrogen fixation and plant production, but reduced seed production only when the stress was very severe, to the point that most of the plants were killed.

Table 1. Number of pods of *Medicago truncatula* at each clipping when grown under different water stress treatments.

Treatments	Days of treatment				
	6	12	18	24	30
C	100	367	627	928	1024
T_1	56	345	633	939	1002
T_2	42	163	529	877	1024
T_3	20	44	274	467	628
T_4	0	0	0	0	66

Table 2. Pod weight (g) of *Medicago truncatula* at each clipping when grown under different water stress treatments.

Treatments	Days of treatment				
	6	12	18	24	30
C	0.30	0.40	5.60	6.50	6.85
T_1	0.29	0.39	4.77	6.30	6.70
T_2	0.28	0.60	3.30	5.30	6.40
T_3	0	0.50	2.91	4.56	6.38
T_4	0	0	0	0	1.36

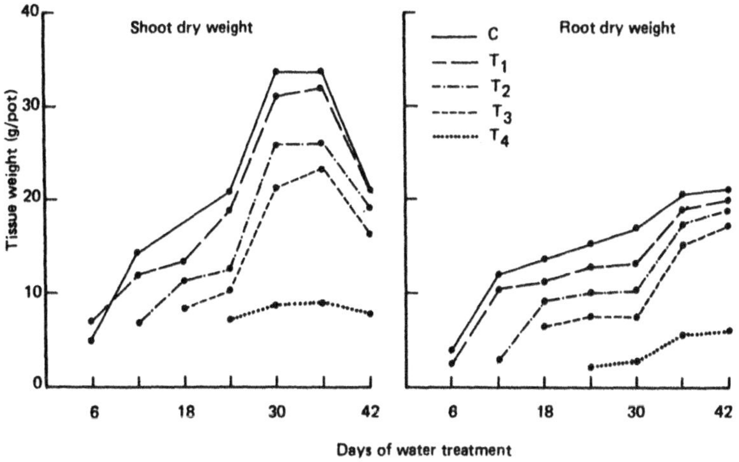

Fig. 2. Biomass in shoots and roots of *Medicago truncatula* grown under different water stress treatments.

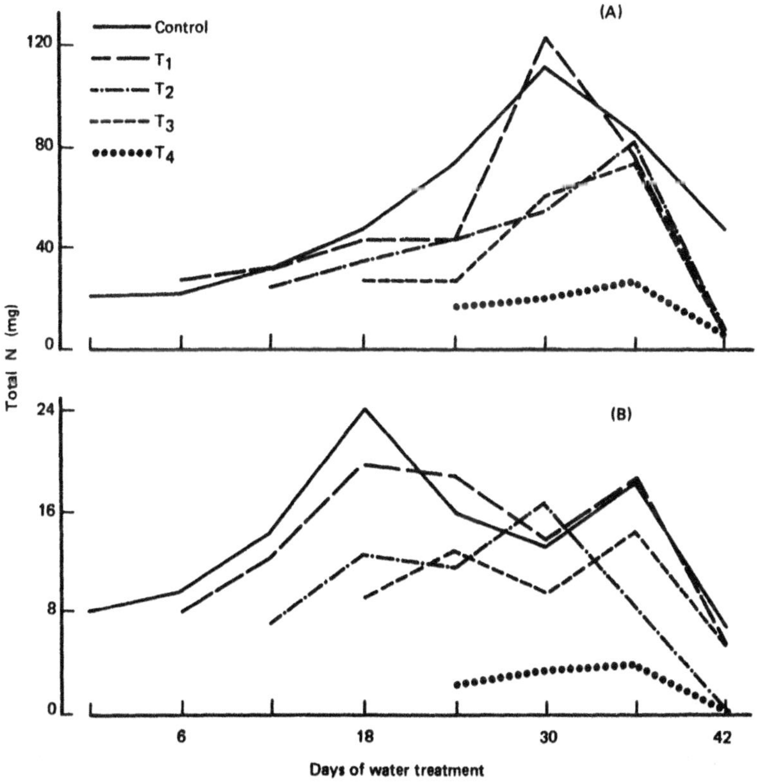

Fig. 3. Total nitrogen of shoots (A) and roots (B) at each clipping of *Medicago truncatula* grown under different water stress treatments.

During 1986, the same experiment will be repeated using two cycles of water stress treatments, one at an early stage of plant development and the other after the plants are well developed.

References

Chatterton, B, and Chatterton, L. 1986. The establishment and management of annual medic pastures in the Near East Region. FAO, Rome, Italy.

Ismaili, M., Briske, D.D. and Weaver, R.W. 1983. Nitrogen fixing activity of water stressed siratro. Agronomy Journal 75: 649–653.

D.P. Beck and L.A. Materon (editors), Nitrogen Fixation by Legumes in Mediterranean Agriculture
 ISBN 90 247 3624 2

Annual Forage Legumes in the Arid and Semi-Arid Regions of Turkey

ESVET ACIKGOZ

Department of Agronomy, Faculty of Agriculture, Uludag University, Bursa, Turkey

Summary. The cultivation of forage legumes, particularly vetches, is described with reference to the conditions of cultivation in Turkey. The uses of vetches are outlined and the problems of cultivation highlighted. Experiments to improve legume yields and those of subsequent cereal crops are also discussed.

Introduction

The total area of Turkey is 776000 km^2 of which 25.4 million ha is arable land. According to FAO statistics (FAO 1981) this area of arable land is the largest in Europe and the Middle East excluding USSR. Official statistics from the State Statistical Institute of Turkey show that the land use can be divided into four types: cropland (24.9 million ha or 32%), permanent crops (3.0 million ha or 3.9%), natural rangelands (21.0 million ha or 27%), and forests and woodlands (20.2 million ha or 26%) (Anon. 1960-84).

Annual rainfall ranges from about 250 mm in the center of the peninsula to 2500 mm in the East Black Sea region. Central and southeastern areas generally receive less than 500 mm average annual precipitation and the distribution pattern is typically continental with 70% of the precipitation falling between December and May. Spring rains are limited and erratic. Plant growth is seriously affected by the amount and distribution of rainfall.

In the arid and semi-arid regions, traditional cereal/fallow cropping systems are used by the majority of the farmers and crop rotations or continuous cropping are not widely practiced. Cereal hectarage has remained almost stable in the past two decades. Every year, about 16.7 million ha are seeded and 8 million ha are fallow. Hence, approximately one third of the cropland lies fallow for 12-14 months.

Definite statistics on the hectarage of forage crops in Turkey are not presently available. Nevertheless, official statistics in 1984 showed a total of 475000 ha under forage crops, which is slightly over 1.9% of the total cropland and 2.84% of the sown area seeded every year. Common vetch, *Vicia sativa* L., is the most important forage crop (197000 ha), followed by alfalfa, *Medicago sativa* L. (166000 ha), sainfoin, *Onobrychis sativa* Lam. (83000 ha), bitter vetch, *Vicia ervilia* L. Willd. (17000 ha), and others (10000 ha). This is equivalent to 1.18, 0.99, 0.49, and 0.10%, respectively of the total area seeded to field crops. It is well known that narbonne vetch, *Vicia narbonensis* L., and Hungarian vetch, *Vicia pannonica* Crantz, are grown in various parts of Turkey but their acreages are not recorded in the statistics. Probably these figures are included with that of common vetch. Fenugreek (*Trigonella foenum-graecum* L.), field pea (*Pisum sativum* ssp. *arvense* L. Poir), grasspea (*Lathyrus sativus* L.), and white lupin (*Lupinus albus* L.) are also grown to some extent but their acreages are not known.

Common vetch, bitter vetch, and peas were cultivated as early as 5000-7000 B.C. in Asia minor and carbonized seeds were found at different places during archaeological excavations (Sencer 1977). Until now, their culture has been continuous for thousands of years.

Vetch

The statistics show an increase in the area under vetch during the past two decades. It is grown in almost all parts of the country, but about 70% of the total acreage is concentrated in the Middle Anatolian region and transitional zones. In Mediterranean and southeastern regions, vetch is grown very occassionally or not at all (Fig. 1), but it is concentrated in some provinces such as Cankiri. There, vetches cover 12.5% of the cropland while in Tokat, Gumushane, Corum, and Erzurum provinces, vetches account for 2-6.3% of the croplands (Anon. 1960-84; Urem 1985). Vetch is used in parts of Europe, the Middle East, and North America as an annual crop for hay, seed, green manure, and pasture, and in Turkey it is widely used as a seed or hay crop. It is sometimes used as green manure, but very little as pasture, and estimates on the use of vetches as forage or seed are not available. In coastal areas, vetches

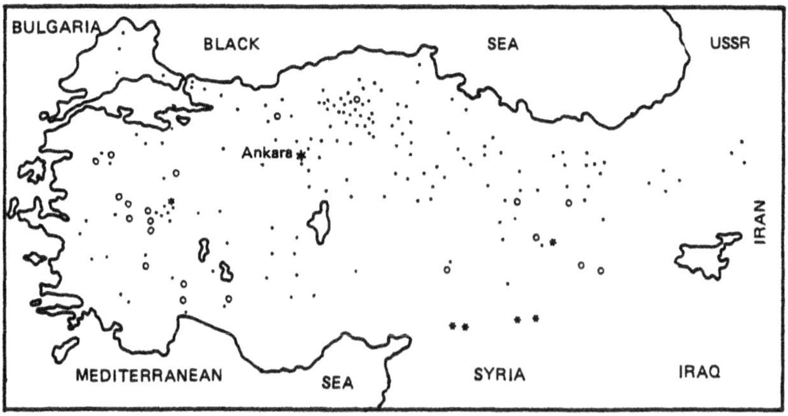

Fig. 1. Distribution of annual forage legumes in Turkey. Each symbol represents 1000 ha.
•: Common vecth including Hungarian vetch and narbonne vecth.
O: Bitter vetch.
*: Other legumes.

are principally used for hay or green herbage, while in central regions, mostly for seed. This pattern is partly due to climatic conditions and partly the animal feeding habits of farmers. In central regions, the dry climate favors the generative development of plants. Also, sheep flocks are fed with cereal straws and some seed concentrates during winter, and roughly ground vetch seed is very popular as a concentrate feed mixed with straws or cereal grains (Acikgoz *et al.* 1985).

Some other vetch species, mainly narbonne vetch and Hungarian vetch, are commercially grown in some parts of the country. Narbonne vetch is grown for seed production in eastern parts of Turkey (Christiansen-Weniger 1973; Tarman 1972). The latter species has been newly introduced but its acreage is expanding rapidly for hay production in Cankiri and Corum provinces (Sahin 1980).

Hay yield of vetches in Turkey varies depending on climatic conditions and soil fertility. Yields of 1.5–2.5 t/ha are common in the arid and semi-arid inner regions (Elci 1975; Tarman 1964) but may reach 7.5 t/ha in the coastal area without irrigation (Cakmakci 1985; Elci 1975).

In the past decade seed yield of vetches varied from 0.75 to 1 t/ha (Anon. 1960-84). In seed production, shattering is the main problem and seeds of the currently used cultivars shatter as soon as the pods are mature. Because of the low, viny growing habit of common vetch, seed harvest is difficult by combine. Recently, over 1000 common vetch strains, mostly from Turkish collections, were screened for several traits including erectness, shatter resistance, and yield under dryland conditions. In this study, some erect and shattering resistant genotypes were identified and selected (Ozkaynak 1981).

In extensive breeding studies conducted at different institutions, some high-yielding common vetch, hairy vetch, and Hungarian vetch cultivars were selected and released (Elci 1975; Urem 1985). However, most of the vetch seed planted in Turkey is still homegrown.

It is generally reported that common vetch is sensitive to cold, while hairy, narbonne, and Hungarian vetches are tolerant (Acikgoz 1982; Erkun and Alinoglu 1960; Schoth and Mckee 1966). Winter killing is a major problem in the regions with continental climate, so common vetch is seeded in spring. However, hay or seed yields of spring-seeded common vetches vary considerably depending on the amount and distribution of spring rains. Certainly fall seedings, if successful, would lead to higher forage and seed yields particularly under dryland conditions. Some studies were established to determine the relative cold tolerance of common vetch cultivars and experimental lines from different origins. Few strains were more tolerant and none of those tested had sufficient cold tolerance for seeding in fall. However, narbonne, Hungarian, and hairy vetches can be seeded in the fall without winter damage in central regions of Turkey (Acikgoz 1982b; Acikgoz 1982c).

The genus *Vicia* contains about 150 species and 59 of these are native to Turkey (Davis 1970). Therefore, it is believed that wild or indigenous *R. leguminosarum* strains are present in the soils of Turkey. Today inoculation is advised, but it is not used in vetches. Although vetch species in general don't respond to inoculation with rhizobia, research has been carried out to investigate the possibility of improving growth and yield using native and introduced strains. Treated plots produced slightly higher seed or hay yields at some sites and in certain years, but differences were not significant in most cases (Ersin 1978; Ulgen 1978).

Other Forage Legumes

Other annual forage legumes have very little importance in Turkish agriculture. Bitter vetch was one of the most important forage legumes until recent years. However as it is low-yielding and produces short straw, its acreage is going down annually. It is grown mainly for seeds in the poor dryland fields of the arid regions. Fenugreek is commercially grown in some provinces but its seed is used for spice rather than concentrate feed for animals. Also, boiled and salted protein-rich white lupin seeds are used as human food in some provinces. Grasspea and field pea are grown in some parts of the country for seed production and their seeds are used like vetch seeds. Very little information is available on grasspea and field pea culture. Garden pea is widely grown as a vegetable crop for green market and canning purposes, while its fresh or dried haulms and damaged seeds are extensively used for livestock feeding, particularly in the vicinity of canning factories (Acikgoz 1982a; Acikgoz 1982c).

Approximately 8 million ha of cropping area in Turkey is under fallow in the arid and semi-arid regions (Fig. 2). Several rotation studies have been established by a number of institutions in attempts to utilize this enormous capacity. In these long-term studies annual and perennial legumes, pulses, some oil crops, and corn grown on fallows drastically reduced the subsequent

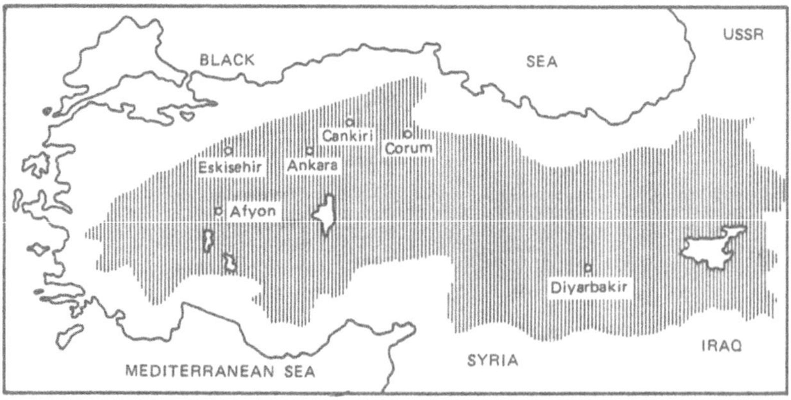

Fig. 2. Fallowlands in Turkey.

Table 1. Effect of legume or legume/cereal mixtures grown on fallows on subsequent wheat and hay yields (t/ha) in Ankara Province (Kurt and Tan 1984).

Rotation	Seeding rate (kg/ha)	Normal fall		Late fall	
		Wheat yield	Hay yield	Wheat yield	Hay yield
Wheat/fallow	160	2.54		2.40	
Wheat/cereal for hay	160/180	2.10	2.92	2.90	2.80
Wheat/common vetch (CV)	160/80	2.57	2.10	2.52	1.87
Wheat/narbonne vetch (NV)	160/160	2.61	2.39	2.61	2.23
Wheat/field pea (FP)	160/100	2.67	2.17	2.64	1.83
Wheat/CV + cereal	160/50 + 90	2.26	2.78	2.23	2.69
Wheat/NV + cereal	160/90 + 90	2.33	2.89	2.27	2.86
Wheat/FP + cereal	160/60 + 90	2.39	2.58	2.34	2.65

wheat yield (Christiansen-Weniger 1973; Kalayci 1981; Kurt and Tan 1984). No fertilizer was applied to the fallow crop or the wheat in these studies.

Recently, more attention has been given to growing annual forage legumes or pulses on fallow land. Several experiments were carried out using annual forage legumes alone or their mixtures with cereals for hay on fallows of arid and semi-arid regions of the country. In these studies, more promising results were obtained than in previous studies.

In Ankara Province, receiving about 375 mm annual rainfall, some cereals and annual legumes were grown on wheat fallows for hay production. All plots were sown at two different dates during fall and late fall, fertilized with 25 kg N/ha and 80 kg P_2O_5/ha as diammonium phosphate. The 6 years of results clearly showed that legumes, or their mixtures, did not greatly affect wheat yield and an average 2.5 t/ha of high quality hay was produced. The highest wheat yields were recorded on plots previously grown with pure legumes. However, these plots yielded less hay than legume + cereal mixtures. Barley and oats grown on fallow and cut for forage negatively affected the subsequent wheat yield (Table 1).

Similar experiments were set up in other provinces. In Corum, receiving 410 mm annual precipitation, barley and its mixture with field pea, hairy vetch, and narbonne vetch grown on fallows did not reduce wheat yield in the following year. In this study, an average 4 t/ha of hay was produced (Table 2).

Table 2. Effect of barley or barley/annual legume mixtures grown on fallow and for forage on wheat and hay yields (t/ha) in Corum Province (Tan 1984).

Rotation	Seeding rate (kg/ha)	Wheat yield	Hay yield
Wheat/fallow	160	2.81	
Wheat/barley	160/180	2.87	4.57
Wheat/barley + field pea	160/90 + 80	2.90	4.74
Wheat/barley + hairy vetch	160/90 + 30	2.85	3.61
Wheat/barley + narbonne vetch	160/90 + 100	2.96	4.89

In another study, fall-seeded pure narbonne vetch, hairy vetch, and field pea cultivars cut for hay did not affect subsequent wheat yield in Eskisehir conditions with 390 mm rainfall. However, when these were grown for seed or green manure, wheat yield was reduced, possibly because moisture and mineral elements in soil were further depleted. Similar results were obtained in Afyon Province where rye mixed with Hungarian vetch, hairy vetch, and field pea produced 2.5–3.0 t/ha hay without adversely affecting subsequent wheat yield (Bakin 1976). In the southeast, there was no clear difference in wheat yield between fallow/wheat and vetch/wheat cropping systems. In the latter an average of 6.5 t/ha green herbage was produced (Atkan 1981).

In addition to poor climatic conditions and early and over grazing have destroyed the desirable plants in the natural rangelands of Turkey (Bakir and Acikgoz 1976; Bakir 1971). Farmers freely feed their sheep on the poor rangelands during most of the year. Recently there has been a trend towards introducing forage legumes or cereal/legume mixtures in place of fallows to provide nutritious grazing in spring and early summer. Although detailed studies are not available, preliminary studies showed promising results. Weaned, 2.5 month-old male lambs were grazed on spring-seeded barley + common vetch or fall-seeded barley + Hungarian vetch mixtures grown on fallows in Ankara. One hectare of mixture maintained 10-20 lambs without supplementation for 42 days in spring (Cangir *et al.* 1984; Elicin *et al.* 1983). When the grazing period was extended, animals lost weight. It was concluded that when lambs are kept for at least 6-7 weeks on the mixtures and average 100-200 kg/ha, liveweight gain may be obtained.

Common vetch is very suitable for grazing or hay making on fallows. It is high-yielding and animals prefer it to other vetch species (Munzur 1982). However, the use of common vetch is being impeded by the seeds' lack of adequate cold tolerance either alone or when mixed with cereals. Unfortunately, none of the common vetch cultivars have enough cold tolerance. Therefore, narbonne vetch and Hungarian vetch are now preferred in most fall seedings. Since there are some difficulties in drying narbonne vetch forage, farmers are advised to use it as green fodder or as silage rather than for hay making (Kurt and Tan 1984; Tan 1984). The use of Hungarian vetch is not widespread despite its good cold and drought tolerance and relatively upright growth habit (Acikgoz 1982b). In several studies, Hungarian vetch did well on fallows alone or in mixtures with cereals (Kurt and Tan 1984; Tan 1984). It also withstands the cold winters in inner Anatolia without damage (Acikgoz 1982b; Aktan 1981; Kurt and Tan 1984; Munzur 1982). Thus, in some regions the acreage of Hungarian vetch for hay production is increasing.

It is estimated that approximately half of the fallowland is suitable for pulse/wheat or forage legume/wheat cropping systems in Turkey. If one quarter of this area was used for hay production, at least 2.5 million tons of hay would be produced. This is three times higher than current forage hay production or one seventh of the total estimated hay production from 26.0 million ha of rangeland and forage crop area recorded in 1971 (Bakir 1971).

References

Acikgoz, E. 1982a. Cold tolerance and its association with seedling morphology and chemical composition in annual forage legumes. I. Field peas (*Pisum sativum* ssp. *arvense* L. Poir). Zietschrift fuer Pflanzenzuhtung 88: 118-126.

Acikgoz, E. 1982b. Cold tolerance and its association with seedling morphology and chemical composition in annual forage legumes. II. Vetch (*Vicia*) species. Zietschrift fuer Pflanzenzuchtung 88: 278-286.

Acikgoz, E. 1982c. Parameters of cold tolerance in common vetch. Euphytica 31: 997-1001.

Acikgoz, E., Katkat, V., Omeroglu, S., Okan, ve B. 1985. Mineral elements and amino acid concentrates in field pea and common vetch herbages and seeds. Zietschrift fuer Acker-und Pflanzenbau 155: 179-185.

Aktan, S. 1981. Using of fallows efficiency in Southeastern Anatolia. Pages 59-70 *in* Symposium on Using Fallowlands in Arid Regions. Publication 593. Scientific and Technical Research Council of Turkey, Agriculture and Forestry Group, Ankara, Turkey.

Anonymous. 1960-84. The summary of agricultural statistics. Rep. of Turkey, Prime Ministry State Institute of Statistics Publication.

FAO (Food and Agriculture Organization). 1981. Production Yearbook. Volume 35. FAO, Rome Italy.

Bakir, O. 1971. Range management in Turkey. Pages 69-78 *in* Proceedings of the CENTO Conference on Development of Feed Resources. Tropical Products Institute, London, England.

Bakir, O. 1976. Developmental research on forage production on fallow land in the western transition zone of Turkey. Final Research Report, Faculty of Agriculture, Ankara, Turkey.

Bakir, O. and Acikgoz, E. 1976. Present situation and development possibilities of forage crops culture in Turkey. Publication No. 61. Grassland and Animal Husbandry Research Institute.

Cangir, S., Elcin, A. Karabulut A., Munzur, M. and Ilaslan, M. 1984. Investigations on fattening performance of weaned lambs grazed on mixtures seeded on fallow. Publication No. 98. Grassland and Animal Husbandry Research Institute.

Christiansen-Weniger, F. 1973. Fundamentals of Field Crops Culture in Turkey. Mentes Publisher, Turkey.

Cakmakci, S. 1985. Effect of sowing season, row spacing and cutting stages on yield of green herbage, dry matter and protein in common vetch (*Vicia sativa* L.). M.Sc. thesis. Uludag University, Bursa, Turkey.

Davis, P.H. 1970. Flora of Turkey and East Aegean Islands. Edinburgh University Press, Edinburgh, UK.

Elci, S. 1975. Vetch (*Vicia*) culture. Agr. Aff. Publication D-167. Department of Food, Agriculture and Animal Husbandry.

Elicin, A., Ilaslan, M., Munzur, M., Cangir, S. and Karabulut, A. 1983. Fattening performance of weaned lambs grazed on vetch + barley mixture seeded on fallow land. Publication No. 84. Grassland and Animal Husbandry Research Institute.

Erkun, V. and Alinoglu 1960. Preliminary results on adaptation studies of forage crops conducted in 1952-1959. Department of Agriculture of Turkey.

Ersin, B. 1978. Effect of inoculation on yield and nitrogen content of vetch hay in Aegean Region. Topraksu Publication No. 62. Republished as No. 37. Menemen, Izmir, Turkey.

Guler, M., Pala, M., Durutan, N., Karaca, M., Avcin, A. and Avci, M. 1981. Criteria used in determination of fallowland borders. Pages 9-17 *in* Symposium on Using Fallowlands in Arid Regions. Publication 593. Scientific and Technical Research Council of Turkey, Agriculture and Forestry Group, Ankara, Turkey.

Kalayci, M. 1981. Studies on reducing of fallows in Eskisehir region. Pages 195-207 *in* Symposium on Using Fallowlands in Arid Regions. Publication 593. Scientific and Technical Research Council of Turkey, Agriculture and Forestry Group, Ankara, Turkey.

Kurt, O. and Tan, A. 1984. Forage production possibilities on fallows by fall or late fall seedings. Publication No. 93. Grassland and Animal Husbandry Research Institute.

Munzur, M. 1982. Seeding rates of cereal plus vetch mixtures and their fitness to grazing and hay production. Ph.D. thesis. Ankara University, Ankara, Turkey.

Ozkaynak, I. 1981. Some important traits of common vetch strains selected from native ecotypes of Turkey. Publication No. 758. Faculty of Agriculture, University of Ankara, Ankara, Turkey.

Sencer, A. 1977. Plant collection studies for long term storage. Pages 183-188 *in* VI Science Congress, Scientific and Technical Research Council of Turkey, Agriculture and Forestry Group, Ankara, Turkey.

Schoth, H.A. and Mckee, R. 1966. The vetches. Pages 205-211 *in* Forages (Hughes, H.D., Heath, M.E. and Metcalfe, D.S., eds.). Iowa State College Press, Ames, Iowa, USA.

Sahin, H. 1981. Reducing of fallows in Corum-Cankiri Development. Pages 263-276 *in* Symposium on Using Fallowlands in Arid Regions. Publication 593. Scientific and Technical Research Council of Turkey, Agriculture and Forestry Group, Ankara, Turkey.

Tan, A. 1984. Hay productivity of some annual legume/barley mixtures under the arid conditions of Corum province by sowing in spring and fall during the fallow years. Publication No. 91. Grassland and Animal Husbandry Research Institute.

Tarman, O. 1954. Growing leguminous forage crops. Department of Agriculture of Turkey.

Tarman, O. 1972. Grassland husbandry and forage crops culture. Publication No. 464. Faculty of Agriculture, University of Ankara, Ankara, Turkey.

Ulgen, H. 1978. Selection of effective strains of *R. leguminosarum* under greenhouse and field conditions. Publication No. 79. Soil and Fertilizer Research Institute.

Urem,A. 1985. Growing of important forage crops and traits of registered cultivars in Turkey. Publication No. 58. Aegean Regional Experimental Station, Ismir, Turkey.

D.P. Beck and L.A. Materon (editors), Nitrogen Fixation by Legumes in Mediterranean Agriculture
© 1988 ICARDA – Printed in the Netherlands. ISBN 90 247 3624 2

The Role of Legumes in Agricultural Production in Cyprus

I. PAPASTYLIANOU

Agricultural Research Institute Nicosia, Cyprus

Summary. Studies in the past 6 years have shown that, in Cyprus, legumes such as common vetch, faba beans, ochrus vetch, and medics which respond to *Rhizobium leguminosarum* and *R. meliloti* can fix up to 80% of their nitrogen without inoculation. Other legumes, such as chickpea, peanut, soybean, and field bean, which respond to other *Rhizobium* species, have nodulation problems and inoculation is needed to ensure good biological nitrogen fixation. In rotation studies in the rainfed areas of Cyprus, it was found that the inclusion of common vetch, which fixes nitrogen, results in higher protein output and grain yield of subsequent cereal crops with less nitrogen fertilizer input compared to a continuous cereal production system.

Introduction

The value of legumes in agricultural production in the Mediterranean region was known in Greece long ago. Theophrastus (370-285 B.C.) pointed out that wheat and barley exhaust the land more than other crops while leguminous crops, such as beans, seem to manure the land. "Wheat exhausts the land more than any other crop, and next to it barley; wherefore the former requires good soil, while barley will bear even on somewhat crumbling soils; and of leguminous plants chickpea is the most exhausting, although this crop is in the ground only a very short time. Beans, as was said, are in other ways not a burdensome crop to the ground, they even seem to manure it, because the plant is of loose growth and rots easily; wherefore the people of Macedonia and Thessaly turn over the ground when it is in flower.'

Rafiq (1982), examining the causes of low crop yields in the Near East and Northrop production. Common practices in the region which help to maintain soil fertility are fallowing, growth of leguminous weeds, application of livestock excreta during grazing of fallow fields and to small-scale forages, or growth of grain legumes in rotation with wheat or barley. Nitrogen fertilizers are not commonly applied, but may be applied in limited amounts in most countries of the region (FAO 1984).

Systematic cultivation of legumes in rotation with cereal is practically non-existent (Gotsch 1980), but the importance of forage legumes in rotation with cereal crops has been demonstrated in Cyprus (Samios 1979) and recently emphasis was given by ICARDA to such studies for the region (Keatinge *et al.*

1986). Legume studies should not be limited to the rotational aspect but should also include plant genetic improvement, *Rhizobium*, nitrogen fixation, and agronomic studies.

This report presents such multidisciplinary studies of legume evaluations in Cyprus.

Soil and Climatic Conditions in Rainfed Areas of Cyprus

Soil in areas where rainfed crops are grown in Cyprus are calcareous, of medium to fine texture, with low nitrogen (total soil nitrogen around 0.1%) and phosphorus (bicarbonate-extractable phosphorus 1-5 ppm), but adequate potassium (exchangeable greater than 250 ppm).

Rainfall occurs during November through April and is rather erratic, both in amount and distribution: most of the rainfall (40%) occurs in December and January. In the cereal-growing areas, average annual rainfall is around 300 mm. Crops may suffer from water stress at any time of their growth cycle but it predominantly occurs in March–April during anthesis and grain-filling, reducing grain yield, and during sowing, resulting in late emergence and thin stands.

Nitrogen Fixation

The amount and percentage of nitrogen fixed by legumes commonly grown in Cyprus are presented in Table 1. These values were estimated using ^{15}N methodology in a cooperative project with the International Atomic Energy Agency (IAEA). In these studies there was no inoculation so the legumes

Table 1. Nitrogen fixed (kg N/ha and %N derived from fixation) by legumes in two seasons in Cyprus, using the A-value method with oats (1982/83) and barley (1983/84) as reference crops.

Crops	1982/83		1983/84	
	kgN/ha	%df [2]	kgN/ha	%df
Common vetch (*Vicia sativa* (L.) [1]	93	73	106	75
Peas (*Pisum sativum* (L.) [1]	23	36	91	63
Medics (*Medicago truncatula* Gaerth				
cv Jemalong (1982/3)	122	67		
cv Cyprus (1983/4)			90	70
Ochrus vetch (*Lathyrus ochrus* (L.) A.P. de Candole)	71	66	105	79
Bitter vetch (*Vicia ervilia* (L.) Willd.)	45	56	105	70
Broad beans (*Vicia faba major*)	122	63	176	80
Tick beans (*Vicia faba minor*)			180	84
Woollypod vetch (*Vicia dasycarpa* Ten.)			151	79
Chickpea (*Cicer arietinum* L.)			25	41

[1] Oat was reference crop in both seasons.

[2] % of total plant nitrogen derived from fixation.

depended on indigenous populations of *Rhizobium* for nitrogen fixation. Rainfall at the experimental site (Dromolaxia) in the 2 years of the study was 234 and 330 mm. Low values of nitrogen fixation (around 60%) were obtained in 1982/83 while in the next season, values were generally higher (70 – 80%). Of the legumes studied, only chickpea had low nitrogen fixation due to poor nodulation.

Most of the rainfed legumes obtained 60 – 80% of their total nitrogen from fixation. This is considered satisfactory under the prevailing conditions, which included moisture stress during the first year and no precautions for pathogen control. *Sitona* weevil, which damages nodules, has been frequently observed in Cyprus, especially on faba beans and common vetch. However, the extent to which nitrogen fixation is reduced by *Sitona* weevil has not yet been determined under our conditions.

Inoculation with *Rhizobium*

As shown in Table 1, an appreciable amount of nitrogen is fixed by many legume species in Cyprus, indicating the effectiveness of the nodules present on these crops. The possibility of improving the indigenous population of *Rhizobium* by selection and introduction is under consideration. However, the need for inoculation is urgent for a few legume species in Cyprus either because no nodules form or those present do not have the characteristic pink color which indicates effectiveness. Crops with such problems are chickpea under rainfed conditions, soybean, peanut, and field bean.

Most of the legumes grown by farmers under rainfed conditions are included in Table 1. Medics are not a cultivated crop but they grow naturally on uncultivated land. With the exception of chickpea, cultivated crops belong to the genera *Lathyrus*, *Vicia*, and *Pisum* which respond to *R. leguminosarum* biovar *viceae*. For crops which respond to *Rhizobium* species other than *R. meliloti* and *R. leguminosarum*, it seems that there are nodulation problems.

The results of the first inoculation studies are presented in Table 2. Inocula were supplied by Nitragin Company (USA) and ICARDA. The data showed that an appreciable yield response could be obtained by inoculating non-nodulated legumes such as soybean and chickpea. Vetch, which can fix 70% of

Table 2. Effect of inoculation on dry matter and nitrogen yield of three crops in Cyprus.

Crop	Yield * (kg/ha)		Nitrogen yield (kg N/ha)	
	Inoculated	Uninoculated	Inoculated	Uninoculated
Soybean	4853 (+)	3677 (−)	302 (+)	212 (−)
Vetch	5020 (+)	4366 (+)	120 (+)	107 (+)
Chickpea	2628 (+)	1683 (−)	71 (+)	41 (−)

* Grain yield for soybean and dry matter yield for vetch and chickpea (+) and (−) denote the presence or absence of nodules, respectively.

Table 3. Performance of vetch grown in Cyprus at different locations and years at two levels of nitrogen fertilizer (Papastylianou 1986).

Year	Location	Fertilizer (kg N/ha)	DM yield (kg/ha)	Nitrogen yield (kg/ha)
1981/82	Laxia	0	4090	112
		60	4320	117
1982/83	Kivisili	0	4609	112
		60	4553	104
1983/84	Athalassa	0	5079	
		60	4305	
1982/83	Dromolaxia	15	4714	127
		90	3105	100
1983/84	Dromolaxia	0	7004	165
		90	7100	171

its nitrogen (Table 1), also showed a yield response with inoculation. Inoculation studies of peanut and field bean have not yet shown any response due to the presence of high mineral nitrogen in the irrigated areas where the experiments were carried out.

Nitrogen Fertilization of Legumes

Although vetch showed a response to inoculation (Table 2), it did not respond to nitrogen fertilization. In studies on vetch carried out over many years and locations in Cyprus, dry matter and nitrogen yields were not increased with nitrogen fertilization (Table 3). In contrast, nitrogen fertilization had a negative effect on nitrogen fixation. In Dromolaxia in 1983 and 1984 the amount of nitrogen fixed by vetch was reduced by 48 and 36%, respectively, when 90 kg N/ha was applied.

In studies with soybean, nitrogen fertilization and inoculation treatments gave equal grain yields, with both treatments yielding 25% higher than the uninoculated, unfertilized control.

Genetic Improvement

Genetic improvement of legumes in Cyprus was attempted through introduction and selection of new cultivars. Such studies were carried out for faba beans, soybeans, chickpeas, field beans, vetch, peanuts, and lentils. The criteria for selection were grain characteristics and yield, disease resistance, and plant characteristics such as height, branching, etc. On no occassion was the nitrogen fixation factor considered.

The main findings of these studies are outlined below.

Chickpeas

Selection of a tall variety suitable for mechanical harvesting and with resistance to ascochyta blight. The local variety is short and is affected by blight (Hadjichristodoulou 1984).

Faba beans

None of the introduced varieties outyielded the local one. However, populations of the local variety were selected with the desired agronomic characteristics (Della, personal communication).

Vetch

Two of the introduced varieties from Greece gave equal yield to the local variety (Droushiotis, personal communication).

Peanuts

Five varieties were selected with pod and nut yields slightly higher than those of the local variety (Hadjichristodoulou, personal communication).

Field beans

One of the introduced varieties gave a yield equal to the local variety but had smaller seeds (Hadjichristodoulou, personal communication).

Lentils

Eighteen varieties introduced from ICARDA performed better and were earlier than the local check (Della, personal communication).

Soybean

Soybean was introduced in Cyprus to study the adaptation, productivity, and cost of production under local conditions. Yields were as high as 4.30 t/ha but cost of production (irrigation) and low world prices render the crop uneconomic (Hadjichristodoulou 1986).

Table 4. Grain yield (kg/ha) of barley grown in vetch-barley for grain or in barley for grain-barley for grain rotations and different levels of nitrogen fertilization at two locations, 1981-85.

Rotation	Fertilizer on barley (kg N/ha)	Laxia				Dromolaxia			Mean
		1981 /82	1982 /83	1983 /84	1984 /85	1982 /83	1983 /84	1984 /85	
	0	1232	664	1434	2550	1187	1030	3014	1586
Vetch * –	30	2188	1016	1367	2237	1426	1277	3476	1855
Barley	60	2121	856	807	1019	1680	1419	3807	1572
	90	2965	829	672	684	1524	1464	3901	1719
	0	239	560	336	736	545	396	2643	779
Barley –	30	986	739	1113	1748	769	537	2922	1259
Barley	60	1845	1299	1269	1569	948	1277	2601	1544
	90	1762	829	634	768	1113	1000	3115	1317
	120	2069	889	358	391	1143	929	3029	962
Rainfall (mm)	226	207	255	265	234	330	487	286	

* Vetch was not fertilized.

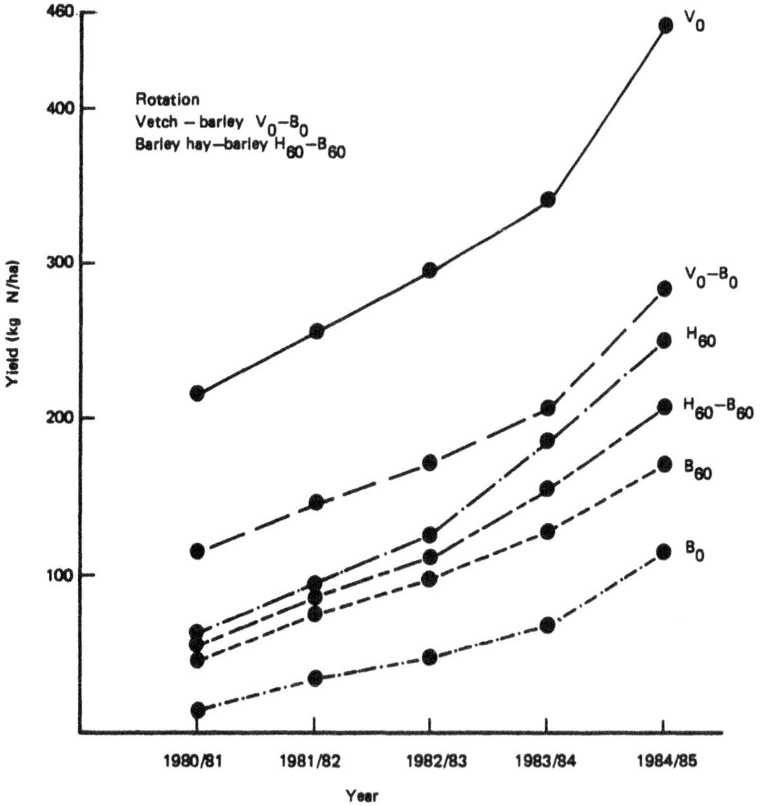

Fig. 1. Cumulative nitrogen yield of crops in vetch-barley for grain rotation without nitrogen fertilization, and in barley for hay-barley for grain rotation with 60 kg N/ha fertilizr in both phases, 1980-85, Laxia, Cyprus.

Legumes in Rotation with Cereals

The residual effect of legumes on subsequent crops is one of the most important factors to be considered when they are introduced into a farming system. Another important factor is their high protein yield which is important for human and animal nutrition. These two factors were studied in rotation experiments carried out during the past 5 years at two rainfed sites in Cyprus. In Table 4, grain yield of barley grown after vetch is compared with grain yield in a continuous barley growing system. The data show that (a) barley after vetch without any nitrogen fertilization can yield as much as continuously cropped barley with 60 kg N/ha, (b) barley fertilized with unlimited nitrogen in continuous cropping cannot yield as much as barley after vetch when in the latter system barley is fertilized with 30 kg N/ha, and (c) high doses of nitrogen fertilizer (above 60 kg N/ha) caused yield reduction in both rotations.

The total nitrogen yields of two rotations are presented in Fig. 1. The rotations were vetch-barley for grain without nitrogen fertilization, and barley for hay-barley for grain with 60 kg N/ha applied in both phases.

These treatments were chosen to demonstrate the importance of legumes in rotations in countries where farmers do not apply any nitrogen fertilizer. Undoubtedly, nitrogen output from crops in the rotation which included legume is higher than in the barley hay-barley grain rotation.

Considering the results in Table 4 and Fig. 1 the best rotation practice should be vetch (without nitrogen fertilization) – barley for grain supplied with 30 kg N/ha, for conditions similar to the experimental sites. These results and conclusions will be tested with future cycles of these rotation studies.

Conclusions

These studies showed that:

a) legumes fixed 60 – 80% of their nitrogen when effective nodules were present;
b) no nitrogen fertilizer is needed for legume production but in some cases the literature recommends small doses of starter nitrogen;
c) legumes have great value in rotation with cereals in rainfed areas for protein yield and increasing yield of the subsequent crop; and
d) inoculation with a suitable strain of *Rhizobium* is important for efficient nitrogen fixation and the final yield of legumes.

All the above advantages of legume over non-legume crops are due to their symbioses with *Rhizobium* sp. which result in biological nitrogen fixation (BNF). This characteristic should be borne in mind in studies aiming to improve legume yields. If BNF is not considered in such studies it is not possible to economically optimise production.

Although the importance of legumes has been known for millenia these crops are not as widely cultivated as might be expected. The main constraints to legume cultivation are:

a) Yield. Productivity of these crops is limited by environmental factors, disease problems, lack of genetic improvement of plants, and presence of proper rhizobial strains.
b) Mechanization and cost of production. This is a very important constraint. Most of the legumes are harvested by hand which increases cost of production. Also, labor cost limits cultivation, and control of broad-leaved weeds is done mainly by hand.
c) Marketing. For cereals there is a well organized world market. For legumes such a market exists only for soybean. What will happen if production of faba beans increases? Most grain legumes are consumed by humans and generally the possibility of increasing local consumption is rather limited. Marketing of hay produced by forage legumes is even more difficult, because of transportation costs.
d) Subsidies. In most countries (like in Cyprus) cereals are subsidized and are covered by insurances for yield loss. This is not the case for legumes.

These are some of the constraints which should be considered in parallel to BNF and show that legume studies should be a joint venture for genetisists, agronomists, microbiologists, physiologists, and soil scientists.

References

FAO (Food and Agriculture Organization) 1984. Fertilizer Yearbook. Volume 34. FAO, Rome, Italy.

Gotsch, G. 1980. Improving dryland agriculture in the Middle East and North Africa. Food Research Institute, Standford University and the Ford Foundation Middle East Regional Office, Cairo, Egypt.

Hadjichristodoulou, A. 1984. New chickpea varieties for winter sowing and mechanical harvesting. Technical Bulletin 5B. Agricultural Research Institute, Nicosia, Cyprus. 8pp.

Hadjichristodoulou, A. 1986. Variety and sowing date trials with soybeans. Technical Bulletin 75. Agricultural Research Institute, Nicosia, Cyprus, 7pp.

Harper, J.E. 1974. Soil and symbiotic nitrogen requirement for optimum soybean production. Crop Science 14: 255-260.

Keatinge, J.D.H., Somel, K., Cooper, P.J.M., and Tully, D. 1986. Experimental design of small plot crop rotation trials in the dry areas of N. Syria. I. The influence of agronomic management. Proceedings of the ICARDA/ACSAD/IDRC Regional Workshop on Crop Rotation Trials in Rainfed Agriculture, Design and Evaluation. March 1985, Damascus, Syria. ICARDA, Aleppo, Syria.

Papastylianou, I. 1986. The contribution of legumes in rotations in the rainfed areas in Cyprus. Proceedings of the ICARDA/ACSAD/IDRC Regional Workshop on Crop Rotation Trials in Rainfed Agriculture, Design and Evaluation. March 1985, Damascus, Syria. ICARDA, Aleppo, Syria.

Rafiq, M. 1982. Crop production management in the agroecological regions of the Near East and North Africa. FAO (Food and Agriculture Organization), Rome, Italy.

Samios, T.H. 1979. Constraints to cereal production and possible solutions in Cyprus. *In* The Gap Between Present Farm Yield and the Potential. Fifth Cereal Workshop, Algiers, Algeria.

Theophrastus. 370-285 BC. Enquiry into plants. Translation by Sir Arthur Hort, The Loeb Classical Library, William Heinemann, London, 1916.

Discussion

Danso

You have shown the potential of legumes in terms of improvement of soil nitrogen, but I guess we have to be a little careful about legumes with a high nitrogen demand for growth, and which are harvested for grains. In this case the smaller proportion of soil nitrogen uptake relative to a cereal may translate into a large quantity of nitrogen and a negative effect on soil nitrogen.

Papastylianou

The actual soil nitrogen balance after legumes, particularly grain legumes, could be negative. However, this has to be measured for a particular grain legume under particular conditions. The negative nitrogen balance, if it occurs, is due to the removal of the total above ground herbage. This is not true for pastures where part of the nitrogen in herbage is returned through animal excreta. In rotation studies of both grain and pasture legumes grown in rotation with cereals, it has been shown that subsequent crops benefit from legumes when compared to continuous cereal growing systems, whatever the soil nitrogen balance. This is because the value of legumes in rotations is not only in fixing nitrogen but also in disease control, soil structure improvement, and moisture storage (the latter in Cyprus shown by the Soil Department). Also in my results, it is shown that the value of legumes is not only due to nitrogen fixation.

Cocks

Do Cypriot farmers own livestock?

Papastylianou

Legumes in the rainfed areas are grown on only 10% of the land, mainly by farmers who have their own livestock. Most of the farmers grow only cereals, so if they are persuaded to grow legumes we have to secure a market for their legumes.

D.P. Beck and L.A. Materon (editors), Nitrogen Fixation by Legumes in Mediterranean Agriculture
© 1988 ICARDA – Printed in the Netherlands. ISBN 90 247 3624 2

Woody Legumes in Arid and Semi-arid Zones: the *Rhizobium-Prosopis chilensis* Symbiosis

J. OLIVARES, M.A. HERRERA and E.J. BEDMAR
Estacion Experimental del Zaidin, CSIC, 18008-Granada, Spain

Summary. Desertification of arid and semi-arid zones is increasing. A number of approaches have been suggested to prevent further desertification, an example being the planting of legume trees in these zones. In this study the effect of inoculation of *Rhizobium* strains from different woody legumes onto *Prosopis chilensis* grown in test tubes and in pots was examined.

Introduction

According to Spears (1980), the area of tropical forest has decreased to half of what it was a century ago, resulting in a negative effect on tropical zones with a dry climate where desertification is increasing.

These arid and semi-arid zones, inhabited by one tenth of the world's population, are located mainly in two large subtropical belts. One is in the north and includes the Sahara and Sahel, the Middle East, Southern Asia, and part of North America, and the other is in the south including part of southern Africa, Australia, and the dry zones of Peru, Chile, Argentina, and Brazil.

Two of the main causes of desertification are overgrazing and the overexploitation of forests for different products. These practices typically cause low productivity, a lack of firewood, and a decrease in soil fertility.

Arid and semi-arid zones cover approximately four-tenths of the earth's land area. In these zones water is the principal limiting factor although, according to several authors (Felker and Bandurski 1979), the low availability of nitrogen and organic matter in the soil seems to be as important as water. Together with these limiting factors, there may be high salinity, high temperature, and low levels of other mineral nutrients. Irregular and scarce rainfall, wide fluctuations between day and night temperatures, and low relative humidity are common characteristics of warm, arid zones. These ecosystems are fragile and the physical environment can be easily destroyed if certain limits are exceeded.

To increase the productivity of these zones, it is necessary to begin with integral planning that takes into account the nature of the soil and the population it supports. In this way, it has been theoretically and practically demonstrated that, under arid and semi-arid conditions, trees are the plants which best take advantage of soil and climate. In addition, perhaps the most important advantage of trees in the developing world is that many combine more than one use or product in a single tree e.g., fuel, fodder, food, fertilizer, fiber, and timber (Ben Salem 1980).

Felker and Bandurski (1979) pointed out that the ideal plant for these areas must have a number of characteristics, such as not altering the soil and having little or no need for irrigation, high productivity, and the ability to fix nitrogen and produce high quality protein. These authors add that it is possible to find plants with these characteristics only among legumes.

Different legume trees have been reported by various authors to grow in arid and semi-arid zones (Felker and Clark 1980; Habish and Khairi 1970; Khudairi 1957; Nakos 1977). *Acacia* and *Prosopis* are two examples. They grow rapidly, can be a good source of green manure, can restore and regenerate soils, and are very effective in preventing soil erosion.

Bacteria isolated from the root nodules of these plants are considered to be slow-growing *Rhizobium* (*Bradyrhizobium*) (Allen and Allen 1981), although different authors disagree with this conclusion (Habish and Khairi 1968; Trinick 1968). Dreyfus and Dommergues (1981) classified *Acacia* species into three groups according to the characteristics of the *Bradyrhizobium* isolated from their nodules. The first group includes species infected by slow-growers, the second group contains those infected by fast-growers, and the third are plants infected by both types of bacteria.

Prosopis spp. are infected by slow- and fast-growing *Bradyrhizobium* spp. which are promiscuous, i.e., they infect species of different genera of legume trees such as *Acacia*, *Desmanthus*, *Leucaena*, and *Mimosa* (Allen and Allen 1981).

According to Felker (1981), the ability to fix nitrogen is the characteristic that has allowed legume trees to survive in the low nitrogen levels of arid soils. It has also contributed to the evolution of drought tolerance.

Many of the valuable characteristics of these plants depend on the formation of an effective nitrogen-fixing symbiosis. In most cases, the effectiveness of the symbiosis is strongly related to the microsymbiont (Trinick 1968). Therefore, identification of ineffective and effective strains of *Rhizobium* is important.

We have tried to study the characteristics of the out-of-group nodulations of some legume trees and the implication of the lack of host specificity in the establishment of an effective symbiosis in these plants. In this study, the *Bradyrhizobium* GRH2–*P. chilensis* symbiosis was used as a model.

Materials and Methods

Microorganisms

Rhizobial strains isolated from the root nodules of different woody legumes growing in Chile and strains of *Rhizobium* belonging to the classic cross inoculation groups are listed in Table 1.

Table 1. Strains of *Rhizobium* and *Bradyrhizobium* isolated from woody legumes and of the classic cross-inoculation groups.

Strain	Isolated from
R. leguminosarum GRO24	*Vicia faba*
R. meliloti GR4	*Medicago sativa*
R. trifolii GRO17	*Trifolium repens*
R. phaseoli GRO12	*Phaseolus vulgaris*
Bradyrhizobium japonicum USDA110	*Glycine max*
Bradyrhizobium sp. "*Lupinus*" GR:46	*Lupinus albus*
Bradyrhizobium sp. (*Vigna*) CB756	*Vigna sinensis*
Bradyrhizobium sp. GRH1	*Acacia cyanophylla*
Bradyrhizobium sp. GRH2	*A. cyanophylla*
Bradyrhizobium sp. GRH3	*A. melanoxylon*
R. leguminosarum GRH5	*Prosopis chilensis*
Bradyrhizobium sp. GRH9	*Sophora microphylla*

Plant cultures

Short-term tests for nodulation and nitrogen fixation were carried out on plants growing in test tubes of different sizes according to Olivares *et al.* (1980). Long-term experiments were carried out using Leonard jars (Leonard 1943) according to Herrera *et al.* (1985), or pots containing unsterile soil.

Physiological assays

Nitrogenase activity was measured on detached root systems excised at the cotyledonary node according to Herrera *et al.* (1985). Dry weight of the whole plants was determined after drying at 80 °C for 24 h.

Results

All the rhizobial strains isolated from the root nodules grew fast except GRH1, which belongs to the slow-growing group.

Results of compatibility cross tests carried out between different bacteria and plant species are presented in Tables 2 and 3. Table 2 shows the results of inoculation of some legume trees with the bacteria shown in Table 1, with respect to nodulation and nitrogen fixation. It can be seen that these plants are infected, although not always effectively, by strains isolated from woody legumes, as well as by *Bradyrhizobium japonicum*, *Bradyrhizobium* sp. "*Lupinus*", and in the case of *P. chilensis* by *R. leguminosarum*. Table 3 shows nodulation and nitrogen fixation results of herbaceous legumes when they were inoculated with bacteria isolated from legume trees. These bacteria

Table 2. Cross inoculation test of three legume trees with different root-nodule bacteria.

Bacterial strain	*Acacia cyanophylla*	*Acacia melanoxylon*	*Prosopis chilensis*
R. meliloti GR4	O	O	O
R. leguminosarum GRO24	O	O	I
R. phaseoli GRO12	O	O	O
R. trifolii GRO17	O	O	O
Bradyrhizobium japonicum USDA110	I	I	I
Bradyrhizobium sp. (*Lupinus*) GRO46	I	I	O
Bradyrhizobium sp. GRH1	I	I	O
Bradyrhizobium sp. GRH2	E	E	E
Bradyrhizobium sp. GRH3	E	E	E
R. leguminosarum GRH5	I	I	I
Bradyrhizobium sp. GRH9	O	I	O
Bradyrhizobium sp. (*Vigna*) CB756	O	O	O

E = Effective nodulation; I = ineffective nodulation; O = no nodulation.

Table 3. Cross inoculation test of herbaceous legumes with bacterial strains isolated from root nodules of legume trees.

Bacterial strain	*Medicago sativa*	*Lupinus albus*	*Phaseolus vulgaris*	*Trifolium repens*	*Vicia sativa*	*Glycine max*	*Vigna sinensis*
GRH1	I	I	I	I	O	O	O
GRH2	I	I	E	I	I	O	O
GRH3	I	O	E	I	O	O	O
GRH5	E	O	E	O	O	O	O
GRH9	I	O	I	I	O	O	O

E = Effective nodulation; I = ineffective nodulation; O = no nodulation.

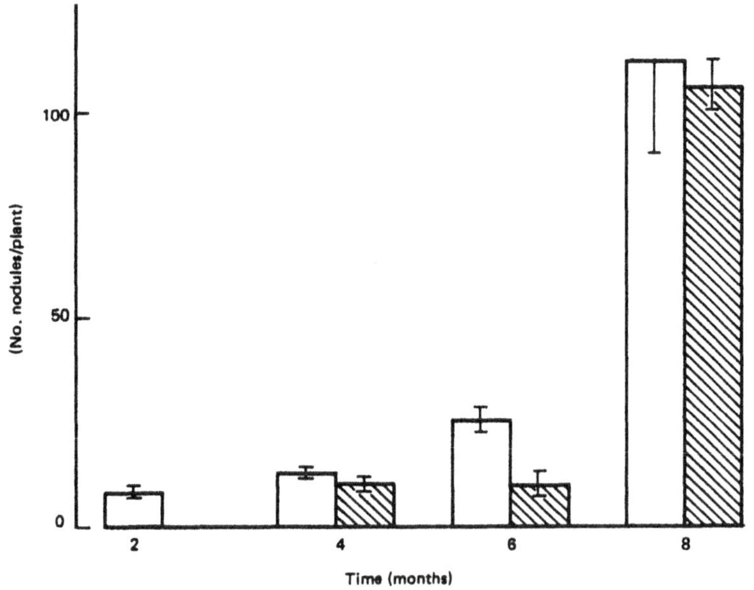

Fig. 1. Number of nodules on roots of *P. chilensis* plants grown in unsterile soil inoculated with *Rhizobium* sp GRH2 (unshaded columns) and uninoculated (hatched columns).

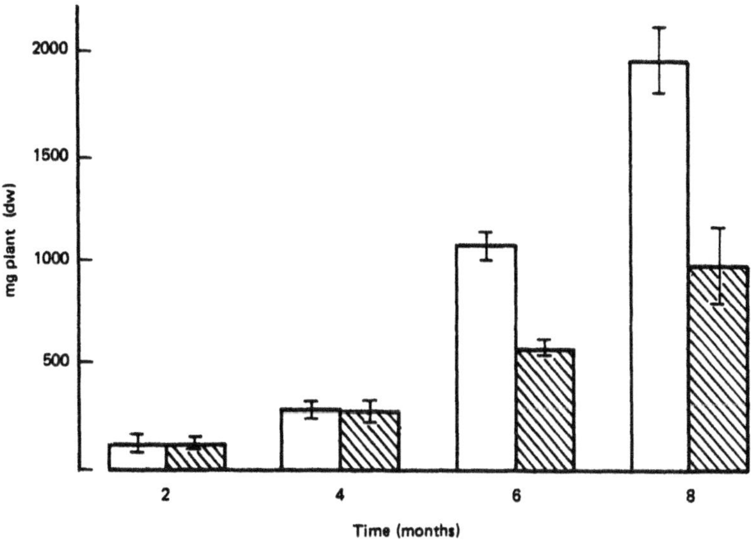

Fig. 2. Dry weight of *P. chilensis* plants grown on unsterile soil inoculated with *Rhizobium* sp GRH2 (unshaded columns) and inoculated (hatched columns).

displayed a broad host specificity, being able to nodulate almost all the plant species assayed.

In long-term inoculation experiments carried out on soil to study the *P. chilensis–Bradyrhizobium* GRH2 symbiosis, the number of nodules per plant at 2, 4, 6, and 8 months after sowing was recorded. The results are presented

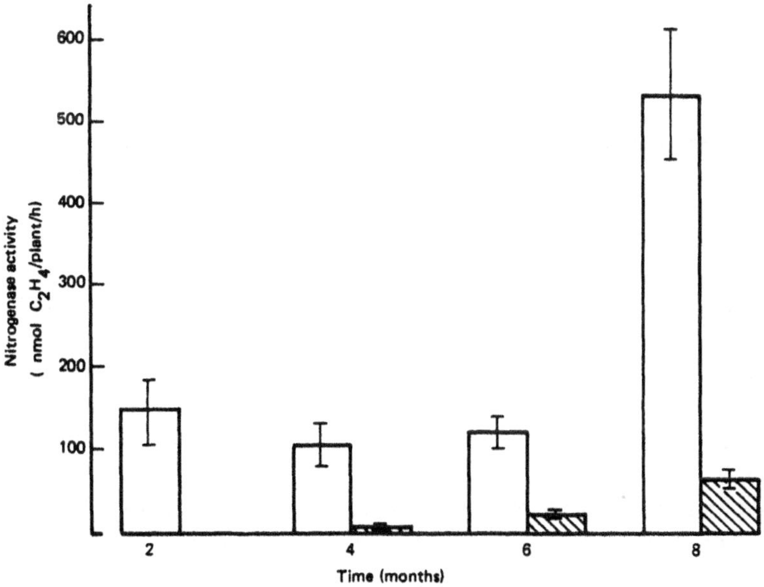

Fig. 3. Specific nitrogenase activity of *P. chilensis* plants grown on unsterile soil inoculated with *Rhizobium* sp GRH2 (unshaded columns) and uninoculated (hatched columns).

in Fig. 1. Uninoculated plants also nodulated due to infection by autochthonous *Bradyrhizobium*. In this case, nodules were not effective and the nodulation rate was slower, although after 8 months inoculated and uninoculated plants had a similar number of nodules. Dry weight data for these plants are presented in Fig. 2. Differences between inoculated and uninoculated plants were clear when plants were older than 4 months.

The corresponding nitrogenase activity values are shown in Fig. 3. A low level of enzyme activity was detected in uninoculated plants because of autochthonous *Bradyrhizobium* present in the soil.

Discussion

Previous work established that rhizobial strains isolated from woody legumes can infect other legume trees (Basak and Goyal 1980; Dreyfus and Dommergues 1981; Trinick 1982), and recently Herrera *et al.* (1985) reported that bacteria isolated from different legume trees can infect and nodulate many herbaceous legumes. A study of the plasmid patterns of some of these rhizobia and of the location of *nif* genes was recently carried out (Toro *et al.* 1984). All strains except GRH1 harbor 2–5 large plasmids and *nif* genes were located in one of these plasmids.

Table 2 shows that several strains of *Rhizobium* and *Bradyrhizobium* species isolated from herbaceous legumes can infect *P. chilensis*, among them *R. leguminosarum*, which is frequently present in soils. It is surprising that *P. chilensis* is nodulated by this bacterium and not by *R. phaseoli* or *R. trifolii*, which are genetically more closely related to it (Beringer 1980). These three *Rhizobium* species under current classification are considered to be biovarieties of only one bacterium, according to genetic analysis. The susceptibility of this *P. chilensis* to infection in such a promiscuous way represents an obstacle in attaining good levels of nitrogen fixation in field culture. A variety of nodule forms have been reported when *P. chilensis* is inoculated with *Bradyrhizobium* sp. (*Vigna*) 176A32 (Virginia *et al.* 1984; Baird *et al.* 1985).

The inoculation of *P. chilensis* with a specific and effective strain of *Rhizobium* increases plant growth, expressed as dry matter weight, by 100% over that in uninoculated plants (Fig. 2). Ineffective rhizobial strains present in the soil also determine the infection of roots and the formation of ineffective nodules (Fig. 1). Ineffective bacteria compete with effective ones (see the low nitrogenase activity values in Fig. 3), hence the necessity of inoculating with an appropriate bacterial strain. Rhizobial strains to be used for the preparation of inocula must be selected according to the host and their respective capacities to adapt themselves to the environmental conditions in which the chosen legume tree will be grown.

The results for the *Bradyrhizobium* GRH2–*p. chilensis* symbiosis reported here are expected to be similar where other legume trees are involved.

Acknowledgement

This work was supported by La Comision Asesora de Investigacion Cientifica y Tecnica, Grant No 1764/82.

References

Allen, O.N. and Allen, E.K. 1981. The Leguminosae: A Source Book of Characteristics, Uses and Nodulation. University of Wisconsin Press, Madison, Wisconsin, USA.

Baird, L.M., Virginia, R.A. and Webster, B.D. 1985. Development of root nodules in a woody legume, *Prosopis glandulosa* Torr. Botanical Gazette 146: 39-43.

Basak, M.K. and Goyal S.K. 1980. Studies on tree legumes. II. Further additions to the list of nodulating tree legumes. Plant and Soil 56: 33-37.

Ben Salem, B. 1980. Forestry in arid zones. Unasylva 32: 16-18.

Beringer, J.E. 1980. The development of *Rhizobium* genetics. Journal of General Microbiology 116: 1-7.

Dreyfus, B.L. and Dommergues, Y. 1981. Nodulation of *Acacia* species by fast- and slow-growing tropical strains of *Rhizobium*. Applied and Environmental Microbiology 41: 97-99.

Felker P. 1981. Uses of tree legumes in semi-arid regions. Economic Botany 35: 174-186.

Felker, P. and Bandurski, R.S. 1979. Uses and potential uses of leguminous trees for minimal energy input agriculture. Economic Botany 33: 172-184.

Felker, P. and Clark, P.R. 1980. Nitrogen fixation (acetylene reduction) and cross inoculation in 12 *Prosopis* (mesquite) species. Plant and Soil 57: 177-186.

Habish, H.A. and Khairi, S.M. 1968. Nodulation of legumes in the Sudan: cross-inoculation groups and the associated *Rhizobium* strains. Experimental Agriculture 4: 277-284.

Habish, H.A. and Khairi, S.M. 1970. Nodulation of legumes in the Sudan. II. Rhizobia strains and cross inoculation of *Acacia* spp. Experimental Agriculture 6: 171-176.

Herrera, M.A., Bedmar, E.J. and Olivares, J. 1985. Host specificity of *Rhizobium* strains isolated from nitrogen-fixing trees and nitrogenase activities of strain GRH2 in symbiosis with *Prosopis chilensis*. Plant Science 42: 177-182.

Khudairi, A.K. 1957. Root-nodule bacteria of *Prosopis stephaniana*. Science 125: 399.

Leonard, L.T. 1943. Method of testing bacterial cultures and results of tests of commercial inoculants. United States Department of Agriculture, Circular Number 703. Washington DC, USA.

Nakos, G. 1977. Acetylene reduction (N{2}-fixation) by nodules of *Acacia cyanophylla*. Soil Biology and Biochemistry 9: 131-133.

Olivares, J., Casadesus, J. and Bedmar, E.J. 1980. Method for testing degree of infectivity of *Rhizobium meliloti* strains. Applied and Environmental Microbiology 39: 967-970.

Spears, J.S. 1980. May agriculture and forestry coexist in tropical regions? Unasylva 32: 2-12.

Toro, N., Herrera, M. A. and Olivares, J. 1984. Location of *nif* genes on large plasmids in *Rhizobium* strains isolated from legume tree root nodules. FEMS Microbiology Letters 24: 113-115.

Trinick, M.J. 1968. Nodulation of tropical legumes. I. Specificity in the *Rhizobium* symbiosis of *Leucaena leucocephala*. Experimental Agriculture 4: 243-253.

Trinick, M.J. 1982. Biology. Pages 76-146 *in*: Nitrogen Fixation. Vol. 2. *Rhizobium* (Broughton, W.J., ed.). Claredon Press, Oxford, England.

Virginia, R.A., Baird, L.M., Lafavre, J.S., Jarrell, W.M., Bryan, B.A. and Shearer, G. 1984. Nitrogen fixation efficiency, natural }15{N abundance, and morphology of mesquite (*Prosopis glandulosa*) root nodules. Plant and Soil 79: 273-284.

Discussion

Broughton

In your experiments to demonstrate plasmids in your isolates, which radioactive probe did you use?

Olivares

pID1 of *Rhizobium meliloti* (*nif* KDH).

Broughton

Did all the fast-growing isolates contain symbiotic plasmids?

Olivares

Yes, but not the single slow-growing isolate (GRH1).

Broughton

Did you test the isolates for free-living nitrogen fixation activity and hydrogen uptake?

Olivares

Yes, we did. They do not fix nitrogen in the free-living stage and they do not show hydrogenase activity.

D.P. Beck and L.A. Materon (editors), Nitrogen Fixation by Legumes in Mediterranean Agriculture
© 1988 ICARDA – Printed in the Netherlands. ISBN 90 247 3624 2

Biological Nitrogen Fixation Research on Grain Legumes in Ethiopia – An Overview

DESTA BEYENE

Department of Soil Science, Institute of Agricultural Research, P.O. Box 2003, Addis Ababa, Ethiopia

Summary. Food legumes are extensively grown in Ethiopia and constitute a major part of the diet of rural and urban populations. In Ethiopia, research on grain legumes has concentrated mainly on improving yields through selection, breeding, and the use of improved agronomic practices. In this paper, past work on the biological nitrogen fixation of grain legumes is reviewed and the need for further, intensive research stressed. Particular emphasis is placed on the introduction of effective *Rhizobium* strains into productive crop systems.

Introduction

Ethiopia is located between 3° and 18° north and 33° and 48° cast. It has a total land area of 122 million ha, of which 84 million ha are suitable for agricultural production. Currently, only 13 million ha are cultivated and food legumes, which are a major part of the diet of rural and urban populations, occupy about 13% of this area.

Agriculture is the mainstay of the Ethiopian economy, providing roughly 56% of the gross national product. Despite this, the majority of farmers still practice traditional production methods and, as a result, grain yields for most of the pulse crops are very low. Commercial fertilizers have helped to improve yields of some pulse crops. However, supplies are limited and the soaring prices of commercial fertilizers in recent years have seriously limited the production potential of grain legumes as many farmers are reluctant to bear the increased cost. Farmers are now looking for alternatives to these fertilizers.

Biological nitrogen fixation (BNF) is an area of research that could be exploited as a source of nitrogen to replace inorganic nitrogen fertilizers. It is well established that the symbiotic relationship between a legume and rhizobia helps to improve the nitrogen status of a soil.

To date, very limited information has been available on BNF research in Ethiopia. This paper, therefore, tries to review the present state of BNF research on grain legumes and stresses the need for more intensive research in this field.

Importance and Distribution of Grain Legumes

Grain legumes are an important source of protein and are the daily food for a majority of the Ethiopian population. The seeds from pulses are mostly boiled and used as part of a vegetable salad, soup, or snack, but they are also eaten fresh as green seed, and the dry seed is used in the preparation of local dishes such as 'Shiro wot' and 'Kit wot'.

Pulses provide a large cash income for farmers and earn foreign exchange for the country. In 1978, Ethiopian export of the various pulse crops was 25409 tons with a value of US$8.6 million.

The four major food legumes commonly grown in Ethiopia are faba bean (*Vicia faba*), chickpea (*Cicer arietinum*), lentil (*Lens culinaris*), and field pea (*Pisum sativum*). Haricot bean (*Phaseolus vulgaris*) and soybean (*Glycine max*) are grown in a few regions in the Rift Valley area, while the major pulses are extensively grown in the central highlands (Fig. 1). Faba bean is the most extensively grown pulse crop. The total area under cultivation, the total

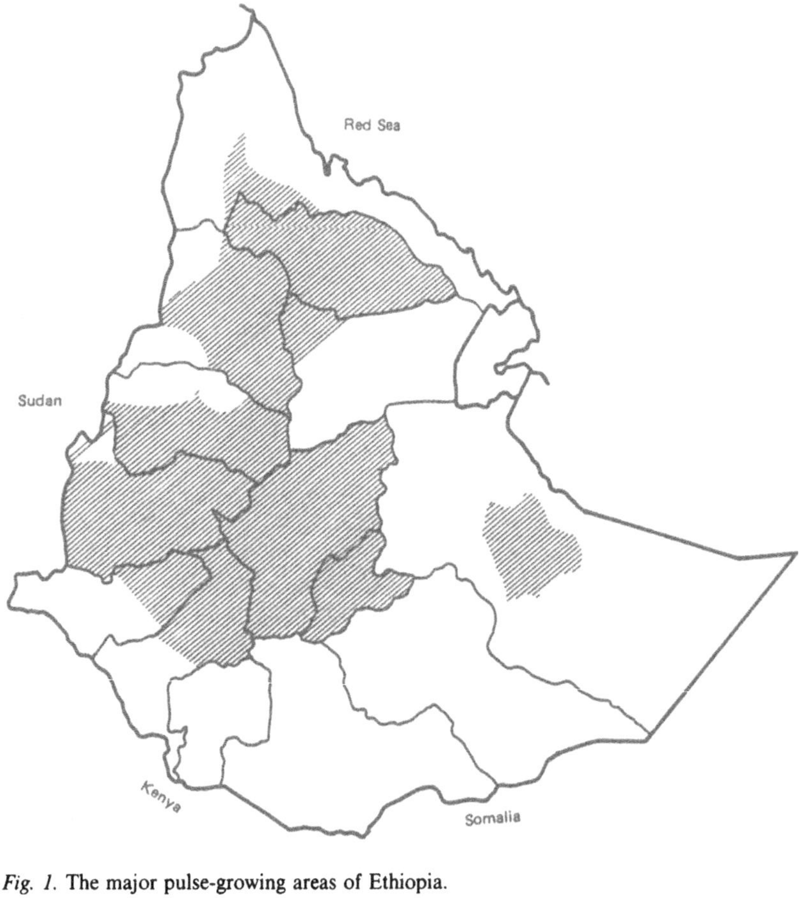

Fig. 1. The major pulse-growing areas of Ethiopia.

Table 1. Area (×1000 ha), average yield (q/ha), and total production (×1000 t)of the major pulse crops in Ethiopia (1978-84) *.

Crop	Year						
	1978	1979	1980	1981	1982	1983	1984
Faba beans							
Area	295.9	304.8	306.5	348.9	380.6	345.6	320.1
Yield	9.0	9.4	15.3	13.5	17.1	11.1	8.3
Prod.	2650.8	2873.9	4699.3	4699.4	5998.6	3848.0	2615.6
Chickpeas							
Area	141.6	148.7	149.1	133.1	129.0	163.6	149.6
Yield	5.5	5.7	7.9	7.6	9.2	8.4	6.2
Prod.	771.9	846.3	1182.9	1011.2	1181.1	1049.4	818.1
Field peas							
Area	127.3	133.7	169.0	171.3	147.7	118.1	139.5
Yield	6.9	7.1	8.7	9.4	8.0	8.9	3.9
Prod.	874.1	953.0	1476.8	1615.0	1319.5	1163.7	899.1
Lentils							
Area	58.5	58.5	53.3	69.1	54.5	38.6	143.0
Yield	4.6	5.2	11.2	7.2	5.2	8.2	3.3
Prod.	269.7	306.6	599.0	497.0	411.2	239.9	143.0

* Tilaye (1985).

production, and the average yield of each of the four major crops are shown in Table 1.

Production Constraints

The major factors contributing to poor yields for most of the pulse crops are:

(a) diseases, insects, and weeds,
(b) poor cultural practices,
(c) poor nodulation and fruit setting,
(d) poor soil fertility status (acidity, low P, and salinity),
(e) lodging,
(f) frost and hail damage, and
(g) lack of high-yielding, stable cultivars.

Intensive research on crop improvement and soil management is needed to overcome these constraints.

Status of Research on Grain Legumes

Breeding and agronomy

In Ethiopia research on pulses was started by the Institute of Agricultural Research (IAR) only in the last decade. Since then, numerous variety and

Table 2. The number of pulse varieties released in Ethiopia, with optimum sowing dates and plant spacings.

Crop	Number of released varieties	Sowing date	Spacing (cm)
Faba bean	1	5 – 15 June	40× 5
Field pea	1	1 – 10 July	20× 5
Haricot bean	2	20 – 30 July	40×15
Soybean	4	10 – 20 June	40× 5
Chickpea	2	25 Aug-8 Sept	30×10
Lentil	2	28 June-10 July	20× 5

selection trials and agronomic studies have been carried out, aiming to improve the crop yields and disease tolerance of the major pulse crops. The results of these trials have been reported in different IAR publications (Tilaye 1985; IAR 1984b; Ohlander 1980). The number of varieties released and cultural practices for each crop are shown in Table 2.

Soil Fertility Studies

Fertilizer trials on pulses carried out at various locations are reviewed and reported elsewhere (Beyene 1986; IAR 1984a). Generally, a fertilizer rate of 10 kg N and 20 kg P/ha is adequate for reasonable yields in most soils.

Rhizobial Inoculation Studies

Some inoculation trials have been carried out at IAR Centers on a range of pulse species. For most pulses, there was a general tendency for nodule size

Table 3. The effect of inoculant and fertilizer application on the yield of faba beans at Holetta, 1981.

Inoculant and fertilizer treatment	Grain yield (t/ha)	Straw yield (t/ha)	Average no. of nodules/plant		Total weight of nodules/plant	
			4 weeks from emergence	at 50% flowering	4 weeks from emergence	at 50% flowering
Control	0.38	0.78	1.07	3.7	2	8
26 kg P/ha	0.70	1.67	0.93	5.3	1	15
60 kg N/ha	1.09	2.37	very small	very small		
Inoculant	0.63	1.32	16.33	48.2	20	122
Inoculant + 26 kg P/ha	1.05	2.30	37.0	86.8	42	291
Inoculant + 60 kg N/ha	1.11	2.33	8.93	28.3	25	60
Inoculant + 60 kg N/ha + 25 kg P/ha	1.97	1.41	33.07	90.2	22	184

and number, and grain yield to increase with rhizobial inoculation (Beyene 1985; Sertsu and Beyene 1986). At Holetta, inoculant application significantly increased faba bean yield and the number and weight of nodules (Table 3), while at other locations, the response to inoculation was high but not significant. Generally, the response of most pulses to seed inoculation was influenced by location variables.

Research Thrust on Biological Nitrogen Fixation

Simple field trials on rhizobial inoculation cannot provide adequate information on the extent of nitrogen fixation in grain legumes. More intensive investigation is required to assess actual bacterial activity and to estimate the level of nitrogen fixation by the different grain legumes. Therefore the establishment of a national BNF research program has been given paramount attention.

The BNF research has the following objectives:

- Collection of nodules from the main food grain legumes grown in Ethiopia.
- Isolation and testing of strains of *Rhizobium* from collected nodules.
- Comparison of the effectiveness of local *Rhizobium* strains with imported inoculants.
- Conducting relevant greenhouse and field experiments to determine the quantity of nitrogen fixed by pulse crops. The acetylene reduction assay and ^{15}N technique will be used to measure nitrogen fixation.
- Production of inoculants from the rhizobial strains selected to be effective and efficient in their ability to fix nitrogen.
- Training researchers from national programs in BNF research.
- Developing closer contact with international institutions involved in BNF research.

To realize these objectives, the Department of Soil Science of the Institute of Agricultural Research has established a soil microbiology unit at the Holetta Agricultural Research Center. This unit will be responsible for directing and coordinating all the research on BNF in Ethiopia. Efforts will be made to set up the necessary working facilities for BNF research at the Center. Links have already been established with the International Atomic Energy Agency (IAEA) in Austria, ICARDA in Syria, and NifTAL in the USA for possible financial and technical support.

Looking Ahead

Studies at Holetta and Nazret Research Centers illustrate the fact that there is considerable scope for increasing food legume yields through improved nitrogen fixation. As relatively little attention has previously been devoted to studying BNF in grain legumes, intensive investigations on nitrogen fixation have now been initiated. Hopefully these investigations will act as a catalyst to accelerate efforts on BNF research in Ethiopia.

Collection of strains of *Rhizobium* from the major pulse crops started recently (Abebe 1986) and will be intensified. Different rhizobial strains of the economically important pulses in Ethiopia have been isolated from nodules of host legumes grown in the major pulse producing areas. These include: 110 isolates of *Rhizobium leguminosarum* from *P. sativum*, *V. faba*, and *L. culinaris*; 20 isolates of *R. phaseoli* from *P. vulgaris*; 328 isolates of *R. japonicum* and cowpea-type *Rhizobium* sp. from *G. max* and *Cicer arietinum*; and 34 isolates of *R. trifolii* from *Trifolium* spp., all of which were collected by the Nazret Research Station microbiology laboratory. Work towards identification of the most suitable rhizobial strains for Ethiopian conditions will be intensified, as will efforts to test appropriate inoculants from neighboring countries.

Many African countries are now seriously involved in BNF research. In fact, countries like Kenya, Egypt, Zimbabwe, Rwanda, and Nigeria have already accumulated the necessary information to be able to produce their own commercial inoculants locally. The prospect for BNF research in Ethiopia is very bright, especially with the initiation of the ICARDA/IFAD Nile Valley Project. Therefore, we expect it will not be very long before inoculants can be produced locally at a commercial level.

In the future, the BNF research program will be extended to include research on forage legumes and will devote an increasing proportion of its effort to more detailed studies on nitrogen fixation and inoculant production.

References

Abebe, A. 1986. Culture collection of *Rhizobium* strains of important pulses of Ethiopia. Paper presented at the Workshop on the Review of Soil Science Research in Ethiopia, 11-14 Feb 1986. Addis Ababa, Ethiopia.

Beyene, D. 1985. The state of research on biological nitrogen fixation in Ethiopia. Pages 72-84 *in* Biological Nitrogen Fixation in Africa: Proceedings of the First Conference of the African Association for Biological Nitrogen Fixation (Ssali, H. and Webb, C., eds.). 23-27 July 1984, Nairobi, Kenya.

Beyene, D. 1986. The response of pulse crops to N and P fertilizers. Paper presented at the Workshop on the Review of Soil Science Research in Ethiopia, 11-14 Feb 1986, Addis Ababa, Ethiopia.

IAR (Institute of Agricultural Research). 1984a. Soil Science Department Progress Reports for the Periods 1980/81 and 1981/82. IAR, Addis Ababa, Ethiopia.

IAR (Institute of Agricultural Research). 1984b. Department of Field Crops Progress Report of Highland Pulse Team 1982/83. IAR, Addis Ababa, Ethiopia.

Ohlander, L.J.R. 1980. Research on haricot bean production in Ethiopia 1972-1976. Swedish University of Agricultural Sciences, Uppsala, Sweden.

Sertsu, S. and Beyene, D. 1986. Review of rhizobial inoculation trials on different pulse crops. Paper presented at the Workshop on the Review of Soil Science Research in Ethiopia, 11-14 Feb 1986, Addis Ababa, Ethiopia.

Tilaye, A. 1985. Review of back-up research on faba beans in Ethiopia. Paper presented at ICARDA/IFAD Nile Valley Project Sixth Annual Coordination Meeting, 9-13 Sept 1985, Cairo, Egypt. ICARDA, Aleppo, Syria.

D.P. Beck and L.A. Materon (editors), Nitrogen Fixation by Legumes in Mediterranean Agriculture
 ISBN 90 247 3624 2

The Status of Biological Nitrogen Fixation Research in Lebanon

MAHMOUD B. SOLH

American University of Beirut, Lebanon

Summary. Although the importance of food and forage legumes in Lebanon is now minor compared to fruit and vegetables, they have been grown here since the beginning of agriculture. Demand is now increasing and efforts are being made to increase production. This paper reports the various aspects of legume research carried out by the American University of Beirut and ICARDA. It includes studies on the effects of inoculation, fertilization, pea weevil, delayed leaf senescence and maintenance of nodule activity beyond pod maturity, and compatibility studies between inocula and pesticides.

Introduction

Food and forage legume crops grown in Lebanon cover about 15000 ha which is 7.5% of the area under agricultural production in an average year. In order of importance, food legume crops include lentils (*Lens culinaris* Medik), chickpeas (*Cicer arietinum* L.), phaseolus beans (*Phaseolus vulgaris*), faba beans (*Vicia faba*), peas (*Pisum sativum*), peanuts (*Arachis hypogaea*), lima bean (*Phaseolus lunatus* L.), and lupin (*Lupinus albus*). Forage legumes include common vetch (*Vicia sativa*), alfalfa (*Medicago sativa*), and berseem (*Trifolium alexandrinum*). Although several of these crops originated elsewhere, most are considered to be endogenous to Lebanon, and lentils, chickpeas, peas, faba beans, and vetch have probably been grown in Lebanon since the beginning of agriculture.

The role of legume crops in the Lebanese agricultural economy is minor compared to that of the high value fruits and vegetables. However, a number of legumes are important in dryland farming, which covers about 75% of the total cultivable land. Production constraints have decreased food legume production which, along with increasing demand, reduced the self-sufficiency ratio from over 60% in the early 1970's to less than 20% in the early 1980's (FAO 1977-83a; 1977-83b). Promotion of modern technology in handling these crops should reduce the gap between supply and demand.

Research

In spite of the ancient history of legume crops in Lebanon, research on nitrogen fixation is limited, and there is no national research program on nitrogen fixation. The collapsed infrastructure of the national research institutes and the lack of qualified personnel have contributed to this deficiency. Though research efforts are fragmented and lack continuity, they may be grouped into the following main areas:

- The effect of inoculation and fertilization on yield, nodulation, and other plant characters.
- The effect of pea weevil (*Sitona* spp.) damage on nodulation and yield of lentil.
- Delayed leaf senescence in soybean and maintenance of nodule activity beyond pod maturity.
- Compatibility studies between inoculum and pesticides.

The institutions involved in this research are the Faculty of Agricultural and Food Sciences of the American University of Beirut (FAFS-AUB) and the International Center for Agricultural Research in the Dry Areas (ICARDA). The two institutions cooperate in joint research on lentils, chickpeas, and faba beans. Some of the field trials and most of the laboratory studies reported in this paper were carried out at ICARDA.

Studies on the Effect of Inoculation and Fertilization on Yield, Nodulation, and Other Plant Characters

Field studies on inoculation and fertilization were done on chickpeas, lentils, lupins, and soybeans in the Beqa'a Plain, where agronomic crops are grown on a relatively large scale. In all these studies, exogenous specific rhizobia were used. In both chickpeas and lentils, inoculation did not increase yield although nodule dry weight was increased significantly in the former (Shad 1975; Ahmad 1976; Islam *et al.* 1986). Apparently the exogenous rhizobia used were not more efficient than the endogenous rhizobia in the local soil. Good nodulation was observed in uninoculated chickpeas, lentils, and faba beans grown in farmers' fields in the Beqa'a Plain where these crops are traditionally grown. Fertilization with 50 kg N/ha had no effect on chickpea and lentil yields although nodule dry weight was significantly reduced in chickpeas. In lentils, fertilization with 100 kg P_2O_5/ha also had no effect on nodulation or yield in the Terbol area where the soil is very fertile.

In lupins, inoculation with commercial strains of *Rhizobium lupini* produced significant increases in seed yield, primary yield components, protein content, and dry weight of nodules (Iqbal 1975). Inoculation with 4.18 g and 20.90 g of commercial inoculant per kilogram of seed increased seed yield by 45 and 75% in the two cultivars Local and Giza II, respectively. Protein content was also increased by 30-34% and 35-38%, respectively. Yield differences between the two inoculation rates were highly significant in Giza II only. The application of 130 and 210 kg N/ha had no effect on seed yield or other plant characters.

Although soybean is not grown commercially in Lebanon, research is carried out due to its increasing importance as an industrial crop in the Near East region. Seed inoculation, using commercial strains of *R. japonicum*, significantly increased yield in two cultivars, Elf and Williams, at research sites where soybeans had not previously been grown (Solh *et al.* 1986a). Apparently, strains of *R. japonicum* are absent in local soils. Supplemental nitrogen (50 kg N/ha) applied as urea at planting and flowering (side-dressed and foliar), had no significant effect on agronomic parameters at two population densities (300000 and 450000 plants/ha). However, an earlier study on the effect of nitrogen and micronutrients on nodulation and seed of soybean using granulated inoculum (13.5 kg/ha) showed that 50 kg N/ha applied before planting resulted in highly significant increases (30%) in seed yield to 5045 kg/ha, seed protein, 100-seed weight, and nodule dry weight. However, 50 kg N/ha applied 4 weeks after planting, and 50 kg N/ha applied preplant plus 100 kg N/ha at the late bloom stage, had no effect on yield. Foliar application of 1.90 kg Fe plus 2.16 kg Mn/ha and 54g Mo/ha improved nodulation, seed yield, and protein content (Khan 1976).

From these studies on nodulation and fertilization it is apparent that, being endogenous to the Beqa'a Plain, both lentils and chickpeas showed no response to inoculation, indicating that native rhizobial strains were at least as effective as the exogenous commercial rhizobial strains. However, both lupin and soybeans, which are not grown in the Beqa'a Plain, responded positively to inoculation. Nitrogen fertilization generally had no effect on these crops except in one study on soybean when nitrogen as a starter increased yield. Iron, manganese, and molybdenum had positive effects on soybeans.

Studies on the Effect of Pea Weevil (*Sitona* spp.) on Nodulation and Yield of Lentils

Pea weevil (*Sitona* spp.) is a serious pest on lentils in Lebanon. The adult feeds on the leaves of seedlings while the larvae attack the nitrogen fixing nodules, which greatly reduces the nitrogen fixing ability of the plant. Larval feeding is confined to the bacteria present in the nodules.

In field studies in the Beqa'a Plain at two locations with different soil fertility, nodule damage by *Sitona* weevil was 69.1-81.8% in a local cultivar and two adapted introductions (Solh *et al.* 1986b). In the location with low soil fertility, treatment with carbofuran reduced nodule damage in the local cultivar and in one introduction to 6.9 and 9.0%, respectively and increased yield by 45% (Tables 1 and 2). The yield of Lebanese local was not affected in spite of the drastic increase in the number of normal nodules per plant as a result of carbofuran treatment. It seems that the few intact nodules (about 20% of the total) in Lebanese local were able to compensate for the damaged nodules. These results indicate genotypic interaction to different levels of nodulation.

Mineral nitrogen fertilization at 80 kg N/ha failed to compensate for the

Table 1. Normal nodules per plant (%) of two microsperma lentil cultivars as affected by nitrogen and insecticides at three sampling dates (Solh *et al.* 1986b).

Treatment	April 8		April 29		May 20	
	Lebanese local	Syrian local	Lebanese local	Syrian local	Lebanese local	Syrian local
Control	50.7b	42.8c	22.6b	29.5b	18.2c	20.4b
Nitrogen (80 kg/ha)	60.5b	58.4bc	29.7b	38.0b	28.0c	22.9b
Carbofuran (3 kg a.i./ha)	89.4a	92.5a	92.5a	82.7a	93.1a	91.0a
Aldicarb (3 kg a.i./ha)	85.3a	83.3a	81.4a	69.4a	64.6b	74.6a

Values for each sampling date followed by the same letter are not significantly different at P = 0.05.

Table 2. Grain yield, biological yield and harvest index of two microsperma lentil cultivars as affected by nitrogen and insecticides (Solh *et al.* 1986b).

Treatment	Grain yield (t/ha)		Biological yield (t/ha)		Harvest index	
	Lebanese local	Syrian local	Lebanese local	Syrian local	Lebanese local	Syrian local
Control	1.49a	1.17b	5.00a	4.61a	0.29a	0.25b
Nitrogen (80 kg/ha)	1.45a	1.51ab	4.75a	5.23a	0.29a	0.28b
Carbofuran (3 kg a.i./ha)	1.55a	2.16a	5.11a	5.21a	0.30a	0.41a
Aldicarb (3 kg a.i./ha)	1.64a	1.76ab	5.18a	5.06a	0.30a	0.34ab

Values for each sampling date followed by the same letter are not significantly different at P = 0.05.

fixed nitrogen lost in the introduced line as a result of nodule damage. The lack of response in some lentil cultivars to nitrogen fertilization may be explained by the fact that lentil landraces, such as those used in these studies, have evolved over hundreds of years on marginal lands of low fertility. At the second location, the lack of response to reduced nodule damage by carbofuran treatment and nitrogen fertilization was attributed to the high soil fertility (Islam *et al.* 1986).

In Lebanon, lentils are grown on marginal land with low fertility, so reducing nodule damage may result in considerable yield increases in certain genotypes. However, carbofuran and other effective chemicals are not economical for lentils due to the generally low yields and the financial status of small farmers. Thus there is a need to develop more economical means to reduce nodule damage, such as breeding resistant cultivars.

Delayed Leaf Senescence in Soybean and Maintenance of Nodule Activity beyond Pod Maturity

Delayed leaf senescence in soybean can extend nitrogen fixation activity beyond the pod-filling stage (Abu-Shakra *et al.* 1978). Thus, more nitrogen is

Table 3. Seasonal nitrogen fixation, specific [1] and total [2] nodule activity, and nodule dry weight [3] for DLS6 and Corsoy soybeans (Habib 1979).

Days after emergence	DLS-6			Remarks	Corsoy			Remarks
	Specific nodule activity	Total nodule activity	Nodule dry weight		Specific nodule activity	Total nodule activity	Nodule dry weight	
28	58.75	47.41	0.81	Initial bloom	59.92	40.81	0.68	Initial bloom
35	27.81	29.09	1.08	Full bloom, terminal growth stopped, multiple branching started	28.45	31.42	1.11	Early podding, full bloom
42	19.19	30.30	1.75	Early pod stage, multiple branching at most nodes	11.61	20.16	1.76	Mid-pod stage
49	11.57	26.16	2.25	Podding and branching continued	9.07	20.90	2.28	Late pod, early seed development
56	7.47	24.67	2.96	Pod and early seed development	5.94	16.34	2.70	Second seed development
63	ND	ND	3.59	Pod and seed development	7.74	16.17	3.10	Rapid seed-fill
70	6.03	22.06	3.47	Pod and seed development	4.59	10.80	2.34	Rapid seed-fill
77	4.35	16.08	4.21	Pod and seed development	0.51	0.75	1.33	Physiological seed maturity
84	1.96	7.83	4.00	Seed-fill and seed maturity	0.0	0.0	ND	Seed maturity
98	2.80	8.45	3.52	Seed-fill and seed maturity				
112	1.47	5.02	3.35	Mature seeds				

[1] umoles C_2H_4/g dry nodule × hr
[2] umoles C_2H_4/plant × hr
[3] g/plant.
ND = No data

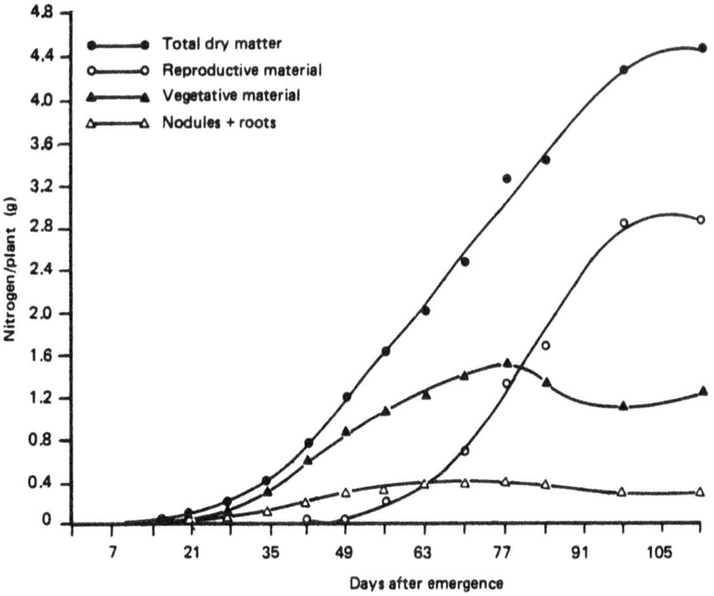

Fig. 1. Seasonal nitrogen accumulation in the various plant components of nodulated DLS-6 soybeans (Habib 1979).

available to plants at this critical stage. However, lines with delayed leaf senescence (DLS) introduced to Lebanon had lower yields than the adapted cultivar Corsoy, although these lines maintained nodule activity far beyond the pod maturity stage (Table 3; Habib 1979).

In a study on nitrogen distribution during plant ontogeny (Habib 1979), DLS-6 accumulated larger quantities of nitrogen and dry matter per plant than did Corsoy (Figs 1 and 2). Also, DLS-6 maintained relatively constant levels of nitrogen and dry matter in its vegetative parts during seed development and maturation. In Corsoy, there was a sharp decline in the nitrogen and dry matter contents of these parts during the reproductive period, due to translocation of food materials from the vegetative parts to the developing seeds. This more efficient partitioning in Corsoy apparently leads to signifi-

Table 4. Percent heterosis for yield over mid- and high- parent values in three soybean crosses(Dewan 1981).

Crosses	Mid-parent value	High-parent value	Hybrid value	Percent heterosis over mid-parent	Percent heterosis over high parent
DLS/WI	35.0	53.9	63.1	80.47 *	17.08
DLS/CL	29.1	42.2	75.4	159.10 *	33.23 *
DLS/CO	32.0	48.3	65.2	33.19 *	16.86

DLS = DLS-6, WI = Williams, CL = Clark-63, CO = Corsoy.
* Significant at P = 0.01.

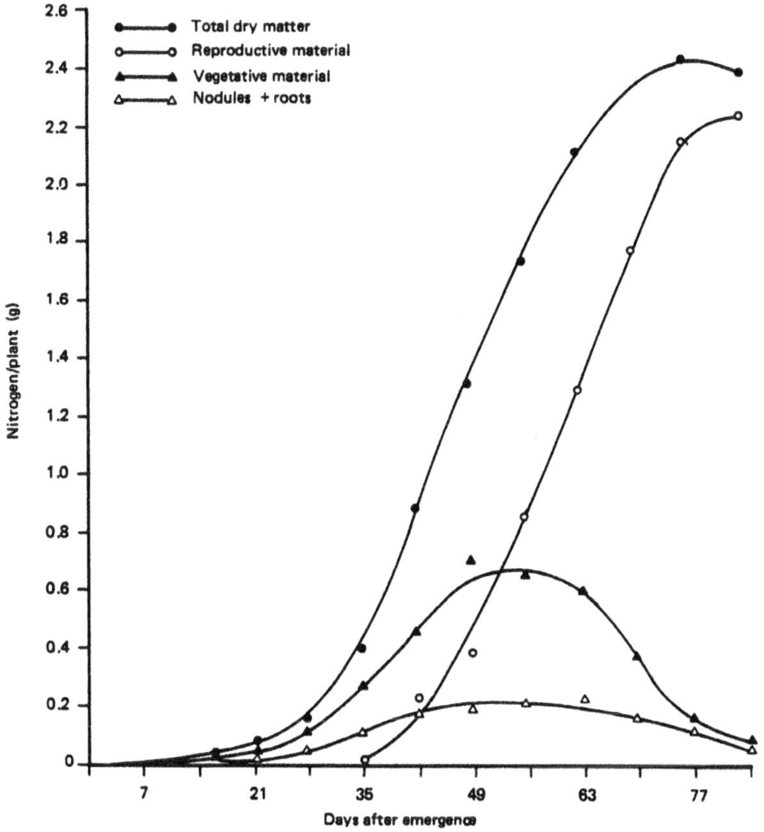

Fig. 2. Seasonal nitrogen accumulation in the various plant components of nodulated Corsoy soybeans (Habib 1979).

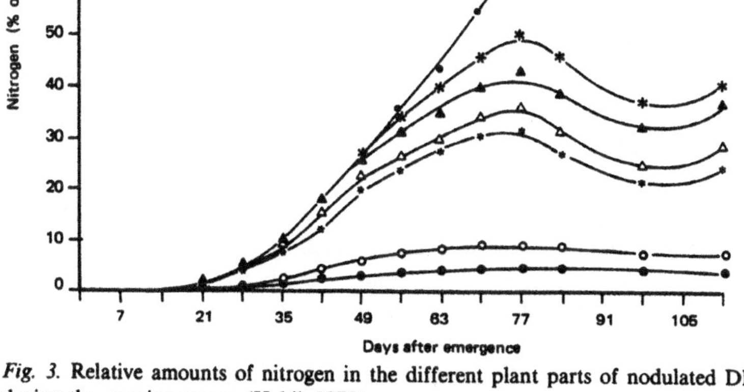

Fig. 3. Relative amounts of nitrogen in the different plant parts of nodulated DLS-6 soybeans during the growing season (Habib 1979).

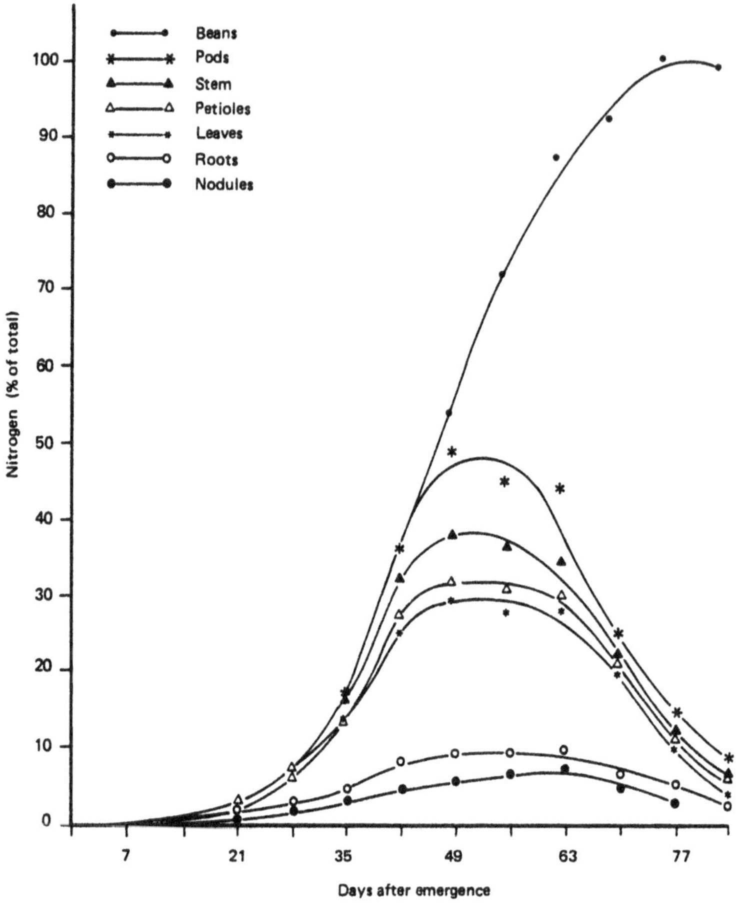

Fig. 4. Relative amounts of nitrogen in the different plant parts of nodulated Corsoy soybeans during the growing season (Habib 1979).

cantly higher seed yield per plant (Figs 3 and 4). However, the low yield of DLS-6 was attributed to a poor yielding and unadapted genetic background which masks the advantage of the delayed leaf senescence character (Habib 1979). Therefore, a study was initiated in 1980 to transfer this character to a high-yielding genetic background adapted to the Beqa'a Plain conditions. DLS-6 was crossed with three adapted cultivars (Clark-63, Corsoy, and Williams) to study combining ability and partition genetic variance, and to recommend an effective breeding methodology (Dewan 1981). Hybrid vigor was apparent in these crosses (Table 4) and crosses of DLS-6 with Clark 63 and Corsoy were promising. Both general and specific combining ability were important and thus family and line selection methods were recommended to develop cultivars with high yield and delayed leaf senescence. Unfortunately, research was interrupted when F_4 selections were lost in 1982. Efforts were resumed last season. Other work at the University of California at Davis on delayed leaf senescence of soybeans supports the hypothesis that maximum yields can only be achieved with plants whose leaves senesce during reproductive development (Phillips *et al.* 1984).

Compatibility Studies between Inoculum and Pesticides

Limited studies on compatibility between inoculum and herbicides, fungicides, and insecticides have been done under field, greenhouse, and laboratory conditions. Work on a few fungicides and insecticides is still underway, using Calaxin-M^{R} (tridemporph and manab), Tecto 60^{R} (thiabendazole), Daconil^{R} (chlorothabonil), and Sportak^{R} (prochloraz) on chickpeas, and Malathion^{R} (malathion), Carbofuran^{R} (furadan), and Actellic 2% (perimiphors) on faba beans, soybeans, and lima beans. Except for Daconil and Sportak, all these chemicals were applied to inoculated seed. Number, size, and total dry weight of nodules as well as acetylene reduction were monitored in pot culture throughout the growing season. Preliminary results of a pot experiment and laboratory studies on fungicides indicated that Tecto 60 (3 g/kg seed) had no toxic effect on strains of *Rhizobium* specific to chickpeas, even when the recommended rate was doubled and the fungicide was in direct contact with the inoculum. Calaxin-M^{R} was safe at all rates provided there was no direct contact with the inoculum.

Work on herbicides is done mainly under field conditions and includes neither acetylene reduction nor incubation studies. In a field and pot experiment on chickpeas, Treflan (trifluralin) applied pre-planting at 0.50-0.75 kg a.i./ha resulted in partial weed control and had no effect on nodulation (Shad 1975). Chickpea yields were slightly increased compared to the unweeded control but were significantly reduced compared to the hand-weeded treatment.

The ICARDA international weed control trials of 1981/82 on lentils, faba beans, and chickpeas were conducted in the Beqa'a Plain and included the following herbicides, applied pre-emergence: Tribunil (methabenthiazuron, 2.5-3.5 kg a.i./ha), Malaron (chlorbromuron, 1.0-3.5 kg a.i./ha), Igran (trebutryne, 2.0-4.0 kg a.i./ha), and Gesagard (prometryne, 1.0-2.0 kg a.i./ha). In faba beans, Tribunil, Malaron, and Igran, applied at various rates, had no effect on number and dry weight of nodules; Igran at 3.0 kg a.i./ha was the most effective for weed control. In chickpeas, similar effects on nodulation were obtained, but Igran at 4.0 kg a.i./ha was the most effective. In lentils, Malaran, Igran, and Gesagard also had no effect on nodulation and Gesagard at 1.5 kg a.i./ha was the most effective in reducing weeds. A new set of more promising herbicides are being tested this year on the same three crops using the ICARDA international trials of 1985/86.

Future Research

Research on nitrogen fixation in Lebanon is insufficient, fragmented, and lacks direction, and as such does not form a solid research program. Due to the importance of food legumes in Lebanon, future efforts must incorporate a systematic research program on nitrogen fixation to tackle the following:

- Assessment of the effectiveness and distribution of native rhizobial strains on major food legume crops such as chickpeas, lentils, faba beans, phaseolus beans, peas, and alfalfa.
- Introduction and identification of highly effective exogenous rhizobial strains compatible with adapted improved cultivars.
- Identification of economical controls for *Sitona* weevil in lentils.
- Compatibility studies between efficient rhizobial strains and effective herbicides and seed dressings.

With the soaring prices of nitrogen fertilizers in Lebanon, investment in research on nitrogen fixation could help to reduce production costs of legumes and subsequent crops, as well as maintain the productivity of the already diminishing agricultural land in the country.

References

Abu-Shakra, S.S., Phillips, D.A. and Huffaker, R.C. 1978. Nitrogen fixation and delayed leaf senescence in soybeans. Science 199: 973-975.

Ahmad, R. 1976. The influence of irrigation, nitrogen and inoculation on chickpea production. MSc. thesis. Faculty of Agricultural Sciences, American University of Beirut, Beirut, Lebanon. 64 pp.

Dewan, S.B.A. 1981. Combining ability and components of genetic variance in a soybean (*Glycine max* L. Merrill) diallel cross involving delayed leaf senescence. MSc. thesis. Faculty of Agricultural Sciences, American University of Beirut, Beirut, Lebanon. 70 pp.

FAO (Food and Agriculture Organization). 1977-1983a. FAO Production Yearbooks. FAO, Rome, Italy.

FAO. 1977-1983b. FAO World Trade Yearbooks. FAO, Rome, Italy.

Habib, G. 1979. Nitrogen distribution during plant ontogeny in nodulating and non-nodulating soybeans. MSc. thesis. Faculty of Agricultural Sciences, American University of Beirut, Beirut, Lebanon. 118 pp.

Iqbal, M. 1975. The influence of inoculation and nitrogen fertilization on lupin production. MSc. thesis. Faculty of Agricultural Sciences, American University of Beirut, Beirut, Lebanon. 101 pp.

Islam, M.S., Solh, M.B. and Saxena, M.C. 1986. Effect of fertilization, inoculation and carbofuran on nodulation, yield and protein content of lentil. LENS Newsletter. (In press.)

Khan, G.S. 1976. Effect of nitrogen and micronutrients on nodulation and seed yield of soybeans. MSc. thesis. Faculty of Agricultural Sciences, American University of Beirut, Beirut, Lebanon.

Phillips, D.A., Pierce, R.O., Edie, S.A., Foster, K.E. and Knowles, P.K. 1984. Delayed leaf senescence in soybean. Crop Science 24: 518-522.

Shad, R.A. 1975. The effect of trifluralin and inoculation on the nodulation and yield in chickpeas. MSc. thesis. Faculty of Agricultural Sciences, American University of Beirut, Beirut, Lebanon. 76 pp.

Solh, M.S., Hamid, S.A.M., Itani, H. and Ryan, J. 1986a. Nitrogen supplementation of inoculated and non-inoculated soybeans at varying plant densities. Dirasat. (In press.)

Solh, M.S., Itani, H.M. and Hawar, M.S. 1986b. The effect of *Sitona* weevil on nodulation and yield of lentils and implication of certain control measures. Lebanese Science Bulletin. (In press.)

Discussion

Moawad

Some researchers could obtain a response to late foliar application of urea as a spray. Could you comment on that in the light of what you mentioned about the early and late leaf senescence?

Solh

Research that showed a positive response to fertilization with a foliar spray during flowering included macro- and micronutrient elements other than nitrogen. Even with other nutrients studies show contradictory results with respect to the yield response in soybean. With respect to delayed and normal leaf senescence there was no difference in response between the two senescing habits.

Engin

Was the use of Aldicarb negative on nodulation?

Solh

Aldicarb reduced nodule damage by *Sitona* spp. significantly in the two cultivars. However, as a result of reducing nodule damage, yield was increased considerably though not significantly in one cultivar.

Engin

Use of Aldicarb for cotton is common in Turkey. What would you say about the residual effects of Aldicarb on the soil rhizobial population?

Solh

No information is available.

Brockwell

Where do you see the value of this particular line with green leaves considering harvesting?

Solh

Green leaves might create a problem with mechanical harvesting and therefore a desiccant would be needed to drop the leaves before harvesting. However, such green leaves may be useful to us as feed if soybean is grown on a small scale where hand labour is readily available. But to be grown economically, soybean probably needs to be grown on a large scale in this part of the world.

Eaglesham

Although it is true that soybeans are usually grown on a large scale requiring mechanized harvesting for which the delayed leaf senescence (DLS) type would be unsuitable, small-scale production of the crop is increasing, for instance in Africa. I believe that scientists working with farming systems which may include small-scale production of soybean would be interested in combining grain production with significant organic nitrogen input to the soil by exploiting the DLS type.

Part II

Genetic and Physiological Aspects of Legumes and Rhizobia

D.P. Beck and L.A. Materon (editors), Nitrogen Fixation by Legumes in Mediterranean Agriculture
 ISBN 90 247 3624 2

Analysis of the Host Genes Controlling the Legume-*Rhizobium* Symbiosis: Some Technical Problems and Pitfalls

T.A. LIE, J. PIJNENBORG and P.C.J.M. TIMMERMANS

Agricultural University, Wageningen, The Netherlands

Summary. A number of factors interfering with the analysis of host genes controlling the legume-*Rhizobium* symbiosis are discussed and methods for the aseptic cultivation of a large number of plants described. The strong effect of temperature on nodulation is illustrated with an example of a pea which has a temperature-sensitive nodulation trait. In a number of crosses, there was a strong maternal effect in the reciprocal F_1 hybrids. A practical solution to circumvent the maternal effect is the use of specific nitrogenase activity, expressed per gram of shoot tissue.

Introduction

A basic knowledge of the genetic background of the legume-*Rhizobium* symbiosis is essential to formulate guidelines for the plant breeder and the microbiologist to improve nitrogen fixation in leguminous crops. The ultimate goal is to introduce desirable genetic traits into the host plant and the associated rhizobial strain and so obtain an effective symbiosis. Our knowledge about the bacterial genes involved in the symbiosis is expanding rapidly, but unfortunately, research on the corresponding plant genes is lagging far behind.

This imbalance is partly due to technical problems inherent in the aseptic cultivation of a large number of plants for genetic analysis. Another reason is the lack of genetic variants in the plants studied, with regard to symbiotic traits. In view of the pronounced influence of the environment on the symbiosis, particularly temperature (Lie 1971), genetic analysis must be performed under environmentally controlled conditions. More recently we observed a strong maternal effect on nodulation and nitrogen fixation (Lie and Timmermans 1983; Lie 1984), which is a complicating factor in the genetic analysis. In this paper, some problems and pitfalls in the analysis of the host genes controlling the symbiosis will be discussed and if possible a practical solution presented.

The Requirement for Aseptic Plant Cultures

The use of bacteriologically controlled conditions in the genetic analysis cannot be sufficiently emphasized. The symbiotic response of the plant is

Table 1. Suppression of nodulation of pea cv Afghanistan with the Turkish *Rhizobium* strain Tom, by the non-nodulating strains PF_2, 310a, and PRE from Europe (Winarno and Lie 1979).

Rhizobial strain(s)	Number of nodules/plant	Inhibition (%)
PF_2	0	
310a	0	
PRE	0	
Tom	90	
Tom + PF_2	0	100
Tom + 31 0a	31	66
Tom + PRE	101	0

highly dependent on the rhizobial strain used. Moreover, the presence of another *Rhizobium* species may change the symbiotic response dramatically. A good example of the specificity of the plant towards rhizobial strains and the effect of another bacterial strain is shown in Table 1. The Afghan pea only forms root nodules with strain Tom from Turkey, but not with the European strains PF_2, PRE, and 310a. These strains, although unable to nodulate the Afghan pea, may compete with the nodulating strain Tom. In the extreme case complete inhibition of nodulation was achieved with strain PF_2 (Winarno and Lie 1979), illustrating the necessity of using aseptic cultures to obtain reliable results from the genetic analysis.

For the classical Mendelian analysis, a large number of plant individuals are required to study the inheritance of symbiotic traits in the host plant. Therefore, simple but reliable methods should be developed to cultivate large numbers of plants under aseptic conditions. With small-seeded plants like clovers or medics, the plants were grown on agar slopes in small tubes. This method is not suitable for pea plants; even in large tubes of 1 l capacity nodulation is erratic and severely delayed (Lie unpublished). For our purpose, it is sufficient to grow the plants with only the roots under aseptic conditions, and in the methods described, nodulation is excellent. For pea plants, hydroponic growth in nitrogen-free nutrient solution has been successfully used in most cases, but in some cases where nodulation is very low, e.g. due to the use of less compatible partners (Lie 1981), a simple modified Leonard jar has been devised. In the case of nutrient solution culture, Erlenmeyer flasks or 250 ml serum bottles can be used. The latter are very convenient as acetylene reduction of the detached root system can be performed in the same bottle by replacing the cotton plug with the original screw cap, which is important if large numbers of plants are to be handled.

For routine analysis 360 ml preservation jars covered with a foam plastic plug instead of a cotton plug were used. The jars were kept overnight at 105 °C and the following day nutrient solution, which had been boiled for at least 1 h, was poured while hot into the jars. The foam plastic was also boiled separately and, after cooling, a pregerminated pea seedling was inserted into

Fig. 1. Modified Leonard jars using plastic coffee beakers

an incision in the foam plastic plug. For climbing plants, a plastic or wooden stick can be inserted into another incision in the plug. Using this method, several hundred jars can be prepared by a single person in a day.

The modified Leonard jar consists of a plastic coffee beaker of about 100 ml capacity filled with perlite. The conical beaker fits into a 360 ml preservation jar, which serves as a reservoir for the nutrient solution. A foam plastic wick was inserted through a slit made in the bottom of the plastic beaker. Before use, the whole system was kept overnight at 60 °C and the following day boiling nutrient solution was poured through the perlite until the jar was completely filled. After sowing, the perlite can be covered with a layer of sterilized, fine gravel to prevent contamination (Fig. 1).

The Need for Temperature Control

The temperature at which the plant is grown may affect several metabolic processes in the plant, and indirectly affect nodulation and nitrogen fixation.

Table 2. The effect of a short exposure to a high (25 ° C) or low (20 ° C) temperature on nodulation of pea cv Ankara inoculated with the European *Rhizobium* strain PF_2.

Plants grown at 20 ° C except where indicated otherwise	Mean no. of nodules/plant	Plants grown at 25 ° C except where indicated otherwise	Mean no. of nodules/plant
Control	0.25	Control	7.6
Day 0-3 at 25 ° C	6.4	Day 0-3 at 20 ° C	2.1
Day 0-5 at 25 ° C	9.5	Day 0-5 at 20 ° C	1.1

But the symbiotic system is much more sensitive to high or low temperatures than the plant itself. In addition, we found that symbiotic systems, consisting of plant and rhizobial strains which are less compatible with each other, are extremely sensitive to small changes in temperature even for a short period. Therefore, temperature fluctuations in a greenhouse which is not well controlled with respect to temperature can have a pronounced effect, particularly on nodulation.

In our previous experiments with pea cv Iran (Lie 1971), no nodulation was obtained when peas were inoculated with most European strains at 18-20 ° C. Nodulation occurred at 25 ° C and the requirement for the higher temperature was confined to the first few days after inoculation (Lie 1971; Lie *et al.* 1976). This phenomenon is due to a temperature-sensitive (ts) gene Sym 1 (Lie 1984) in pea cv Iran. The effect is strain specific, obviously not or less observable when compatible strains from the Middle East were used.

Temperature-sensitive nodulation is not restricted to pea cv Iran but is also found in pea-*Rhizobium* systems which are not so compatible with each other (Lie 1981). Recently, another wild pea line from Turkey was obtained which behaved similarly to pea cv Iran when a European strain was used (Table 2). At present , the analysis of a cross between this Turkish pea line (cv Ankara) and cv Iran is not completed and it is not yet possible to decide whether we have to deal with the same gene (Sym 1) or with another one. Again, as in the case of the Iran pea, the Middle East rhizobial strains can overcome the ts barrier, and nodulation proceeds normally at 20 ° C and at 25 ° C.

Maternal Effects

The formation of root nodules and their functioning are integrated into plant metabolic processes and certainly depend on photosynthesis. It is therefore not so surprising that the plant background can have a strong effect on nodulation and nitrogen fixation. This was certainly the case when plants of different seed size were compared (Tables 3 and 5),for example the cultivated pea lines cv Rondo and cv L110 which have large seeds and the primitive landrace cv Afghanistan. But it is also observable between cv Afghanistan and cv Iran, which have almost the same seed size.

Table 3. Nodulation of pea cvs Rondo and Afghanistan and their reciprocal F_1 hybrids with *Rhizobium* strains PRE and PF_2, respectively (Lie 1984).

Rhizobium strain	*Pisum sativum* L.			
	cv Rondo (R)	F_1 (RxA)*	F_1 (AxR)*	cv Afghanistan (A)
PRE	68±26**	5.6±2.3	0	0
PF_2	169±45	123±57	15±11	0.3±0.8

* In the F_1 hybrids the first letter denotes the female.
** Number of nodules per plant.

Significant maternal effects on nodule number were found when the reciprocal F_1 hybrids were compared (Table 3) and the effect can be accentuated by the strain of *Rhizobium* used. These results demonstrate the problem we encountered in classifying the F_2 individuals. It is advisable to also compare backcrosses with the same maternal plant background.

In the cross between cv Afghanistan and cv Iran, which have almost the same seed size, the mother had a strong effect on nitrogen fixation (Table 4). Both pea lines produce a large number of nodules with the Middle East strain F_{13} but the size and color of the nodules are different (Lie and Timmermans 1983). Our results showed that due to maternal effects, shoot weight and nitrogen fixed per plant are not suitable parameters for genetic analysis. Fortunately, as shown in Table 4, nitrogenase activity is a good parameter when expressed as per gram of shoot, since in this case, the maternal effect can be eliminated.

The usefulness of this parameter is also shown in Table 5. In this experiment, the large pea cv L110 formed ineffective nodules, whereas the small Afghan pea formed effective nodules with strain Tom. The F_1 (L110xA) inherits the large size of the mother and the nitrogen fixing activity of the father, resulting in a strong heterosis. Again, when calculated on the basis of shoot weight, the nitrogenase activity can be used.

Table 4. Growth and nodulation of pea cvs Afghanistan and Iran and their reciprocal F_1 hybrids with *Rhizobium* strain F_{13} (Lie and Timmermans 1983).

	Pisum sativum L.			
	cv Afghanistan (A)	F_1 (AxI)*	F_1 (IxA)*	cv Iran (I)
Fresh weight				
Shoot (g/plant)	1.17	1.98	3.42	4.21
Nodules (mg/plant)	97	168	237	362
N_2 fixed				
mg N/plant	2.7	7.8	13.3	19.8
μmoles C_2H_4/plant/h	0.43	2.43	4.13	6.20
μmoles C_2H_4/g shoot/h	0.37	1.23	1.27	1.47
% N in shoot	1.57	3.12	3.24	3.96

* In the F_1 hybrids the first letter denotes the female.

Table 5. Nitrogen fixation in pea cv L 110 and cv Afghanistan and their F_1 hybrid when inoculated with *Rhizobium* strain Tom (Lie unpublished).

	Pisum sativum L.		
	cv L 110	F_1 (L110xA)	cv Afghanistan (A)
Shoot fresh wt (g)	1.79	2.45	1.01
μmoles C_2H_4/plant/hour	0.19	1.90	0.78
μmoles C_2H_4/g shoot/hour	0.11	0.78	0.77

In a genetic analysis where large numbers of plants are needed, it is easy to determine the fresh weight and the nitrogenase activity using the acetylene reduction method. Good results can also be obtained using the nitrogenase activity or total nitrogen fixed per unit of nodule tissue. These methods, however, require much more labor, particularly for harvesting root nodules.

Concluding Remarks

The examples shown here clearly demonstrate the need for standardized methods for genetic analysis. It is essential to carry out the experiments with aseptic plant cultures and to use a single, well defined rhizobial strain. Several environmental factors can influence the symbiosis, but temperature especially should be controlled. Temperature fluctuations when the symbiosis is in the critical nodulation phase can have a pronounced effect. In the case of pea cv Iran, the temperature of the root system is decisive for nodulation, and the use of a temperature regulated waterbath is a good method when the plants are grown in a less well-controlled environment like a normal greenhouse.

We stress the use of simple, but reliable methods in the analysis, so that large numbers of plants can be handled in a short time. A number of methods are suggested here, and by improving the methods and studying some of these systems, we can obtain a better insight into genetic control by the host plant.

References

Lie, T.A. 1971. Symbiotic fixation under stress conditions. Plant and Soil, Special Volume: 127-137.

Lie, T.A., Hille Ris Lambers, D. and Houwers, A. 1976. Symbiotic specialisation in pea plants: some environmental effects on nodulation and nitrogen fixation. Pages 319-333 *in* Symbiotic Nitrogen Fixation in Plants (Nutman, P.S., ed.). Cambridge University Press, Cambridge, UK.

Lie, T.A. 1981. Gene centres, a source for genetic variants in symbiotic nitrogen fixation; host-induced ineffectivity in *Pisum sativum* ecotype fulvum. Plant and Soil 61: 125-134.

Lie, T.A. and Timmermans, P.C.J.M. 1983. Host genetic control of nitrogen fixation in the legume-*Rhizobium* symbiosis: complication in the genetic analysis due to maternal effects. Plant and Soil 75: 449-453.

Lie, T.A. 1984. Host genes in *Pisum sativum* L. conferring resistance to European *Rhizobium leguminosarum* strains. Plant and Soil 82: 415-425.
Winarno, R. and Lie, T.A. 1979. Competition between *Rhizobium* strains in nodule formation: interaction between nodulating and non-nodulating strains. Plant and Soil 51: 135-142.

Discussion

Broughton

At what stage do plants that are resistant to nodulation by various *R. leguminosarum* strains actually terminate the nodulation process?

Lie

Pea cv Afghanistan can be infected by European *R. leguminosarum* strains. We found many curled root hairs with initial infections and in some cases also infection threads. The blocking stage is obviously after infection has taken place.

Brockwell

Is it possible to cross Afghan and European peas and, if so, is it known how the nodulating characteristic of Afghan pea is inherited?

Lie

Yes, we have crossed the Afghan pea with several pea lines including the Dutch pea cv Rondo. From F_2-analysis and backcrosses we conclude that the resistance is due to one gene, sym 2. However, we have some evidence that there is another gene (or genes) which determines the number of root nodules. There is also a technical problem, due to maternal effects.

Solh

In your opinion what would be a better strategy to improve symbiosis: to introduce new *Rhizobium* strains or to introduce new genotypes of the host?

Lie

It depends on the situation. The easiest way is to introduce new *Rhizobium* strains, if the native *Rhizobium* population is low and not so competitive. But I think that there is also room for the plant breeder. It is perhaps worthwhile to introduce resistance genes into the plant against the local, and perhaps not so effective, *Rhizobium* population, and specific susceptibility to a superior *Rhizobium* strain or strains to be used as an inoculant. The pea cv Afghanistan and the Middle East *Rhizobium* strain Tom may be a good model system.

Smith

The persistance of a population of legumes which have been growing in an agricultural situation for a long time is a result of their adaptation to the environment. To introduce a more efficient variety, the variety must be changed in some way or protected (let us ignore geographical isolation for the present).

It is difficult to change the environment below ground in order to introduce a new rhizobial strain. It is also difficult to protect the strain from competition with indigenous ones.

Do you consider that major aspects of environment, as experienced by *Rhizobium*, are those produced by the plant and thus can be controlled? Where the plant is an annual, and the bacteria have to be free-living for a long time, will it be possible to produce the necessary change in conditions below ground?

Lie

It is very difficult to control the environment during the saprophytic free-living stage. Little is known about this subject. Perhaps the *Rhizobium* inoculant can be protected on the seed by pelleting the seed with a layer of material which is favorable for the growth of this bacterium.

Materon

Although it is not possible to change the environment in the wide ecological sense, the use of sprays, chemical fertilizers, cultivation, and novel agricultural methods inevitably will affect soil microclimate, and so in certain situations it may be commercially profitable to protect the rhizobia.

D.P. Beck and L.A. Materon (editors), Nitrogen Fixation by Legumes in Mediterranean Agriculture
 ISBN 90 247 3624 2

Cloning Gene(s) for Competitive Nodulation Blocking on Afghanistan Peas from *Rhizobium leguminosarum*

D.N. DOWLING, U. SAMREY, and W.J. BROUGHTON

University of Geneva, Geneva, Switzerland

Summary. The primitive pea cultivar Afghanistan can be nodulated by the *Rhizobium leguminosarum* strain TOM but fails to nodulate with strain PF_2. The gene(s) involved in competitive nodulation blocking were identified on the Sym plasmid of PF_2. They were successfully isolated and cloned into cosmid vectors.

Introduction

Competition amongst rhizobia for nodulation of legumes is a common phenomenon in soils (Bushby 1982; Schmidt *et al.* 1985; Dowling and Broughton 1986). This may be of considerable agronomic importance, but the mechanisms of competition are poorly understood. Competition also occurs when non-nodulating strains in the soil prevent the nodulation of a given host (Diatloff and Brockwell 1976; Lie 1978). One of the best studied is the Afghanistan pea system described by Winarno and Lie (1979). This primitive pea cultivar, which generally fails to nodulate with European isolates of *R. leguminosarum* such as PF_2, can be nodulated by strains from the Middle East such as TOM. The inability of the Afghanistan pea to nodulate with European *R. leguminosarum* strains was attributed to a single recessive gene (Sym-2) in the host plant (Holl and LaRue 1976; Kneen and LaRue 1984), while the ability to nodulate with TOM is due to a host range gene (Hombrecher *et al.* 1984) carried on the Sym plasmid (Brewin *et al.* 1980) of that strain. Although PF_2 does not form nodules on the Afghanistan cultivar, it does cause root-hair curling and infection thread formation (Lie 1978; Winarno and Lie 1979; Dowling unpublished). However, these appear to be aborted at an early stage of infection.

Nodulation of Afghanistan peas with TOM can be completely inhibited in the presence of some European *R. leguminosarum* strains, including PF_2 (Winarno and Lie 1979), and this system was used as a model to study the dynamics and mechanisms of competition (Broughton *et al.* 1980; 1982). These studies revealed that there was no bacterial antagonism, preferential growth rates of either strain, or specific plant factors which enhanced or inhibited either PF_2 or TOM. There was, however, a requirement for living cells. Dead PF_2 cells could bind to root hairs of Afghanistan peas but did not block nodulation by TOM.

This paper describes the isolation, by molecular cloning, and location of gene(s) involved in competitive nodulation blocking of TOM by *R. leguminosarum* PF_2 on Afghanistan pea plants.

Methods

Competition tests were carried out in liquid nitrogen-free media in Erlenmeyer flask assemblies as described by Broughton *et al.* (1980). *Rhizobium* plasmids were labelled and mobilized with Tn5-Mob by the method of Simon (1984) and bacterial matings were carried out on sterile nitrocellulose membranes (Broughton *et al.* 1986).

Large plasmids were prepared by the method of Hirsch *et al.* (1980) followed by purification through CsCl/EtBr density gradients. These plasmids were also analysed by a modified "in gel" lysis procedure (Eckhardt 1978). Total DNA was prepared according to the method of Broughton *et al.* (1985) and cosmid DNA by the method of Maniatis *et al.* (1982).

To prepare a PF_2 cosmid gene bank, the target DNA was restricted and sized (Maniatis *et al.* 1982), cloned, and *in vitro* packaged into the wide host range cosmid pMMB33 according to Frey *et al.* (1983).

DNA hybridizations were carried out in 50% formamide at 42° C (Broughton *et al.* 1985) and colony filters of the cosmid bank were prepared by the method of Maas (1983).

Results

Identification of the symbiotic plasmid in PF_2

As plasmids play a major role in the symbiotic process in fast-growing rhizobia, the role of plasmid DNA was investigated in competitive nodulation

Table 1. Competitive nodulation blocking of TOM on cv Afghanistanby different *R. loti* NZP4010 transconjugants harboring PF_2 plasmids.

Strain	Plasmids present	Nodulation of *Pisum sativum*		Competitive blocking ability
		cv Rondo	cv Afghan	
R.leg.	PF_2(donor)	All	Nod^+,Fix^+ Nod^-	Yes
R.loti(recipient)	None	Nod^-	Nod^-	No
R.loti(233)	$pSYMPF_2$	Nod^+,Fix^- Nod^-	Yes	
R.loti(220)	pPF_{2b}	Nod^-	Nod^-	No
R.loti(237)	$pSYMPF_2$, pPF_{2b},pPF_{2c}	Nod^+,Fix^-	Nod^-	Yes

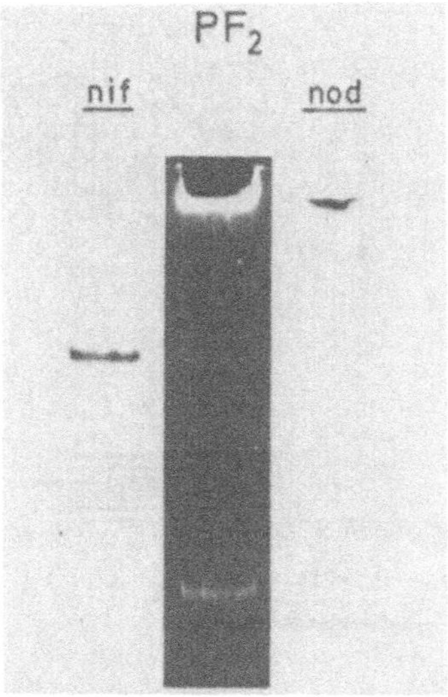

Fig. 1. Identification of the *R. leguminosarum* PF_2 symbiotic plasmid by hybridization of plasmids separated on Eckhardt gels to *R. meliloti nif* and *nod* probes.

blocking by PF_2. The Sym plasmid of PF_2 was identified as the smallest plasmid (250 kbp) in the strain by using *R. meliloti nif* and *nod* probes (Fig. 1). Subsequently, the use of random Tn5-Mob insertions into the genome of PF_2 and the mobilization of kanamycin resistance by RP4.4 into a plasmid-free (competitive nodulation blocking negative) *R. loti* strain (NZP4010) led to the isolation of a number of transconjugants. These were characterized with respect to their plasmid content, symbiotic ability, and competitive nodulation blocking ability (Table 1). Transconjugants harbouring $pSYMPF_2$ could nodulate (but not fix nitrogen) on cultivar Rondo and could block nodulation of TOM on cultivar Afghanistan.

Construction of a Cosmid Gene Bank of *R. leguminosarum* PF_2

A gene bank was constructed in the wide host range cosmid pMMB33. Plasmid mobilization experiments suggested that the genes for competitive nodulation blocking might be encoded on the Sym plasmid. By using plasmid DNA isolated from PF_2 as a hybridization probe against the gene bank, we succeeded in "fishing out" colonies containing plasmid sequences.

Cosmids which had homology to PF_2 plasmid DNA, were mobilized by tri-parental matings into rhizobial strains which did not block nodulation of TOM on Afghanistan. These transconjugants were tested for their competitive effects in mixed inoculations with TOM on Afghanistan.

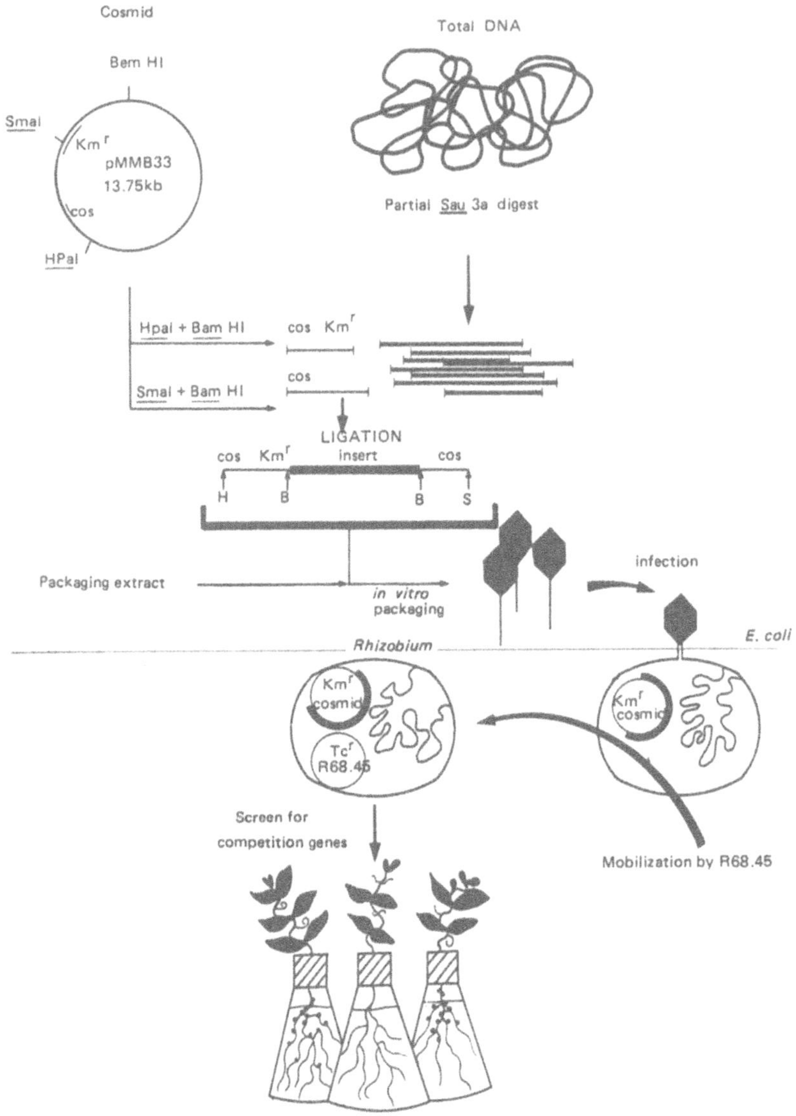

Fig. 2. Diagrammatic outline of the strategy used to clone and screen for competitive nodulation blicking genes from *R. leguminosarum* PF_2.

By using this strategy (Fig. 2) three overlapping cosmids were isolated which blocked nodulation of TOM on Afghanistan peas.

Discussion

The overlapping cosmids isolated have approximately 10-15 kbp of DNA in common, (presumably the location of the nodulation blocking gene(s)) and hybridize only to the Sym plasmid of PF_2 confirming the plasmid mobilization data. The 10-15 kbp common region is currently being sub-cloned for more detailed analysis and mutagenesis.

In conclusion, gene(s) for competitive nodulation blocking on Afghanistan peas have been located in the Sym plasmid of *R. leguminosarum* PF_2 and have been cloned in cosmid vectors.

References

Brewin, N.J., Beringer, J.E. and Johnston, A.W.B. 1980. Plasmid-mediated transfer of host range specificity between two strains of *Rhizobium leguminosarum*. Journal of General Microbiology 120: 413-420.

Broughton, W.J., van Egeraat, A.W.S.M. and Lie, T.A. 1980. Dynamics of *Rhizobium* competition for nodulation of *Pisum sativum* cv Afghanistan. Canadian Journal of Microbiology 26: 562-565.

Broughton, W.J., Samery, U. and Bohlool, B.B. 1982. Competition for nodulation of *Pisum sativum* cv Afghanistan requires live rhizobia and a plant component. Canadian Journal of Microbiology 28: 162-168.

Broughton, W.J., Bohlool, B.B., Shaw, C.H., Bohnert, H.J., and Pankhurst, C.E.. 1985. Conserved plasmid/chromosome sequences in fast- and slow-growing rhizobia that nodulate the same plant. Archives of Microbiology 141: 14-21.

Broughton, W.J., Wong, C.H., Lewin, A., Samery, U., Myint, H., Meyer, Z.A., Dowling, D.N. and Simon, R. 1986. Identification of *Rhizobium* plasmid sequences involved in recognition of *Psophocarpus*, *Vigna*, and other legumes. Journal of Cell Biology 102. In press.

Bushby, H.V.A. 1982. Ecology. Pages 35-75 *in* Nitrogen Fixation. II. *Rhizobium* (Broughton, W.J., ed.). Oxford University Press, Oxford, England.

Diatloff, A., and Brockwell, J. 1976. Symbiotic properties of *Rhizobium japonicum* and competitive success in nodulation of two *Glycine max* cultivars by effective and ineffective strains. Australian Journal of Experimental Animal Husbandry 16: 514-521.

Dowling, D.N. and Broughton, W.J. 1986. Competition for nodulation of legumes. Annual Review of Microbiology. In press.

Eckhardt, T. 1978. A rapid method for the identification of plasmid deoxyribonucleic acid in bacteria. Plasmid 1: 584-588.

Frey, J.M. Bagdasarian, Feiss, D., Franklin, C.H. and Deshusses, J. 1983. Stable cosmid vectors that enable the introduction of cloned fragments into a wide range of Gram-negative bacteria. Gene 24: 299-308.

Hirsch, P.R., van Montagu, M., Johnston, A.W.B., Brewin, N.J. and Schell, J. 1980. Physical identification of bacteriocinogenic nodulation and other plasmids in strains of *Rhizobium leguminosarum*. Journal of General Microbiology 120: 403-412.

Holl, F.B. and LaRue, T.A. 1976. Genetics of plant hosts. Pages 391-399 *in* Proceedings of the First International Symposium on Nitrogen Fixation, volume 2 (Newton, W.E. and Numan, C.J., eds.). Washington State University Press, Pullman, USA.

Hombrecher, G.R. Gotz, Dibb, N.J., Downie, J.A., Johnston, A.W.B. and Brewin, N.J. 1984. Cloning and mutagenesis of nodulation genes from *Rhizobium leguminosarum* TOM, a strain with extended host range. Molecular General Genetics 194: 293-298.

Kneen, B.E. and LaRue, T.A. 1984. Peas (*Pisum sativum* L.) with strain specificity for *Rhizobium leguminosarum*. Heredity 52: 383-389.

Lie, T.H. 1978. Symbiotic specialisation in pea plants. The requirement of specific *Rhizobium* strains for peas from Afghanistan. Annals of Applied Biology 88: 462-465.

Maas, R. 1983. An improved colony hybridization method which significantly increases sensitivity for detection of single genes. Plasmid 10: 296-298.

Maniatis, T., Fritsch, E.F. and Sambrook, J. 1982. Molecular Cloning: A Laboratory Manual. Cold Spring Harbor Laboratory, New York, USA. 545 pp.
Schmidt, E.L. and Robert, F.M. 1985. Recent advances in the ecology of *Rhizobium*. Pages 379-385 *in* Nitrogen Fixation Research Progress (Evans, H.J., Bottomley, P.J., and Newton, W.E., eds.). Martinus Nijhoff, Dordrecht, The Netherlands.
Simon, R. 1984. High frequency mobilization of Gram-negative bacterial replicons by the *in vitro* constructed Tn5-Mob transposon. Molecular General Genetics 196: 413-420.
Winarno, R., and Lie, T.H. 1979. Competition between *Rhizobium* strains in nodule formation. Interactions between nodulating and non-nodulating strains. Plant and Soil 51: 131-142..lm 10;

Discussion

Eaglesham

You seem to be saying that genetic exchange between rhizobia in soil is common. If so, would cross-infection between isolates from e.g. clover, peas, beans, and lucerne not be commonplace?

Broughton

Possibly, but I doubt whether definitive experiments have been done to test your point. There is however, circumstantial evidence that suggests that such events have occurred (Dowling, D.N. and Broughton, W.S., Annual Review of Microbiology, 1986).

Roughley

Would you please comment on the occurrence of genetic transfer in slow-growing rhizobia in soil cf fast-growing strains with sym plasmids.

Broughton

1. *Pseudomonas* plasmids are capable of mobilising plasmids and chromosomes of both fast- and slow-growing rhizobia.

2. Host specificity genes can be transferred from fast-growing rhizobia into slow-growing ones where they are expressed.

3. Not all symbiotic plasmids are self transmissible (or even curable e.g. *psym* of *R. meliloti*).

4. Taken together, these data suggest that there is no reason why such exchange could not occur. It is not possible to predict frequencies however.

Materon

Based on your results on genetic exchange of rhizobia, do you have any information on frequency of transfer between antibiotic-resistant mutants applied to soil as seed inoculants and soil rhizobia? And, would you then recommend the use of antibiotic-resistant mutants for field studies?

Broughton
As far as I am aware, there are no specific data relating to your question.

Since it seems likely that chromosomal genes will be transferred at lower frequencies than plasmid-borne genes, chromosomal antibiotic resistances would be preferred as markers (e.g. streptomycin). Of course it is better to use multiple markers (antibiotic resistance, serology, etc.).

Amarger
Do you have an idea of the frequency of plasmid transfer between your *R. meliloti* and *R. leguminosarum* strains in the soil experiment?

Broughton
We attempted to answer this question by inoculating sterile soil with Nod- *R. meliloti* (WL 113) and Nod + *R. leguminosarum* (T83K3). Transconjugants arose at the frequencies that were 100-fold higher in the presence than in the absence of a *M. sativa* rhizosphere. Furthermore these data suggest that under field conditions several transconjugations could occur per plant per year.

D.P. Beck and L.A. Materon (editors), Nitrogen Fixation by Legumes in Mediterranean Agriculture
 ISBN 90 247 3624 2

Physiology and Genetics of *Rhizobium* Nodulating Chickpeas (*Cicer arietinum* L.)

TOMAS RUIZ-ARGÜESO, ESTHER CADAHIA and ANTONIO LEYVA
Universidad Politecnica de Madrid, Madrid, Spain

Summary. Twenty-two strains of chickpea rhizobia of different geographical origins were examined for the cultural, physiological, and genetic properties currently used to separate fast- and slow-growing groups of rhizobia. These strains were diverse with regard to their growth rates on yeast extract-mannitol medium but all of them acidified this medium and the majority produced an acid reaction with serum zone formation on litmus milk. Low activities of nicotinamide adenine dinucleotide phosphate-6-phosphogluconate dehydrogenase were detected in six of the strains studied. This activity was not correlated with growth rate. One to three plasmids of molecular weight ranging from 36 to over 380 Mdal were reproducibly resolved in 16 strains, but no plasmids were observed in the six remaining strains. Structural *nif* genes, homologous to *nif* DH of *R. meliloti*, were highly conserved and were not located on detectable plasmids in 21 of the 22 strains examined. Strain IC-72M showed a different organization of *nif* genes. On the basis of these results, it is concluded that the chickpea rhizobial group is intermediate between classical fast- and slow-growing groups of rhizobia.

Introduction

Cicer arietinum L., commonly known as chickpeas, garbanzo, Bengal-gram, or gram, is a large-seeded pulse legume widely cultivated in India, the Mediterranean area, the Middle East, and Ethiopia since antiquity. It was brought to the New World, and is now cultivated on a large scale in Mexico, Argentina, Chile, and Peru (Duke 1981). The high seed protein content (up to 30%) and the ability of the crop to perform well under hot, dry conditions, make chickpeas an attractive pulse crop for introduction into new areas (Corbin *et al.* 1977).

Initial host specificity studies by Simon (1914) erroneously placed strains of *Rhizobium* nodulating chickpeas in the pea cross-inoculation group. However, after the work of Raju (1936) chickpea rhizobia are considered to be highly host-specific, nodulating only *Cicer* species and rarely and non-reciprocally to *Sesbania bispinosa* and *S. sesbania* (Gaur and Sen 1979). Based on cross-inoculation studies, these authors suggested that *Cicer arietinum* and its nodule bacteria should be considered a separate cross-inoculation group. In the field, chickpeas are generally nodulated by native strains and respond poorly to inoculation (Subba Rao 1976; Pareek 1979), possibly because the inoculant strains must face severe competition for nodule formation. However, positive responses to inoculation with selected strains have been reported in certain soils in India (Sen 1966; Patil and Medhave 1974) and in other countries (Okon *et al.* 1972).

Relatively little information on the physiological and genetic properties of chickpea rhizobia or on the diversity of strains able to nodulate this legume is available in the literature. The existence of strains with very different growth rates has been reported (Raju 1936; Okon *et al.* 1972). Consequently, chickpea rhizobia have been included both in fast- and slow-growing groups of rhizobia and have been assigned to both *Rhizobium* and *Bradyrhizobium* genera (Jordan 1984). The genetics of nitrogen fixation have lately been the subject of intensive research. We considered it important to also carry out genetic studies on chickpea rhizobia and started with the determination of plasmid content and the location of nitrogen fixation genes. Both nodulation and nitrogen fixation genes have been shown to be carried on large plasmids in fast-growing bacteria (Nuti *et al.* 1979; Hombrecker *et al.* 1981; Banfalvi *et al.* 1981) but slow-growing *B. japonicum* strains possibly have *nif* genes on the chromosome (Haugland and Verma 1981; Masterson *et al.* 1982).

To better understand the taxonomic position of rhizobia nodulating chickpeas, we examined a group of chickpea rhizobial strains of different geographical origin for the cultural and physiological properties currently used to separate classical fast- and slow-growing rhizobia. Also, to initiate genetic studies on this agriculturally important bacterium, we report in this paper on the plasmid profiles and location and conservation of *nif* genes.

Materials and Methods

Bacteria and growth conditions

The strains of *Rhizobium* used in this study were obtained from J. Brockwell, CSIRO, Canberra, Australia (CB1189, CC1192); O.P. Rupela, ICRISAT, Patancheru, India (strains designated IC); H.H. Keyser, USDA, Beltsville, USA (strains designated 3HOa); J.C. Burton, Nitragin Co., Milwaukee, USA (strains designated 27A); and O.A. Brestetsky, AUSIAM, Leningrad, USSR (522). Strains designated UPM were isolated from nodules of chickpea plants grown in Spanish soils. *R. meliloti* Rme 41 and L5.30 were received from J. Olivares, CSIC, Granada, Spain and *R. japonicum* 3I1b143 was obtained from H.H. Keyser.

The rhizobia were isolated and cultivated on yeast extract-mannitol agar (YMA) (Vincent 1970) at 30 °C and maintained on slopes of YMA at 4 °C.

Cultural and biochemical characteristics

Growth rates were calculated from the exponential growth phase of cultures grown in yeast extract-mannitol broth (YMB) (Vincent 1970). Changes in the pH of YMB were determined after 5 days' incubation. Reaction with litmus milk was followed at 26 °C according to Skerman (1967). Nicotinamide adenine dinucleotide phosphate (NADP)-linked glucose-6-phosphate dehydrogenase (NADP-G6PD) and NADP-linked 6-phosphogluconate dehydrogenase (NADP-6PGD) activities were measured as described by Keele *et al.*(1969).

Plasmid screening

Plasmid DNA screening was carried out in cells grown in TY medium (Beringer 1974) using the method of Eckhardt (1978). The molecular weight of plasmids was estimated from their relative mobilities in agarose gels by using pR68-45 (Riess *et al.* 1980), pRme L5-30, pRme 41 (Casse *et al.* 1979), and pAt-C58 (Rosenberg *et al.* 1982) as molecular weight standards.

Isolation, restriction, and hybridization of DNA

Total DNA was isolated as described by Corbin *et al.* (1982) from cultures on TY medium, restricted with EcoR1 endonuclease (Boehringer Mannheim, FRG), and the resulting DNA fragments were separated in 1% agarose gels, stained with ethidium bromide, and photographed under ultraviolet light. The agarose gels were then blotted according to Southern (1975) onto nitrocellulose filters.

DNA from plasmid pID1 containing *nif* DH genes from *R. meliloti* (Banfalvi *et al.* 1981) was nick-translated (Rigby *et al.* 1977) and used as a *nif* probe.

The Eckhardt agarose gels for visualization of plasmids were dried by the procedure of Purrello and Balazs (1983). The dried Eckhardt gels and the nitrocellulose filters were hybridized at 42 °C for 48 h with the *nif* probe using the procedure described by Corbin *et al.* (1982).

Results

Cultural and physiological properties

Three characteristics currently used to separate fast- and slow-growing rhizobia (growth rate and reactions in yeast extract-mannitol broth (YMB) and litmus milk) were examined in 22 strains of chickpea *Rhizobium* and the results are shown in Table 1. These strains were found to be quite diverse in

Table 1. Growth rates and changes produced in yeast extract- mannitol broth and litmus milk media by 22 strains of chickpea rhizobia.

Doubling times (h) [a]	No. of strains	pH on YMB [a]	Changes in litmus milk [b]
4.0–4.9	6	4.7–6.5	Acid, serum zone (6)
5.0–5.9	7	4.9–6.5	Acid, serum zone (6)
			No change (1)
6.0–6.9	3	4.6–6.2	Acid, serum zone (1)
			No change (2)
7.0–7.9	3	5.1–5.9	Acid, serum zone (2)
			No change (1)
8.0 or more	3	4.7–6.6	Acid, serum zone (2)
			No change (1)

[a] Figures are means of at least three determinations.
[b] Figures in parentheses are number of strains giving the indicated reaction.

their growth rates, with doubling times which ranged from 4.0 to 8.5 h. Although most of them had doubling times of less than 6 h, slow-growing strains with doubling times of 8 h or longer, and strains with intermediate growth rates were also identified. All the chickpea rhizobial strains examined acidified the YMB media after 5 days' growth, and there was no correlation between pH change and growth rate. Most strains caused an acid reaction on litmus milk media with the production of a serum zone.

Since differences in carbohydrate utilization between fast- and slow-growing rhizobia have been related to the presence of an operative pentose phosphate cycle (Martinez de Drets and Arias 1972), we have examined the activities of the two enzymes of this pathway (NADP-G6PD and NADP-6PGD) in six strains of chickpea *Rhizobium* with a wide range of growth rates, and in two strains representative of the fast- and slow-growing rhizobial

Table 2. Glucose-6-phosphate dehydrogenase (G6PD) and 6-phosphogluconate dehydrogenase (6PGD) activities in chickpea rhizobial strains compared with *Rhizobium meliloti* and *Bradyrhizobium japonicum*.

Strains	Doubling time (h)	Activities [a]	
		G6PD	6PGD
R. meliloti RME 41	2.5	243.0	70.1
B. japonicum 3ILb143	13.5	9.5	< 0.5
Chickpea rhizobia			
27a2	4.7	132.0	5.3
CC1192	5.0	58.4	11.0
UPM-Ca7	5.7	87.2	20.7
UPM-Ca116	7.5	64.1	6.9
UPM-Ca36	8.0	28.5	12.5
IC-94M	8.5	81.3	26.1

[a] Values are expressed as nmoles of NADP reduced/min/mg protein and are means of two separate measurements.

groups. Results are shown in Table 2. NADP-6PGD activity was detected in cell-free extracts of all the strains examined. However, the activity of this key enzyme in the pentose phosphate cycle was lower in chickpea rhizobia than in the fast-growing *R. meliloti* Rme 41 strain included as control. No detectable NADP-6PGD was observed in cell-free extracts of the slow-growing *B. japonicum* strain 3I1b143. NADP-G6PD activity was present in cell-free extracts of all the rhizobial strains studied.

Plasmid DNA profiles

By using the in-well cell lysis and electrophoresis procedure of Eckhardt (1978), plasmids were reproducibly detected in 16 of the 22 strains of *Rhizobium* examined. The first tracks of (b) and (c) in Fig.1 are representative examples of plasmid-containing strains. No detectable plasmids were observed in the remaining strains. The megaplasmid of *R. meliloti* (molecular weight about 1000 megadaltons (Mdal)) was consistently resolved by the procedure used to visualize plasmids (first track of (a) in Fig. 1). Nine different plasmid profiles were identified among the rhizobial strains on the basis of the number and size of plasmids. A diagrammatic representation of the different plasmid profiles is shown in Fig. 2. The number of plasmids varied from zero to three, with sizes ranging from 36 to over 380 Mdal. Most of the chickpea rhizobial strains grouped in plasmid profiles 1, 2, 3, and 4 contained one or no plasmids.

*Location and conservation of **nif** genes*

To determine if *nif* genes are located on plasmids in strains of *Rhizobium* from chickpeas, the Eckhardt gels used to resolve the plasmid bands were hybridized with the *R. meliloti nif* DH probe. There was one weak hybridization band corresponding to a plasmid of approximately 140 Mdal in strain IC-72M (second track of (c) in Fig. 1) but no hybridization signal was detected corresponding to plasmid bands of any of the remaining 21 strains examined (second track of (b) in Fig. 1 is an example of negative results). All autoradiograms showed specific hybridization to chromosomal DNA retained in the wells of the agarose gels, and a hybridization signal was consistently obtained to the megaplasmid DNA of *R. meliloti* strains included as control (second track of (a) in Fig. 1).

Total DNA from each of the 22 strains was isolated and restricted with EcoR1. The resulting fragments, isolated by electrophoresis in agarose gels, were blotted onto nitrocellulose filters and hybridized to *R. meliloti nif* DH probe. Most strains exhibited a unique electrophoretic banding pattern of EcoR1-generated fragments (tracks (b), (c), (d), and (e) in Fig. 3). In contrast, the pattern of fragments with strong homology to *R. meliloti nif* DNA was identical in 20 of the 22 strains examined (tracks (c) and (d) in Fig. 3 are

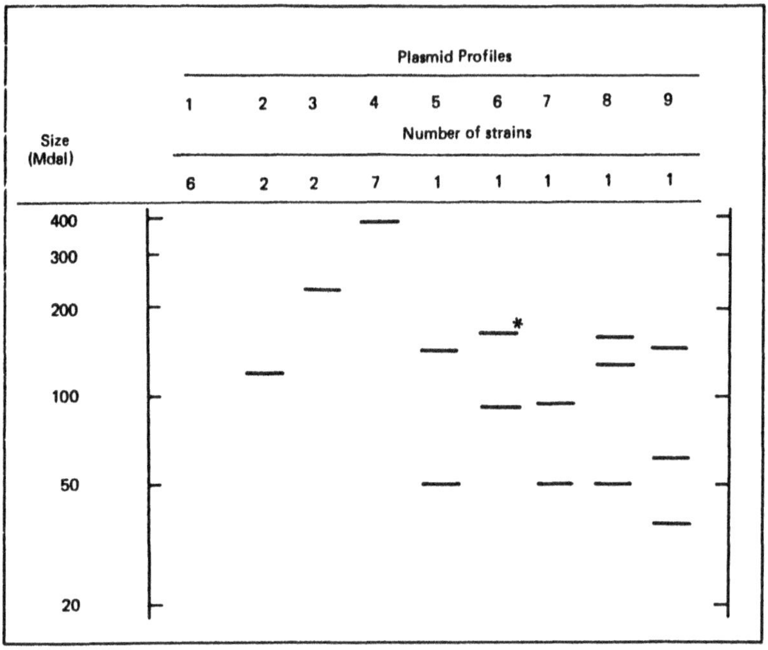

Fig. 2. Diagrammatic representation of plasmids present in 22 strains of chickpea rhizobia after electrophoresis in agarose gels like those in Fig. 1. The strains were distributed according to their plasmid profiles: 1 (UPM–Ca116, UPM–Ca126, 3H0a9, IC–41, IC–60, 522); 2 (UPM–Ca142, IC–2008); 3 (UPM–Ca7, 3H0a10); 4 (27a2, 27a9, 3H0a1, 3h0a5, CB1189, CC1192, IC–2002); 5 (IC–94M); 6 (IC–72M); 7 (IC–76); 8 (UPM–Ca36); 9 (IC–6). The asterisk-tagged line in strain IC–72M of profile 6 indicates that this band is hybridized to *nif* probe.

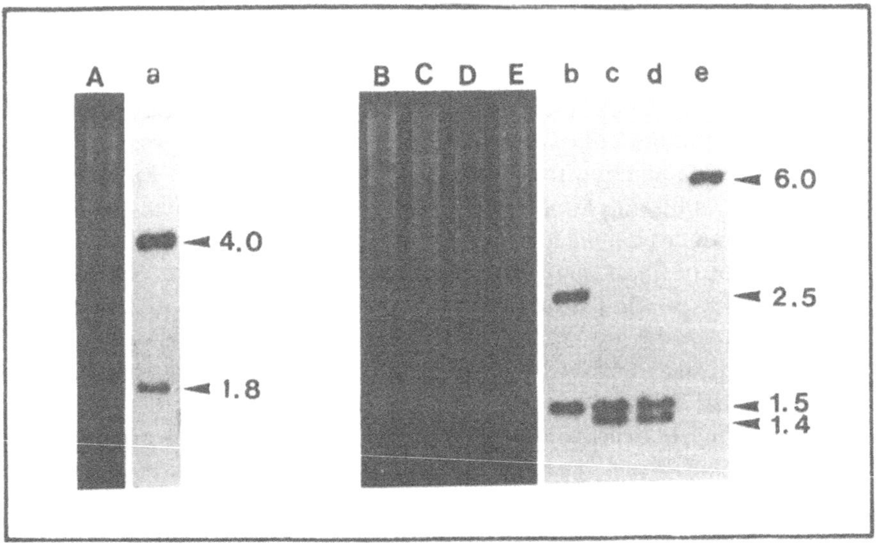

Fig. 3. EcoR1 restriction endonuclease analysis of total genomic DNA (lanes marked with capital letters) and autoradiograms of the corresponding southern blots after hybridization with α^{32}P–labelled pID1 (lanes marked with small letters). Lanes Aa) *R. Meliloti* Rme 41. Chickpea rhizobial strains are shown in lanes Bb (3H0a9), Cc (3H0a1), Dd (CC1192), and Ee (IC-72M). Figures are sizes (Kb) of fragments with strong homology to pID1 probe. The sizes were calculated by using HindIII restriction fragments of λ–DNA as molecular weight standards.

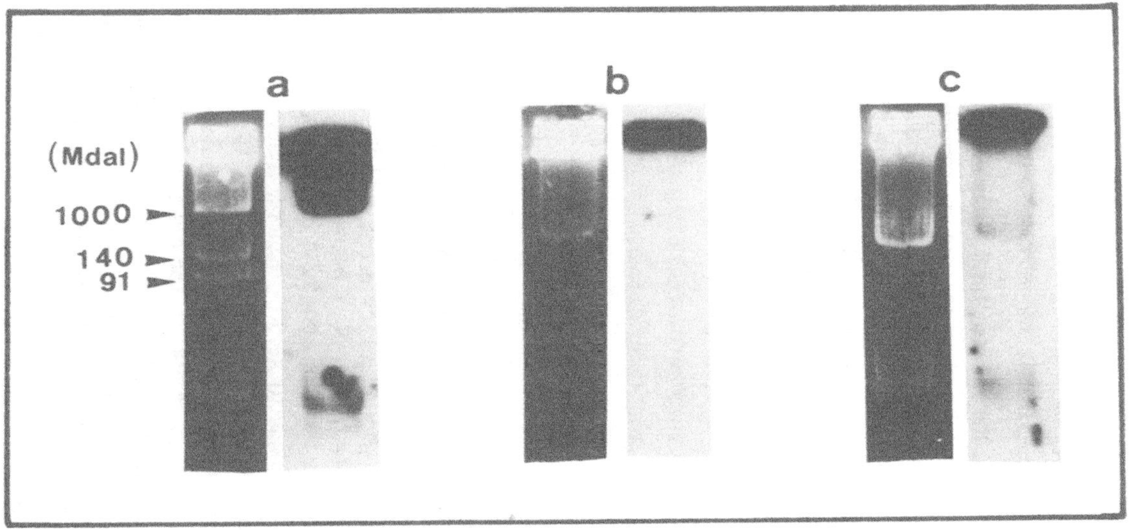

Fig. 1. Composite photograph showing ethidium bromide-stained gels (first of each pair of tracks) and the corresponding autoradiogram (second track) of the dried gels after hybridization to DNA from pID1 *nif* probe. a) A 1:1 mixture of *R. Meliloti* strains Rme 41 and L5.30 containing plasmids of known molecular weight (Casse *et al.* 1979); b) Chickpea rhizobail strain UPM-Ca7; c) Chickpea rhizobial strain IC-72M.

Table 3. Physiological and genetic characteristics of *Cicer-Rhizobium* compared with fast- and slow-growing groups of rhizobia.

Characteristics rhizobia	Chickpea rhizobia	Fast-growing rhizobia	Slow-growing
Growth rate (dt, h)	4–8.5	2–4 [a]	> 6 [a]
pH on YMB	Acidification	Acidification [a]	Alkalinization [a]
Reaction on litmus milk	Acid	Acid or alkaline [a]	Alkaline [a]
	serum zone	serum zone	no serum zone
6-PG dehydrogenase	(+)	+ [b]	– [b]
nif gene location	Chromosome	Plasmids [c]	Chromosome [d]

[a] Vincent *et al.* 1979; [b] Martinez de Drets and Arias 1972; [c] Nuti *et al.* 1979; Rosenberg *et al.* 1981; [d] Haugland and Verma 1981; Masterson *et al.* 1982.

examples). In these strains, the *nif* structural genes appeared to be contained in two EcoR1 fragments of 1.5 and 1.4 kb. The two remaining strains showed different hybridization patterns. Strain 3HOa9 had a minor change affecting the 1.4 kb fragment (track (b) in Fig. 3), while strain IC-72M exhibited a completely different pattern of homology to the *nif* probe, hybridizing only to one EcoR1 fragment of 6.0 kb (track (e) in Fig. 3). These results indicate a strong conservation of the *nif* DH genes within strains of *Rhizobium* from chickpeas.

Discussion

Twenty-two strains of chickpea *Rhizobium* from world-wide collections were examined for the cultural, physiological, and genetic characteristics currently used to separate fast- and slow-growing groups of rhizobia. The results are summarized in Table 3. The strains examined were highly diverse in their growth rates in YMB but very similar on the basis of reactions in YMB and on litmus milk media, and on the location and conservation of *nif* genes.

The growth rate limits used to separate fast- and slow-growing groups of rhizobia have not been clearly defined. Vincent *et al.* (1979) established that fast-growers had doubling times of about 2-4 h, whereas doubling times for slow-growers were more likely to be 6 h or more. Kennedy and Greenwood (1982) reported doubling times of 5.5 h or less for fast-growers and 8.5 h or more for slow-growers. According to these authors, fast- and slow-growing strains exist within chickpea rhizobia (Table 1), thus confirming previous reports (Raju 1936; Okon *et al.* 1972). However, we did not find a clear separation of chickpea rhizobia into two distinct groups on the basis of growth rates since strains that had intermediate doubling times were also observed.

Unlike slow-growers which alkalinize YMB medium and produce an alkaline reaction without the formation of a serum zone on litmus milk, all the strains of chickpea rhizobia studied produced acid in YMB and most of them caused an acid reaction with serum zone formation in litmus milk. With regard to these characteristics, chickpea rhizobia resemble fast-growing rhizobia, particularly *R. meliloti* (Vincent *et al.* 1979; Graham and Parker 1964). Further similarities with the fast-growing group are evident from the enzymatic analysis data. NADP-linked 6-phosphogluconate dehydrogenase (NADP-6PGD) is a key enzyme in the pentose phosphate pathway and has been reported to be absent in cell-free extracts of the slow-growing group of rhizobia (Martinez de Drets and Arias 1972). NADP-6PGD activity was detected in all six rhizobial strains studied in this work (Table 2). In all cases, the observed rates were much lower than those reported for fast-growing rhizobia (Martinez de Drets and Arias 1972). No correlation was found between the growth rate and NADP-6PGD activity of chickpea rhizobia. Although this study should be extended to a larger number of strains, this lack of correlation indicates that there is no evidence to separate chickpea rhizobia into two distinct physiological groups as could be expected from the differences in growth rates.

The plasmid analysis shows that most strains contain only one plasmid or no detectable plasmids. This is in contrast to the multiple plasmids found in most fast-growing *Rhizobium* species by Casse *et al.*(1979), Hirsch *et al.* (1980), Beynon *et al.* (1981), and Djordjevic *et al.* (1982). Furthermore, unlike plasmids of fast-growing rhizobia (Nuti *et al.* 1979; Prakash *et al.* 1981; Hombrecker *et al.* 1981), plasmids of chickpea rhizobia have not shown detectable nitrogen fixation (*nif*) genes. Strain IC-72M may be an exception, since it contains a plasmid of approximately 140 Mdal with homology to *nif*DH genes from *R. meliloti*. However, it should be examined to determine whether or not this homology is due to authentic *nif* genes. The possibility that *nif* genes were located on a megaplasmid that we are not able to resolve cannot be ignored, although it is improbable since megaplasmids of *R. meliloti* of approximately 1000 Mdal were consistently resolved by the plasmid visualization procedure used by Burkardt and Burkardt (1984). Therefore, it is very likely that *nif* genes in chickpea rhizobia are only on the chromosome, as has been reported in slow-growing *Bradyrhizobium japonicum* (Haugland and Verma 1981; Masterson *et al.* 1982).

The *nif*DH homology data show that *nif* genes are highly conserved within chickpea rhizobia. This is substantiated by the fact that EcoR1-restricted total DNA from 21 of the 22 strains examined exhibited an almost identical pattern of fragments homologous to *R. meliloti nif*DH probe. The high degree of sequence conservation is more remarkable considering the different geographical origins of the strains studied.

In conclusion, it is apparent from the summary presented in Table 3 that rhizobia nodulating chickpea form a relatively homogeneous group of rhizobia that do not conform to the criteria used for separation of classical fast- and slow-growing rhizobia. Since, in addition, these endosymbiotic bacteria are highly specific for *Cicer arietinum* we consider that, taxonomically, they should not be split into the genera *Rhizobium* and *Bradyrhizobium*. To clarify the taxonomic position of *Rhizobium*, more extensive biochemical and genetic analyses are needed.

Acknowledgements

Thanks to J.C. Burton, H.H. Keyser, J. Brockwell, O.P. Rupela, O.A. Berestesky and J. Olivares for providing strains of *Rhizobium*. This work was supported by grants from the Comision Asesora de Investigacion Cientifica y Tecnica and from the Ministerio de Agricultura, Pesca y Alimentacion.

References

Banfalvi, Z., Sakanyan, V., Koncz, C., Kiss, A., Dusha, I. and Kondorosi, A. 1981. Location of nodulation and nitrogen fixation genes on a high molecular weight plasmid of *R. meliloti*. Molecular General Genetics 184: 318-325.

Beringer, J.E. 1974. R-factor transfer in *Rhizobium leguminosarum*. Journal of General Microbiology 84: 188-198.

Beynon, J.L., Beringer, J.E. and Johnston, A.W.B. 1980. Plasmid and host-range in *R. leguminosarum* and *R. phaseoli*. Journal of General Microbiology 120: 421-429.

Burkardt, T. and Burkardt, H.J. 1984. Visualization and exact molecular weight determination of a *Rhizobium meliloti* megaplasmid. Journal of Molecular Biology 175: 213-218.

Casse, F., Boucher, C., Julliot, J.S., Michel, M. and Denarie, J. 1979. Identification and characterization of large plasmids in *Rhizobium meliloti* using agarose gel electrophoresis. Journal of General Microbiology 113: 229-242.

Corbin, E.J., Brockwell, J. and Gault, R.R. 1977. Nodulation studies on chickpea (*Cicer arietinum*). Australian Journal of Experimental Agriculture and Animal Husbandry 17: 126-134.

Corbin, D., Ditta, G. and Helinski, D.R. 1982. Clustering of nitrogen fixation (*nif*) genes in *Rhizobium meliloti*. Journal of Bacteriology 149: 221-228.

Djordjevic, M.A., Zurkowski, W. and Rolfe, B.G. 1982. Plasmid and stability of symbiotic properties of *Rhizobium trifolii*. Journal of Bacteriology 151: 560-568.

Duke, A.J. 1981. Pages 50-56 *in* Handbook of Legumes of World Economic Importance. Plenum Press, New York, USA.

Eckhardt, T. 1978. A rapid method for the identification of plasmid deoxyribonucleic acid in bacteria. Plasmid 1: 584-588.

Gaur, Y.D. and Sen, A.N. 1979. Cross inoculation group specificity in *Cicer-Rhizobium* symbiosis. New Phytology 83: 745-754.

Graham, P.H. and Parker, C.A. 1964. Diagnostic features in the characterization of the root-nodule bacteria of legumes. Plant and Soil 20: 383-396.

Haugland, R. and Verma, D.P.S. 1981. Interspecific plasmid and genomic DNA sequence homologies and location of *nif* genes in effective and ineffective strains of *Rhizobium japonicum*. Journal of Molecular and Applied Genetics 1: 205-217.

Hirsh, P.R., van Montagu, M., Johnston, A.W.B., Berwin, N.J. and Shell, J. 1980. Physical identification of bacteriocinogenic nodulation and other plasmids in strains of *R. leguminosarum*. Journal of General Microbiology 120: 403-412.

Hombrecker, G., Brewin, G. and Johnston, A.W.B. 1981. Linkage of genes for nitrogenase and nodulation ability on plasmids in *Rhizobium leguminosarum* and *R. phaseoli*. Molecular General Genetics 182: 133-136.

Jordan, D.C. 1984. Family III. Rhizobiaceae Conn, 1938. Pages 234-266 *in* Bergey's Manual of Systematic Bacteriology, 9th edition, Vol. 1. (Krieg, N.R. and Holt, J.G., eds.). Williams and Wilkins Co., Baltimore,USA.

Keel, B.B., Hamilton, P.B. and Elkan, G.H. 1969. Glucose catabolism in *Rhizobium japonicum*. Journal of Bacteriology 97: 1184-1194.

Kennedy, L.D. and Greenwood, R.M. 1982. 6-phosphogluconate and glucose-6-phosphate dehydrogenase activities, growth rate and acid production as taxonomic criteria for *Rhizobium*. New Zealand Journal of Science 25: 361-366.

Martinez de Drets, G. and Arias, A. 1972. Enzymatic basis for the differentiation of *Rhizobium* into fast- and slow-growing groups. Journal of Bacteriology 109: 467-470.

Masterson, R.V., Russi, P.R. and Atherly, A.G. 1982. Nitrogen fixation (*nif*) genes and large plasmids of *Rhizobium japonicum*. Journal of Bacteriology 152: 928-931.

Nuti, M.P., Lepidi, A.A., Prakash, R.K., Schilperoot, R.A. and Cannon, C. 1979. Evidence for nitrogen (*nif*) genes on indigenous *Rhizobium* plasmids. Nature 282: 533-535.

Okon, Y., Eshel, Y. and Henis, Y. 1972. Cultural and symbiotic properties of *Rhizobium* strains isolated from nodules of *Cicer arietinum* L. Soil Biology and Biochemistry 4: 165-170.

Pareek, R.P. 1979. Studies of effectiveness of different strains of chickpea (*Cicer arietinum* L.). Indian Journal of Microbiology 19: 123-129.

Patil, P.L. and Medhave, N.S. 1974. Seed inoculation studies in gram (*Cicer arietinum*) with different strains of *Rhizobium* sp. Plant and Soil 40: 221-223.

Prakash, R.K. Shilperoot, R.A. and Nuti, M.P. 1981. Large plasmids of fast-growing rhizobia: homology studies and location of structure nitrogen fixation (*nif*) genes. Journal of Bacteriology 145: 1129-1136.

Purello, M. and Balazs, I. 1983. Direct hybridization of labelled DNA to DNA in agarose gels. Analytical Biochemistry 128: 393-397.

Raju, M.S. 1936. Studies on bacterial-plant groups of cowpea, *Cicer* and dhaincha. I. Classification. Zentralblatt fur Bakteriologie Parasitenkunde infektionskrankheiten und Hygiene, Abteilung II, 94: 249-262.

Riess, G., Holloway, B.W. and Puhler, A. 1980. R68. 45, a plasmid with chromosome mobilizing ability (cma) carries a tanden duplication. Genetic Research 36: 99-109.

Rigby, P.W.J., Dieckmann, M., Rodes, C. and Berg, P. 1977. Labelling deoxyribonucleic acid to high specific activity *in vitro* by nick translation with DNA polymerase I. Journal of Molecular Biology 113: 237-251.

Rosenberg, C., Biostard, P., Denarie, J. and Casse-Delbard, F. 1981. Genes controlling early and late functions in symbiosis are located on a megaplasmid in *Rhizobium meliloti*. Molecular General Genetics 184: 326-333.

Rosenberg, C., Casse-Delbard, F., Dusha, I., David, M. and Boucher, C. 1982. Megaplasmids in the plant-associated bacteria *Rhizobium meliloti* and *Pseudomonas solanacearum*. Journal of Bacteriology 150: 402-406.

Sen, A.N. 1966. Inoculation of legumes as influenced by soil and climatic conditions. Indian Journal of Agricultural Science 36: 1.

Simon, J. 1914. Uber die verwandtschaftsverhaltnisse der leguminosen-wurzelbakterien. Zentralblatt fur Bakteriologie Parasitenkunde Infektionskrankheiten und Hygiene, Abteilung II, 41: 470.

Skerman, W.B.D. 1967. Pages 246-247 *in* A Guide to Identification of the Genera of Bacteria. Williams and Wilkins Co., Baltimore, USA.

Subba Rao, N.S. 1976. Field response of legumes in India to inoculation and fertilizer applications. Pages 255-268 *in* Symbiotic Nitrogen Fixation in Plants. International Biological Program No 7 (Nutman, P.S., ed.). Cambridge University Press, London, England.

Vincent, J.M. 1970. Pages 3-4 *in* A Manual for the Practical Study of Root Nodule Bacteria. International Biological Program No 15. Blackwell Scientific Publications, Oxford, England.

Vincent, J.M., Nutman, P.S. and Skinner, F.A. 1979. Pages 49-69 *in* Identification Methods for Microbiologists (Skinner, F.A. and Lovelock, D.W., eds.). Academic Press Inc., New York, USA.

D.P. Beck and L.A. Materon (editors), Nitrogen Fixation by Legumes in Mediterranean Agriculture
© 1988 ICARDA – Printed in the Netherlands. ISBN 90 247 3624 2

Rhizobium Strains from Wild and Primitive Legumes: A Nuisance or a Valuable Gene Pool?

T.A. LIE [1], D. GOKTAN [2] and M. ENGIN [3]

[1] *Dept. of Microbiology, Agricultural University, Wageningen, The Netherlands*
[2] *Engineering Faculty, Ege University, Izmir, Turkey*
[3] *Agricultural Faculty, Cukurova University, Adana, Turkey*

Summary. Soils from several European countries contain strains of *Rhizobium* capable of forming effective nodules on the cultivated pea cultivar Rondo. The range of effectivity, as expressed on cultivar Rondo, is not so great and almost the same range can be found within each country.

The pea cultivar Afghanistan, which is not nodulated by European rhizobial strains, is nodulated by strains which are only present in soils of several countries in the Middle East. Adana, a wild pea cultivar from South Turkey, is even more specific, and only forms an effective symbiosis with strains present in a restricted region in South Turkey.

The majority of the Middle East strains are poor nitrogen fixers when in symbiosis with the cultivated pea, but there are a few strains which are as effective as the European strains. We suggest that, due to selection, the European strains originated from these few effective Middle East strains.

Introduction

Bacteria of the genus *Rhizobium* are soil-borne microorganisms capable of inducing root nodules on plants belonging to the family Leguminosae. In the absence of the host plant, these bacteria grow as a common saprophyte on a wide range of substrates. It is not possible to isolate rhizobia directly from the soil but they can be isolated indirectly from the induced root nodules, which can be regarded as an enrichment culture of *Rhizobium*. However, in view of the high specificity, it is likely that the plant acts as a specific trap plant, capturing only those bacteria with which it is compatible. Therefore the majority of rhizobial strains studied and kept in our bacterial collections are isolated from cultivated crop plants. These plants, in contrast to wild plants and primitive landraces, in general show little genetic variation, since only a limited number of plant genotypes are used to create our modern high-yielding crop plants. Therefore, it is possible that by using these genetically related plants as a trap host, a specific group of rhizobial strains has been selected.

The major area of agricultural production is usually far removed from the regions where the progenitors of crops are supposed to have originated. These regions, designated as centres of origin, centres of diversity, or gene centres are characterised by a great diversity of genetically variable plants (Vavilov 1951). We (Lie 1978; 1981) showed that this is also true for rhizobial strains. We assume that, in the early days of agriculture, the introduction of leguminous plants into a new area was accompanied by a casual transfer of a limited number of rhizobial genotypes as a soil contaminant of the pods or seeds. Under conditions of low soil nitrogen, which is common in primitive agriculture, the bacteria may act as a selective agent favouring those plants which form effective nodules with these bacteria. At the same time, the plants will also favour bacterial growth and survival.

An important characteristic of soil under natural conditions or primitive agriculture is the occurrence of large differences in soil properties within a small area, which may provide specific niches for the different types of rhizobia. The use of modern machines for soil tillage and the wide application of fertilizers tend to reduce or even eliminate these differences, allowing the cultivation of a single plant line over large areas.

In this paper, we explore whether the above considerations may affect the genetic variation of rhizobial populations. Essentially, we compare strains from cultivated plants in agricultural soils with those from wild and primitive legumes in the plant gene centers.

Genetic Variation of Rhizobial Strains in European Soils

Wild and primitive forms of *Pisum sativum* are found in the Middle East, Central Asia, Ethiopia, and the Mediterranean. In the Middle Ages, the pea became a prominent crop in central and north Europe, where the climate is favourable for its growth. Most of the commercial varieties available today originated from the varieties developed in Europe (Marx 1977). In our study, using peas from different parts of the world, we found that all the cultivated pea lines can form an effective symbiosis with a Dutch strain of *Rhizobium* specially selected for peas in the Netherlands (Lie *et al.* 1976). This is also the case for a pea line now being grown in the tropics, where it has been propagated for a long period in the absence of its specific strain (Lie *et al* 1984).

We observed that most agricultural soils in the Netherlands contain effective rhizobial strains for *Pisum sativum* and *Vicia faba*, except in the newly reclaimed soils of the polders. This is in accordance with the experience of farmers in the Netherlands that inoculation of peas and faba beans is not essential.

To obtain an impression of the rhizobial population in European soils, the symbiotic response of the Dutch cultivated pea cultivar Rondo to a number of pure cultures isolated from Dutch and Swedish soils, was investigated. In addition, the response of the pea to inoculation with soils from some other European countries was studied (Table 1). In all cases, effective nodules were

Table 1. The effect of inoculation with pure cultures of *R. leguminosarum* strains or with soils from Europe on the yield of pea cv Rondo from the Netherlands.

Rhizobium strain from	No. of strains or soil samples	Strain yield (% of strain PF_2 *)	
		Mean	Range
Pure cultures			
The Netherlands	7	90.4	81–100
Sweden	10	89.3	65–106
Soil samples			
Greece	4	79.0	76– 84
Italy	5	90.6	72–101
Spain	2	87.5	85– 90

* The best *Rhizobium* strain selected in the Netherlands.

induced on the Dutch pea cultivar. There was some variation in the amount of nitrogen fixed and it is interesting that very effective strains for the Dutch pea, comparable to the Dutch strain PF_2, can be found in Sweden and Italy. These strains are about 10% more effective than the mean of strains from the Netherlands and Sweden (Table 1), indicating that selection of rhizobial strains is worthwhile. However, as shown later, the variation observed among the European strains was much lower than that found among the strains from the Middle East.

The Presence of Specific Rhizobial Strains in Soils of the Middle East

To demonstrate the presence of specific rhizobial strains in the soil, we used a number of pea lines which require highly specific strains. The primitive pea landrace, cultivar Afghanistan, and a wild form from south Turkey, cultivar Adana, were used. These pea lines cannot be nodulated by European strains (Lie 1978) but will nodulate with strains from the Middle East.

Our earlier studies (Lie 1971; 1978) and those from Ohlendorf (1983a; 1983b; 1985) and Young *et al.* (1982) showed that many pea lines from Afghanistan are "nodulation-resistant" to European strains. The cultivar Afghanistan contains a gene, sym 2, which confers general resistance to European strains (Lie 1984; Lie *et al.* 1976). The results in Table 2 show that soils from Europe and America do not contain rhizobial strains capable of nodulating the Afghan pea. In contrast, Afghanistan-specific strains seem to be very widespread in soils of the Middle East. These strains, termed Middle East strains, are also capable of nodulating the pea cultivar Rondo but, as shown later, many of the strains are not very effective with the cultivated pea.

A much more specific pea line was recently obtained near Adana in south Turkey. No nodules were formed on cultivar Adana by strain PF_2 (and many other European strains) and a strain from Anatolia, which are capable of

Table 2. Geographical distribution of *R. leguminosarum* strains capable of nodulating the modern pea cv Rondo and the primitive landrace cv Afghanistan.

Country	No. of soil samples	No. of soils containing *Rhizobium* for	
		cv Rondo	cv Afghanistan
NW and S. Europe	55	52	0
N and S. America	3	3	0
Turkey	145	140	140
Iran	6	6	6
Yemen	3	3	3
Syria	10	nd	7

nd = not determined.

Table 3. Nodulation and nitrogen fixation of pea cv Afghanistan and pea cv Adana in symbiosis with *Rhizobium* strains from the Netherlands (PF_2), central Anatolia (Tom), and South Turkey (Ad^k).

Rhizobium strain	Nodulation		Nitrogen fixation (mg N/plant)	
	Afghanistan	Adana	Afghanistan	Adana
PF_2	None	None	None	None
Tom	Large red nodules	None	11.8	None
Ad^k	Large red nodules	Large red nodules	22.5	26.2

nodulating the Afghan pea (Table 3). However, a highly effective symbiosis was formed with a strain isolated from the same field where this plant was growing.

Table 4. Geographical distribution of *R. leguminosarum* strains capable of nodulating the wild pea cv Adana from south Turkey.

Country	No. of soil samples	Nodulation
The Netherlands	5	None
Sweden	1	None
Italy	3	None
Greece	3	None
Turkey		
West coast (Izmir, Mulga, Fethiye)	32 *	With most samples no nodulation. In some cases a few small white nodules.
Central Anatolia (Konya, Ankara)	5 *	With most samples no nodulation. In one case few, white nodules.
South Turkey (Cukurova plain around Adana and eastwards to Antakya)	18 * 17	With the exception of 1 sample all the other samples induced large, red, very effective root nodules.

* All these soil contain *Rhizobium* capable of nodulating pea cv Afghanistan.

Using cultivar Adana, the distribution of rhizobial strains in Turkey was studied (Table 4). In general, no nodules were formed by strains present in European soils or in soils from the west coast and central Anatolian regions of Turkey. In some cases, a few small, white nodules of an ineffective type were formed. Highly effective nodules were only formed by strains from a restricted area in the Cukurova plain surrounding Adana and in the area eastwards towards Antakya. In the north, the Taurus mountains seem to be a barrier since no Adana-specific strains were present in soils beyond this mountain range. Recently, nodulation was obtained with soils from Syria, north of Aleppo.

The results presented here clearly demonstrate that highly specific strains of *Rhizobium* are present in soils from the Middle East which are not found in European soils. To detect and isolate these specific strains, wild and primitive plants are required with a high symbiotic specificity.

The Symbiotic Response of the Cultivated Pea to Middle Eastern Rhizobial Strains

As shown in Table 2, the strains from the Middle East are also capable of nodulating cultivar Rondo. However, there is extremely large variation in the N_2-fixing capacity, ranging from completely ineffective to effective. The majority of the strains belong to the partially effective type (Fig. 1; Table 5). These results are in sharp contrast to the results obtained with rhizobial populations in European soils, which all belong to the effective group (Fig. 1). This suggests that the genetic variability of strains in Middle Eastern soils is far greater than that from European soils.

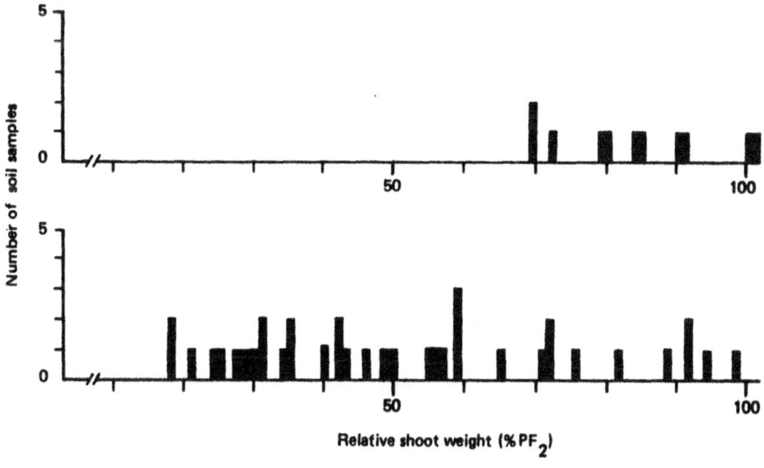

Fig. 1. Frequency distribution of soil samples from Europe (top) and Turkey (bottom) containing rhizobial populations of different symbiotic effectiveness as expreseed on a pea cv Rondo from the Netherlands. Plants inoculated with strain PF_2 serve as a control (100%).

Table 5. Classification of *R. leguminosarum* populations in soils from Europe and Turkey, based on nitrogen fixation, in association with pea cv Rondo *.

Nitrogen fixation (Class) **	*Rhizobium leguminosarum* populations in soils from	
	Europe (28 samples)	Turkey (42 samples)
I	0	19
E/I	10	65
E	90	16

* Plants were grown in a N-free nutrient solution and inoculated with 2-3 g of soil. The data are compiled from several experiments and given as % of the total number of soil samples investigated.

** I, ineffective; E/I, partially effective; E; effective

It is understandable that few studies are devoted to those strains, which are the normal symbionts of wild and primitive legumes. To the agronomist, these strains, being ineffective with our modern high-yielding leguminous crops, are a nuisance.

Concluding Remarks

These results support our hypothesis that rhizobial strains from agricultural soils in Europe have a rather narrow genetic base. This is in accordance with the fact that our modern leguminous crops are derived from a limited number of plant genotypes. When tested on cultivar Rondo, all the European strains (or populations) are capable of forming an effective symbiosis, which is presumably the result of selection in soils low in combined nitrogen.

A much greater genetic variability can be found among the Middle Eastern strains. In this group, strains can be isolated which have symbiotic properties not found in any of the European strains. However, an exception was recently found in a field in Denmark where the pea cultivar Afghanistan was cultivated some years before. From this soil strains were isolated which are capable of nodulating the Afghan pea (Jensen and Engvild 1985). The range of effectivity of the Middle East strains, as expressed on Rondo, is much wider than that observed in European soils. Since the majority of these strains form ineffective or partially effective nodules on the cultivated pea, these bacteria are a problem for the agronomist.

Among the Middle East strains, there are a few which can form an effective symbiosis with the cultivated legume. It is likely that the European strains originated from these bacterial genotypes and were further distributed in agricultural soils along with the cultivated legume.

So far, attention has focussed on the N_2-fixing capacity of *Rhizobium* strains. However, if a strain is to be used as an inoculant, other traits such as competitive ability and survival under stress conditions are also important. We assume that since genetic variability among the Middle Eastern strains is high, the chances of selection among these strains are greater than among the "cultivated" strains.

Acknowledgements

This research is partly financed by OECD. Thanks are due to Ir. J.J. Kessler for providing the soil samples from Yemen.

References

Jensen, E.S. and Engvild, K.C. 1985. Danish *Rhizobium leguminosarum* strains nodulating "Afghanistan" pea (*Pisum sativum* L.). Page 35 *in* Nitrogen Fixation Research Progress (Evans, H.J., Bottomley, P.J. and Newton, W.E., eds.). Martinus Nijhoff, Dordrecht, The Netherlands.

Lie, T.A. 1971. Symbiotic nitrogen fixation under stress conditions. Plant and Soil, Special Volume: 127-137.

Lie, T.A. 1978. Symbiotic specialisation in pea plants: The requirement of specific *Rhizobium* strains for peas from Afghanistan. Annals of Applied Biology 88: 462-465.

Lie, T.A. 1981. Gene centres, a source for genetic variants in symbiotic nitrogen fixation: Host-induced ineffectivity in *Pisum sativum* ecotype fulvum. Plant and Soil 61: 125-134.

Lie, T.A. 1984. Host genes in *Pisum sativum* L. conferring resistance to European *Rhizobium leguminosarum* strains. Plant and Soil 82: 415-425.

Lie, T.A. and Goktan, D. 1984. Gene centres, a source for genetic variants in symbiotic nitrogen fixation: the symbiotic response of the cultivated pea to *Rhizobium leguminosarum* strains from Europe and the Middle East. Plant and Soil 82: 359-367.

Lie, T.A., Hille Ris Lambers, D. and Houwers, A. 1976. Symbiotic specialisation in pea plants: some environmental effects on nodulation and nitrogen fixation. Pages 319-333 *in* Symbiotic Nitrogen Fixation in Plants (Nutman, P.S., ed.). University Press, Cambridge, England.

Lie, T.A. , Akkermans, A.D.L. and van Egeraat, A.W.S.M. 1984. Natural variation in symbiotic nitrogen-fixing *Rhizobium* and *Frankia* spp. Antonie van Leeuwenhoek 50: 489-503.

Marx, G.A. 1977. Classification, genetics and breeding. Pages 21-43 *in* The Physiology of the Garden Pea (Sutcliffe, J.F. and Pate, J.S., eds.). Academic Press, London, England.

Ohlendorf, H. 1983a. Selektion auf Resistenz von *Pisum sativum* gegen *Rhizobium leguminosarum* Stamm 311d. Zeitschrift fuer Pflanzenzuechtung 90: 204-221.

Ohlendorf, H. 1983b. Untersuchungen zur Vererbung der Resistenz von *Pisum sativum* gegen *Rhizobium leguminosarum* Stamm 311d. Zeitschrift fuer Pflanzenzuechtung 91: 13-24.

Ohlendorf, H. 1985. Kontrollmechanismen der Leguminosen-*Rhizobium* Symbiose im Zusammenhang mit dem Infektionsprocess. Angewandte Botanik 59: 279-303.

Vavilov, N.I. 1951. The origin, variation, immunity and breeding of cultivated plants (English translation). The Ronald Press Co., New York, USA.

Young, J.P.W., Johnston, A.W.B. and Brewin, N.J. 1982. A search for peas (*Pisum sativum* L.) showing strain specificity for symbiotic *Rhizobium leguminosarum*. Heredity 48: 197-201.

Discussion

Mytton

Have you grown the Afghan pea for several seasons in a soil containing European populations of *R. leguminosarum* to see if *Rhizobium* variants arise which can overcome its resistance to nodulation?

Lie

No, we have not done these experiments. We did introduce a Middle Eastern *Rhizobium* strain Tom, which is serologically identifiable and also resistant to streptomycin, into some Dutch soils. This strain can be recovered after 3 years. There is an interesting report from Danish researchers, who found *Rhizobium* strains capable of nodulating pea cv Afghanistan in one plot, where pea cv Afghanistan was grown 7 years before. These strains seem to be different from the Middle Eastern strain Tom which we use as a standard strain.

Solh

Is the amount of diversity in *Rhizobium* strains different in self-pollinated and cross-pollinated plants?

Lie

I have no specific data, but since genetic variability is greater in cross-pollinated than in self-pollinated plants, this may well be the case.

Papastylianou

We have heard in the past 3 days and in your studies that *R. leguminosarum* is very efficient and there is no need to introduce new strains in the Mediterranean. What will happen if one of the strains which we introduce for our studies is highly competitive and ineffective and it is released in this region?

Lie

One should be careful if a new *Rhizobium* strain from abroad is to be introduced for a local leguminous crop. It may be better to select among the local rhizobial strains, which are presumably adapted to local conditions. For a new plant variety, of course, it is necessary to introduce the associated rhizobial strain. Very little is known about the effect of an introduced strain of *Rhizobium* on the local rhizobial population, and I think this is an important, but technically very difficult study. It is essential to differentiate between the introduced and the local rhizobial strains and to estimate the numbers for some years in the presence of the crop plant or the local indigenous plants. We found in our studies with pea cv Afghanistan grown in water culture, that nodulation of this pea by the Middle Eastern strain Tom is completely suppressed by a Dutch strain, PF_2. We do not know if this will also take place if we grow the pea in soils from the Middle East and inoculate with the highly competitive strain PF_2. We also do not know what will happen to the local *Rhizobium* population after several years of repeatedly introducing a highly competitive strain. This is important in the soils of the plant gene centres, where there is also high genetic variability in *Rhizobium* strains.

Smith

In view of the enormous capacity of living organisms to produce strains by genetic recombination, can it be argued as unnecessary to take steps to conserve strains of rhizobia?

Lie

Genetic recombination implies the availability of different types of genetic material, which is presumably the case in the gene centers. We have some indications that in cultivated soils the genetic variability of *Rhizobium* is relatively low, and we do not know what might happen if we introduce an aggressive rhizobial strain into soils of the gene centres. A real danger is the rapid disappearance of wild and primitive plants, the natural hosts of the indigenous strains. The absence of these plants may also lead to the disappearance of the specific rhizobial strains. Another practical problem is that we need these wild leguminous plants, which have specific *Rhizobium* requirements, for the specific isolation of a certain group of strains.

I am aware of the large problems we have to face in order to conserve strains, i.e., to make a *Rhizobium* gene bank. I am not satisfied with bacterial culture collections, since these are dependent on the plant used for isolation of strains and thus represent only a fraction of the gene pool in the soil. I prefer to collect and store soil samples, from which we can isolate the bacteria by using the correct plant, even after storage for 15 years at 4° C.

What we need is more studies of strains from the gene centers in order to know the genetic potentialities and how to conserve soil samples so that the genetic variability of *Rhizobium* remains stable during storage.

D.P. Beck and L.A. Materon (editors), Nitrogen Fixation by Legumes in Mediterranean Agriculture
© 1988 ICARDA – Printed in the Netherlands. ISBN 90 247 3624 2

Host-*Rhizobium* Relationships and their Implications for Legume Breeding

R.L. MYTTON [1], D.M. HUGHES [1] and J. KAHURANANGA [2]

[1] *Welsh Plant Breeding Station, Aberystwyth, UK*
[2] *International Livestock Centre for Africa, Addis Ababa, Ethiopia*

Summary. The legume-*Rhizobium* symbiosis is subject to genetic interactions which result in large and unpredictable effects on nitrogen fixation. This paper discusses possible strategies to improve the symbiosis either generally, through the plant, or specifically, through selection for specific adaptation between restricted symbionts.

Introduction

Gene expression is recognized as being environmentally dependent, and subtle alterations in growing conditions can produce large changes in individual phenotypes. Many of these effects are unpredictable and result either in alterations to the ranked ordering of genotypes over time and from one environment to another, or to changes in absolute differences between genotypes without altering ranking. These effects make it difficult to define genotypic excellence in relation to a range of environments such as those encountered in practical agriculture.

Engledow (1925) was, perhaps, the first to express the implications of such effects when he wrote, "we can no longer hope, as breeders once did, for the new form which everywhere and in all years will excel". Over the intervening years much has been achieved in the identification of genes, in predicting their behaviour during reproduction and hybridization, and in explaining the processes by which the genotype becomes phenotype. Such understanding has often led to predictive methods of genetic improvement, but despite advances in biometrical analysis of genotype-environment interactions (Finlay and Wilkinson 1963), they frequently defy prediction and must still be dealt with by selecting plants which have been tested in a range of environments and over a number of seasons.

Quantitative variations in nitrogen fixation can be substantially determined by plant–*Rhizobium* interactions (Sherwood and Masterson 1974; Mytton 1975; Mytton *et al.* 1977; Mytton and Livsey 1983; Mytton *et al.* 1984) and this has the same consequences for improving the legume–*Rhizobium* symbiosis as does genotype-environment interaction for general crop improvement. It is theoretically possible to eliminate undesirable effects of reactions and interactions with indigenous rhizobia by selecting elite strains for use as inoculants. This is relatively simple to achieve if *Rhizobium* is not present in the soil, but where it is present, selected strains often fail to influence the legume phenotype (Brockwell 1981). Practical experience suggests that there are a number of undefined factors determining the success of *Rhizobium* genotypes in soil. The ability to colonise roots and produce nodules in competitive situations is also little understood. As a consequence, we are unable to reliably predict the nature of a successful inoculant strain or to ameliorate the environment to favour particular genotypes. In the long term, yields of nitrogen from legume cultivars are dependent upon the forces determining the population structure of rhizobia. Improvements in fixation which have agricultural significance depend upon minimising the effects of less desirable indigenous rhizobia on the legume phenotype. For the future, a growing understanding of the function of genes determining the formation and productivity of the symbiosis should lead to more precise methods of manipulating them and predicting their expression. In the interim, the objectives of crop improvement are too important to wait upon possible future developments and must be persued by more empirical methods. The purpose of this paper is to consider ways in which existing methods of plant breeding can be adapted to avoid the formation of poor symbiotic associations.

When dealing with the interrelationship of several factors, the effect of each on the others can only be measured accurately where all combinations are measured simultaneously. Such factorial experiments provide a wide inductive basis to the main effect comparisons and they reveal and quantify interactions between the various factors. In legume-*Rhizobium* research the best known examples of this approach are probably the early experiments concerned with the transfer of nodule isolates between species. These revealed limitations to exchange and led to the cross inoculation taxonomic groupings of rhizobia (Fred *et al.* 1932).

In common with many attempts to fit biological systems into "neat boxes", it has been found that the boxes are more analogous to a "series of sieves" with barriers to reciprocal exchange being quantitative rather than qualitative (Wilson 1944). *Rhizobium* species have therefore been a subject of controversy ever since they were first defined. There can be no doubt that the original concept of the groupings identifies plant and *Rhizobium* genomes which are generally so strongly interactive as to preclude frequent and fully compatible exchange of partners. This indicates the existence of genetically determined discriminatory recognition phenomena between symbionts. On this basis, a possible strategy for improvement would be to further enhance host-strain specificity leading to the development of a co-adapted, highly effective plant-*Rhizobium* inoculation package. Alternatively, because the degree of specificity differs between species, it should be possible to "buffer" some plant populations rendering them less sensitive to variations in the *Rhizobium* genotype. There are two methods of doing this: a) genetic buffering, in which all individuals in a population have wide spectrum effectiveness, and b) population buffering, in which a genetically diverse population is constructed to contain individuals compatible with different genotypes of *Rhizobium*. Both methods would be aimed at producing populations with improved general symbiotic competence.

Before considering these strategies in more detail, it is important to emphasize the need to ensure that the legumes being considered for improvement are limited by symbiotic nitrogen fixation. It will be difficult to increase supply unless the demand exists.

Improving General Symbiotic Competence

In theory, it would be possible to improve general symbiotic competence in both the plant and the *Rhizobium* genotype, but at present the most practical target is the plant. Methods for establishing and maintaining new plant populations as stable components of agricultural systems are well defined: this advantage cannot reliably be conferred on *Rhizobium* populations in soil.

An essential test in any programme of improvement would be to screen plants against a range of rhizobia but this creates problems in defining a sample which will be representative of rhizobia occurring in general agriculture. To study plant-*Rhizobium* interactions, Mytton *et al.* (1984) have recently adapted a method of joint regression analysis commonly used to interpret genotype–environment interactions in plant breeding programmes (Finlay and Wilkinson 1963). This has allowed a degree of interpretation of host–*Rhizobium* interactions and has identified plant genotypes which have superior symbiotic competence with a range of *Rhizobium* strains.

More recent experiments with *Trifolium* species of African origin confirm that plants differ in their general symbiotic competence. A number of these species are known to have specialised *Rhizobium* requirements (Norris and 't Mannetje 1964). Table 1 shows whole plant dry weight of six African *Trifolium* species grown with all combinations of two *Rhizobium* isolates taken from each host. An analysis of variance is presented in Table 2. All species exchanged rhizobia and most exchanges resulted in reasonable levels of nitrogen fixation and plant growth. Plants were grown in test tube culture under conditions free from combined nitrogen (Rys and Mytton 1985). Nitrogen accumulated by fixation correlates strongly with yield under these conditions (Mytton 1984) and yield can therefore be used as a reliable measure of fixation. In this experiment, acetylene reduction also correlated well with yield ($R = 0.938$ at $P = 0.001$).

Some interactions occurred between strains and species. For example, *T. tembense* did badly with the isolates from *quartinianum* but these strains were generally good with most of the other species. No *Rhizobium* isolate gave consistently high yields with all species, but some plant species had much better broad spectrum effectiveness than others. In particular, *T. quartinianum* gave consistently high yields with the majority of isolates. A further experi-

Table 1. Whole plant dry weight (mg/plant) of six African *Trifolium* species grown with all combinations of two *Rhizobium* isolates from each host (Experiment 1, February 1985, WPBS). Plant species effects shown in columns, *Rhizobium* isolates in rows.

Rhizobium isolate from		*Trifolium* species						
		T. ruepellianum ILCA No (R.6218)	*T. steudneri* (R.6261)	*T. quartinianum* (R.6301)	*T. tembense* (R.8501)	*T. decorum* (R.9447)	*T. simense* (R.9470)	Mean
T. ruepellianum	Ra	23	28	29	8	40	10 (49/5)	27
	Rb	43	7	53	46	31 (202/7)	10	
T. steudneri	Sa	34	19	48	41	33	6	29
	Sb	28	21	55	42	16	7	
T. quartinianum	Qa	40	25	42	13	33	9	27
	Qb	45	25	50 (399/11)	11 (25/2)	30	8	
T. tembense	Ta	35	30	55	10	31	14	
	Tb	35	28	58	13	31	8	30
T. decorum	Da	43	24	55	48	18	10	
	Db	35	10	46	39	23	7	30
T. simense	Sia	33 (347/11) *	13 (127/10)	39	50	21	12	
	Sib	38	13	32	42	33	11	28
Mean		36	20	47	30	28	9	

* Figures in parenthesis refer to acetylene reduction (nmol C_2H_4/plant/h) and specific activity nmol C_2H_4/mg DM/h eg. 347/11 = 347 nmol C_2H_4/plant/h with a specific activity of 11 nmol/mg DM/h. Correlation between plant dry weight and AR = 0.938.

Table 2. Analysis of variance of *Trifolium* species- *Rhizobium* isolate factorial experiment data shown in Table 1.

Sources of variation	df	Mean squares variation
Plant species	5	9966.4 ***
Rhizobium isolates	11	189.0 ***
Species x isolates	55	510.3 *
Residual	288	79.3

ment was carried out in which a larger sample of rhizobial isolates (6/host) was tested onto a reduced number of species (3). Isolates were also tested against *Trifolium repens* and 20 strains of *R. trifolii* were inoculated onto the African clovers. *T. repens* failed to nodulate with the African rhizobia and the African species were not nodulated by any of the *R. trifolii* strains tested, thus confirming the symbiotic specialisation of these groups.

The larger sample of African rhizobia generally confirmed the results of the first experiment with all symbionts exchanging partners reasonably well. However, in this experiment, correlation between acetylene reduction and dry weight ($r = 0.742$, $P = 0.001$) was much lower than in the earlier test. The plant material was from the same sample as in the previous experiment and environmental conditions were identical. This reduction in correlation is very unsatisfactory because the coefficient of determination (r^2) suggests that less than 50% of differences in yield are accounted for by differences in fixation as measured by acetylene reduction. This is clearly erroneous and points to the need to treat the acetylene reduction method with caution (see Witty, this workshop).

We also examined the reaction of these species to soil populations of rhizobia, to supplement the information gathered from nodule isolates. Soil samples were provided by the International Livestock Centre for Africa (ILCA) and used to inoculate plants grown in test tubes. Results are shown in Table 3 with an analysis of variance in Table 4. All plants failed to nodulate in the Sodo and Kitulo soils and inoculation would clearly be essential to correct these deficiencies. Rhizobia in the Shola soil generally allowed most plant species to approach their yield potential as judged against controls given nitrate (Table 3, penultimate column), but there were still significant differences between species. For example, *T. decorum* and *T. quartinianum* grown with rhizobia produced 75 and 110%, respectively of their yield with nitrate.

In general, the yield of plants in the Uyole soil was much poorer than with the Shola soil and interactions occurred between species and soil sources. Notably, *T. quartinianum* and *T. tembense* gave comparable yields with the Shola rhizobia, but with the Uyole sample, *T. quartinianum* increased its yield whilst that of *T. tembense* was more than halved. *T. quartinianum* was the only species to give consistently high yield with both *Rhizobium* populations and

Table 3. Whole plant dry weight (mg) of five Ethiopian *Trifolium* species grown with sample populations of rhizobia taken from four sites in Africa. (Plants are dependent upon nitrogen fixation for growth.)

Species	Site					Controls *		Yield (% of N control)	
	Shola	Sodo	Uyole	Kitulo	Mean	no R/N	+ R/N	Sh **	Uy
T. rupellianum	32	5	20	5	16	9	33	97	60
T. steudneri	20	5	16	6	12	5	24	83	67
T. quartinianum	33	8	37	8	22	9	30	110	123
T. tembense	32	10	14	7	16	9	38	84	37
T. decorum	18	5	14	5	11	4	24	75	58
Mean	27	7	20	6	15	7	30	90	67

* Controls: no R/N = no *Rhizobium*, no nitrogen; + R/N = with *Rhizobium* with nitrogen.
** Sh = Shola, Uy = Uyole.

this supports the results of the initial experiment which indicated that it had good general symbiotic competence. A range of plant-to-plant variation was observed within these species and some of this is likely to be genetically based, indicating that desirable traits could be developed further through plant breeding. The results presented here correlated closely with ILCA's field trial data.

This information, together with that generated on *Medicago sativa* (Mytton and Livsey 1983), can be seen as an encouraging indication that plant breeding methods, if more closely coupled to appropriate reference populations of rhizobia, would allow further genetic improvements in the development of broader-spectrum effectiveness in some species.

The other approach to reducing cultivar sensitivity to different rhizobia would be to develop improved population buffering in legumes. This could be achieved by developing mixtures of differentially adapted plants so that contrasting genotypes within the mixture would be capable of forming highly effective symbiotic associations over a range of *Rhizobium* populations. The method would probably be most relevant to pasture legumes because seedling populations far exceed those of mature plants. Therefore, poor phenotypes should be eliminated and have no impact on production.

Table 4. Analysis of variance of *Trifolium* species-*Rhizobium* population factorial experiment data shown in Table 3.

Sources of variation	df	Mean squares
Plant species	4	456.7 ***
Soils (*Rhizobium* pops.)	3	2979.4 ***
Species X soils	12	164.8 ***
Reps.	6	7.4
Residual	114	8.9

Improving Specific Symbiotic Competence

The second method of improvement is to select for specific adaptation between restricted symbionts. This would normally carry the penalty of reduced relevance to general agriculture but in this case the possibilities of coincidental selection of plant and *Rhizobium* genotypes aimed at producing a co-adapted inoculation package could reduce the undesirable effects of indigenous rhizobia. However, this approach can only succeed if the selected rhizobia preferentially form nodules in the presence of local rhizobia. It is probable that the best hope for achieving this objective rests with the development of legume-*Rhizobium* recognition phenomena. A number of investigations have shown that there is variation for this character within legume species so that individual plants show differential preference for nodulation by particular components of *Rhizobium* strain mixtures (Caldwell and Vest 1968; Sherwood and Masterson 1974; Russel and Jones 1975; Roughley *et al.* 1976; Mytton and De Felice 1977; Caldwell and Vest 1977). These effects are genetically co-determined by the symbionts and are broadly heritable (Mytton and De Felice 1977; Hardarson and Jones 1979).

This method of improvement is currently being developed in *Trifolium repens* at the Welsh Plant Breeding Station. Preliminary results (Table 5) show that in some selection lines, yield and total nitrogen in F_2 and F_2 generations is 33% and 51% greater, respectively, than in the base population. The same material grown with non-limiting nitrate was 9% better in both generations than the base population. This indicates some general improvement in overall yield regardless of nitrogen source but a substantial improvement in fixation is also clearly evident. "Back selection" for yield with high mineral nitrogen has

Table 5. Response to selection for increased yield and nitrogen fixation in white clover (g DM/plant).

1. Lines selected for increased yield with *Rhizobium*

Generation	Fixed N	Growing conditions		
		% increase	Combined N	% increase
Base pop.	5.59		7.76	
F_2	7.44	33.1	8.50	9.5
Base pop.	5.76		7.90	
F_2	8.75	51.9	8.64	9.4

2. Lines selected for increased yield with combined nitrogen

Generation	Fixed N	Growing conditions		
		% increase	Combined N	% increase
Base pop.	5.59		7.76	
F_2	6.80	21.7	8.73	12.5
Base pop.	5.76		7.90	
F_2	7.36	27.7	8.98	13.7

been carried out as a control. The F_2 generation of this material shows a 28% improvement in yield and nitrogen fixation when grown without nitrate, confirming that improving plant vigour and thus the demand for nitrogen can bring about reasonable improvements in fixation, but this was not as large as that achieved in the F_2 selected for high fixation.

It has become evident in these experiments that the base population contains two types of potentially high-yielding plants; those which are primarily limited by nitrogen fixation and those limited by carbon fixation. Carbon-limited plants impede the rate of genetic advance under selection and must be eliminated by screening under high nitrate. This experimental breeding programme is carried out by growing plants in soil with a natural population of rhizobia. Both plant and *Rhizobium* genotypes are selected and carried forward for further cycles of hybridization and reselection.

At present the approach remains experimental but on a limited scale the potential of specifically matched cultivar–*Rhizobium* combinations has been demonstrated in hill land inoculation trials. Young and Mytton (1983) have shown that in situations where populations of indigenous rhizobia are absent or small, inoculation of the white clover cultivars Huia and S184 with WPBS *Rhizobium* strain A allowed Huia to outyield S184 by 600 kg DM/ha; inoculation with strain B reversed the rank order and S184 outyielded Huia by 1200 kg DM/ha. Subsequent trials confirmed the value of this approach and also demonstrated that environmental factors alter the symbiotic competence of the combinations tested to date. This is a factor which must be considered in the selection procedures.

Environmental Effects

Precise definitions of the effects of environmental variables on the survival, nodulating ability, and effectiveness of host–*Rhizobium* associations under field conditions are not easily determined but it is an important area of research which must accompany attempts at genetic improvement.

Of the many factors known to affect the symbiosis, soil nitrogen is particularly important and high levels can mask fixation effects. We have recently examined these effects in relation to establishment of clover in UK upland grazing systems. Grass–clover swards were established using both full cultivation and surface seeding techniques on contrasting soil types which previous observation suggests have either high or low mineralisation potential. A summary of results is shown in Table 6. On soil A the uncultivated treatment (no N) yielded 3.5 t dry matter/ha and this was increased by only 5.5% in the cultivated treatment (3.8 t/ha). Results were different on soil B where cultivation increased yield from 5.6 to 7.8 t/ha (39% improvement, $P < 0.01$). On soil A, additions of 450 kg N/ha improved yield substantially in both the uncultivated and cultivated treatments, with 36% and 60% increases, respectively ($P < 0.01$). On soil B, nitrogen fertilizer increased yield only in the

Table 6. Effect of cultivation and applied nitrogen (450 kg N/ha) on grass yield (t/ha) on two different upland soils.

	Soil A		Soil B	
	−N	+N	−N	+N
Uncultivated	3.6	4.9	5.6	6.8
Cultivated	3.8	6.1	7.8	7.6

uncultivated treatment (21% improvement, $P < 0.05$). In the cultivated treatment, it resulted in a small, but non-significant decrease in yield.

Clearly, in their uncultivated state both soils were strongly nitrogen–limited. Following cultivation, this limitation remained in soil A but it was removed in soil B. Clover was established in these experiments with the aid of inoculation, using strains of rhizobia specifically adapted to the cultivar (S184) and the environment (Mytton and Hughes 1984). A summary of clover yields is shown in Table 7. The legume was slow to establish in the swards and yield data were only obtained from a single, late harvest. Without inoculation, clover yields were very low (9-26 kg/ha) and differences between other treatments were generally non-significant. As expected, nitrogen fertilizer almost eliminated clover from all treatments. Clover responded to inoculation on both soils but the effects of cultivation produced strongly contrasting differences on the two soils. Thus, on soil A, now shown to be strongly nitrogen–limited in both cultivated and uncultivated states, inoculation was most beneficial on the cultivated treatment (144 kg clover DM/ha compared with 69 kg/ha on the cultivated). On soil B, now shown to be nitrogen–limited only when uncultivated, inoculation was least beneficial on the cultivated treatment (82 kg/ha increase compared with 220 kg/ha uncultivated).

Differences between soils in mineralisation of nitrogen following cultivation are clearly large and critical in determining both clover establishment and response to inoculation.

Further evidence implicating environmental factors in modifying the behaviour of symbionts comes from recent experiments with *Vicia faba*. Sorwli and Mytton (1986) have shown that populations of *R. leguminosarum* in Britain are generally not able to fix sufficient nitrogen to meet plant potential demand. Glasshouse experiments suggested that the effectiveness of field populations of rhizobia in soils at the Welsh Plant Breeding Station fell short

Table 7. Effect of cultivation, applied nitrogen, and inoculation on clover yield (kg/ha) on two different upland soils.

	Soil A			Soil B		
	Uninocu-lated	Inoculated	Inoculated +N	Uninocu-lated	Inoculated	Inoculated +N
Uncultivated	9	68	3	8	220	13
Cultivated	17	144	1	26	82	2

Table 8. Response of field beans (cvs. Minor and Maris Bead) to inoculation with *R. leguminosarum* (strains R1 and 29d) and to large applications of combined nitrogen (data taken at flowering).

Treatment	Dry matter yield (t/ha)	Nitrogen yield [1] (kg/ha)	Nitrogen [2] (%)
Minor + R1	1.8 cd*	54 c	3.09 d
Maris Bead + R1	1.6 d	53 c	3.33 cd
Minor + 29d	2.0 bc	68 bc	3.37 cd
Maris Bead + 29d	2.0 bc	63 bc	3.21 d
Minor uninoculated	2.1 b	77 b	3.65 c
Maris Bead uninoculated	2.1 b	71 b	3.44 cd
Minor uninoculated + 560 kg N/ha	2.8 a	110 a	3.98 ab
Maris Bead uninoculated 560 kg N/ha	2.7 a	115 a	4.21 a
S.E. ±		5.1	0.134

[1] Data taken at 10 weeks from sowing
[2] Data taken at 12 weeks from sowing
* Values within a column followed by the same letter do not differ (P = 0.05) according to Duncan's Multiple Range Test.

of plant demand by about 40%. Host-*Rhizobium* genotypes were matched to produce combinations of differing effectiveness. *Rhizobium* strain 29d, in combination with the cultivar *Minor*, was the best from a large sample of the field population, whilst Minor-R_2 was one of the poorest (Mytton *et al.* 1977). These were used in a field inoculation trial (Sorwli and Mytton (in press)). Data taken at flowering (Table 8) show that the selected combinations generally performed as expected with cv. Minor-29d producing 27% more dry matter than Maris Bead-R_2 ($P < 0.05$). However, both cultivars did better with the indigenous rhizobia (over 2t DM/ha) but average yield was increased a further 32% ($P < 0.05$) by adding high levels of fertilizer nitrogen (560 kg N/ha). At grain harvest (Table 9) the differences between *Rhizobium*–depen-

Table 9. Response of field beans (cvs. Minor and Maris Bead) to inoculation with *R. leguminosarum* (strains R1 and 29d) and to large applications of combined nitrogen (data taken at ripe harvest).

Treatment	Seed yield (t/ha)	Total N in seeds (kg/ha)	% N in seeds
Minor + R1	5.4 c*	214 c	4.00 b
Maris Bead + R1	6.0 bc	245 bc	4.06 b
Minor + 29d	5.9 bc	241 bc	4.11 b
Maris Bead + 29d	6.3 bc	257 bc	4.04 b
Minor uninoculated	6.7 b	277 b	4.07 b
Maris Bead uninoculated	6.8 b	272 b	3.95 b
Minor uninoculated + 560 kg N/ha	8.5 a	414 a	4.89 a
Maris Bead uninoculated + 560 kg N/ha	8.3 a	401 a	4.82 a
S.E. ±	0.37	15.9	0.081

* Values within a column followed by the same letter do not differ (P = 0.05) according to Duncan's Multiple Range Test.

dent treatments had diminished so that most gave similar yields. However, plants receiving fertilizer nitrogen retained their advantage, averaging 8.4 t of grain/ha, a 24% improvement over treatments fixing nitrogen.

The improvement in grain yield from treatments receiving non-limiting fertilizer nitrogen suggests a potential for improving yield through better fixation. Thus, the failure of the high potential combination of Minor-29d to increase yield is disappointing but perhaps not surprising since it merely re-emphasizes the fact that we know little about the factors which determine successful nodulation and high levels of fixation in the field. It must also be acknowledged that strain 29d, although native to the site, occurred at a low frequency and must therefore be at some selective disadvantage. There is no reason to suppose that simply by selecting it and re-introducing it as an inoculant this disadvantage can be overcome. It is evident that appropriate changes are also required in either the plant genotype or the environment, or more probably in both, to favour the success of such an elite combination.

Conclusions

All crops are limited in the range of agricultural environments in which they have economic value. As a result, the essence of crop improvement consists of placing the "right" genotype in the "right" environment. It is evident that progress can occur either by modifying genotypes to increase their productivity in particular environments, or by altering the environment to make it better suited to the genotype. In practice, both approaches are important and are interrelated since each element needs to be constantly cross-referenced to variable samples of the other for the whole to succeed. The need to cross-reference is brought about by genotype-environment interactions. If the reaction of plants to environmental variables was always additive, then broad referencing would be unnecessary. Changes in productivity would be proportional to alterations in limiting factors and a single environment would then have predictive value for all others. Experience shows that this is rarely so.

Legume-*Rhizobium* associations can be subject to genetic interactions which result in large and unpredictable effects on nitrogen fixation. The symbiotic phenotype is further subject to environmental interactions. Consequently, practical improvements demand that there be broadly based cross-referencing between plant genotypes, *Rhizobium* populations, and agricultural environments. In this respect, the screening methods developed at CIAT for selecting *Rhizobium* strains to use as inoculants are an excellent example. Both plants and rhizobia are tested in systems which provide reference to appropriate environments and local soil rhizobia (Sylvester-Bradley, this workshop). This is undoubtedly an important component of the programme's success.

The methods and approaches outlined here provide some further examples of limited successes and alternative methods of enhancing fixation and therefore offer the promise of future improvements. But they also point to important gaps in present knowledge. In this respect, more information on the environmental variables which, in situations where there is competition for nodule sites, allow some plant-*Rhizobium* combinations to succeed in nodulation where others fail is likely to be of particular importance in maximizing symbiotic nitrogen fixation.

References

Brockwell, J. 1981. Can inoculant strains ever compete successfully with established soil populations? *In* Current Perspectives in Nitrogen Fixation. Proceedings of the Fourth International Symposium on Nitrogen Fixation, Canberra, Australia, 1980 (Gibson, A.H. and Newton, W.E., eds.). 277pp.

Caldwell, B.E. and Vest, M.G. 1968. Nodulation interactions between soybean genotypes and serogroups of *Rhizobium japonicum*. Crop Science 8: 680-689.

Caldwell, B.E. and Vest, M.G. 1977. Genetic aspects of nodulation and dinitrogen fixation by legumes: the macrosymbiont. Pages 557-576 *in* A Treatise on Dinitrogen Fixation. III. Biology. (Hardy, R.W.F. and Silvers, W.S., eds.). John Wiley and Sons, New York, USA.

Engledow, F.L. 1925. The economic possibilities of plant breeding. Pages 31-40 *in* Report of the Proceedings of the Imperial Botanical Conference (Brooks, F.T., ed.).

Finlay, K.W. and Wilkinson, G.N. 1963. The analysis of adaptation in a plant breeding programme. Australian Journal of Agricultural Research 19: 739-747.

Fred, E.B., Baldwin, I.L. and McCoy, E. 1932. Root Nodule Bacteria and Leguminous Plants. Studies in Science No. 5. University of Wisconsin, Madison, Wisconsin, USA.

Hardarson, G. and Jones, D.G. 1979. The inheritance of preference for strains of *Rhizobium trifolii* by white clover (*Trifolium repens*). Annals of Applied Biology 92: 329-333.

Mytton, L.R. 1975. Plant genotype-*Rhizobium* strain interactions in white clover. Annals of Applied Biology 80: 103-107.

Mytton, L.R. 1984. Developing a breeding strategy to exploit quantitative variation in symbiotic nitrogen fixation. Plant and Soil 82: 329-335.

Mytton, L.R., Brockwell, J. and Gibson, A.H. 1984. The potential for breeding an improved lucerne-*Rhizobium* symbiosis. 1. Assessment of genetic variation. Euphytica 33: 401-410.

Mytton, L.R. and De Felice, J. 1977. The effects of mixtures of *Rhizobium* strains on the dry matter production of white clover grown in agar. Annals of Applied Biology 87: 83-93.

Mytton, L.R., El-Sherbeeny, M.H. and Lawes, D.A. 1977. Symbiotic variability in *Vicia faba*. 3. Genetic effects of host plant, *Rhizobium* strain and host strain interaction. Euphytica 26: 785-791.

Mytton, L.R. and Hughes, D.M. 1984. Inoculation of white clover with different strains of *Rhizobium trifolii* on a mineral hill soil. Journal of Agricultural Science, Cambridge 102: 455-459.

Mytton, L.R. and Livsey, C. 1983. Specific and general effectiveness of *Rhizobium trifolii* populations from different agricultural locations. Plant and Soil 73: 299-305.

Norris, D.O. and 't Mannetje, L. 1964. The symbiotic specialization of African *Trifolium* spp. in relation to their taxonomy and their agronomic use. East African Agriculture and Forestry Journal 29: 214-235.

Roughley, R.J., Blowes, W.M. and Herridge, D.F. 1976. Nodulation of *Trifolium subterraneum* by introduced rhizobia in competition with naturalized strains. Soil Biology and Biochemistry 8: 403-407.

Russel, P.E. and Jones, D.G. 1975. Variation in the selection of *Rhizobium trifolii* by varieties of red and white clover. Soil Biology and Biochemistry 7: 15-18.

Rys, G.J. and Mytton, L.R. 1985. The potential for breeding white clover (*Trifolium repens* L.) with improved nodulation and nitrogen fixation when grown with combined nitrogen. 1. The effects of different amounts of nitrate nitrogen on phenotypic variation. Plant and Soil 88: 181-195.

Sherwood, M.T. and Masterson, C.L. 1974. Importance of using the correct test host in assessing the effectiveness of indigenous populations of *Rhizobium trifolii*. Irish Journal of Agricultural Research 13: 101-108.

Sorwli, F.K. and Mytton, L.R. Nitrogen limitations to field bean productivity: a comparison of combined nitrogen applications with *Rhizobium* inoculation. Plant and Soil. In press.

Sorwli, F.K. and Mytton, L.R. 1986. The nitrogen fixing potential of *Vicia faba* rhizobia from different agricultural locations. Plant and Soil 92: 249-254.

Wilson, J.K. 1944. Over five hundred reasons for abandoning the cross-inoculation groups of the legumes. Soil Science 58: 61-69.

Young, N.R. and Mytton, L.R. 1983. The response of white clover to different strains of *Rhizobium trifolii* in hill land reseeding. Grass and Forage Science 38: 13-19.

D.P. Beck and L.A. Materon (editors), Nitrogen Fixation by Legumes in Mediterranean Agriculture
© 1988 ICARDA – Printed in the Netherlands. ISBN 90 247 3624 2

Infection, Development and Functioning of Nodules Under Drought and Salinity

J.I. SPRENT and H.H. ZAHRAN
University of Dundee, Dundee DD1 4HN, Scotland

Summary. This paper considers the major factors limiting nitrogen fixation by legumes in arid and saline areas, with particular reference to infection, and nodule development and functioning.

Introduction

For reasons which have been discussed elsewhere (Griffin 1981; Skujins 1984; Sprent 1984), microbial cells including rhizobia, are better able to withstand low water potentials (as a result of drought and/or salinity) than most higher plant cells, including those of legumes. Recent studies suggest that it is possible to select rhizobia tolerant to salinity (Singleton *et al.* 1982) and drought (Shoushtasi and Pepper 1985a) and some of these strains are very effective (Shoushtasi and Pepper 1985b). With the current rapid advances in genetic engineering of rhizobia, it should soon be possible to construct rhizobial strains with desired characteristics.

The major factors limiting nitrogen fixation by legumes in arid and saline areas are thus likely to lie in the host plant, from the time of infection by rhizobia to the functioning of the differentiated nodules. This paper considers these problems, drawing from both the literature and work in our own laboratory.

Infection of Legumes by Rhizobia

Most crop legumes which have been studied have root hairs which are outgrowths of epidermal cells and through which rhizobial entry is effected. Two variants of the general pattern are observed (Fig. 1): (a) in plants such as soybean which have only one infectible period, close to the beginning of hair growth, and (b) in others such as clovers which have a second infectible period later in hair development. These two types of legume may also differ in the response of their root hairs to growth and infection by rhizobia at low water potentials (Table 1).

Zahran and Sprent (1986) carried out a detailed study of the effects of sodium chloride salinity and polyethylene glycol (average molecular mass 6000

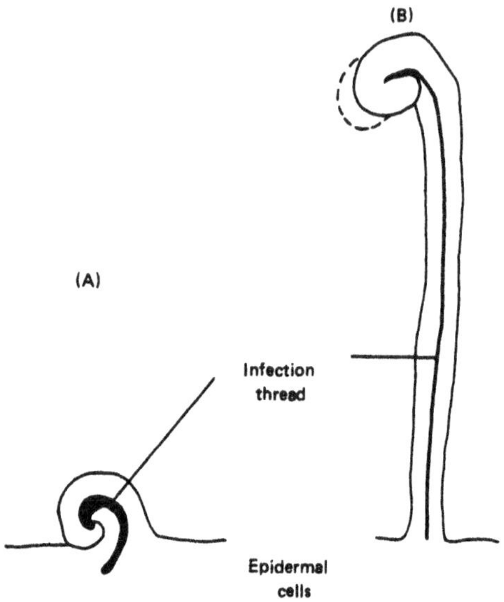

Fig. 1. Hairs with one (A) and two (B) infectible stages.

(PEG 6000) to mimic dehydration stress) on the infection of *Vicia faba* root hairs by rhizobia. The principle findings were:

1. root hair colonization by rhizobia was little affected,
2. NaCl delayed and PEG inhibited the lateral expansion of root hairs,
3. both stress treatments reduced the number of root hairs showing marked curling,
4. the number of root hairs containing visible infection threads was reduced by 38% in plants given 100 mol/m^3 NaCl and by 67% in plants given 200 mol/m^3 PEG 6000, and
5. the number of nodules was approximately halved by each stress treatment.

Further studies (Zahran and Sprent unpublished) showed host cultivar x rhizobial strain interactions in these responses. Generally, combinations giving early infection/nodulation were better able to withstand stress. This could have important practical implications in the timing and use of fresh and slightly saline irrigation waters.

Recent work (Faria *et al.* 1984) suggests that infection other than via root hairs, usually via fissures where lateral roots emerge, may be more common

Table 1. Possible variations in response of different groups of legumes to drought and salinity at the time of rhizobial infection. (Detailed references in Sprent 1984.)

	Soybean type	Clover type
Period of infection of root hairs	early in growth	early and later in growth
Hair growth under water stress	?	may be reduced or stimulated
Hair growth under salinity stress	impaired	impaired
Interaction between calcium and salinity on hair infection	positive (aggravated)	negative (alleviated)

Table 2. Effects of NaCl on growth and nodulation of *Arachis hypogaea* after 21 days. Plants were inoculated with *Rhizobium* strain 3824.

	NaCl in sand (mol/M^3)		
	0	50	100
Shoot fresh weight (g)	1.55 ± 0.90	1.25 ± 0.11	0.92 ± 0.15
Root fresh weight (g)	1.40 ± 0.11	0.82 ± 0.11	0.40 ± 0.13
Number of nodules	14.0 ± 1.2	16.0 ± 1.6	1.8 ± 0.9

than once thought. Amongst crop plants this has been established for *Arachis* (Chandler 1978) and *Stylosanthes* (Chandler *et al.* 1982). These species do not normally produce root hairs. Preliminary studies with *Arachis* (Sprent and McInroy 1984 and Table 2) suggest that this mode of infection may be less sensitive to salinity stress than hair infection. Salt stress in *Arachis* also leads to the production of epidermal root hairs (Sprent and McInroy 1984) whilst decreasing the number and length of the multicellular root hairs around the lateral root junctions (Wishart and Sprent unpublished). We have not observed infection threads in root hairs produced under salinity stress: their role in nutrient acquisition is under consideration. Further studies of the effect of drought and salinity on root hair growth and rhizobial infection are needed, especially in genotypes with some tolerance to these forms of stress.

Nodule Development

Nodule growth is sometimes enhanced under stress, in an apparent attempt by the plant to compensate for reduced nodule efficiency. This has been found for example in *V. faba* grown at low temperature (Fyson and Sprent 1982) and with 100 mol/m^3 NaCl (Yousef and Sprent 1983). Drought, on the other hand, reduces nodule growth (Engin and Sprent 1973; Gallacher and Sprent 1978). As part of a more detailed study of the structure of nodules formed under NaCl and PEG 6000 stress, Zahran and Sprent (unpublished) found that:

1. mitoses in the nodule apical meristem were reduced in number,
2. infected cells showed accelerated senescence, and
3. bacteroids did not differentiate completely into the pleomorphic forms typical of this symbiosis.

All these effects were more pronounced under PEG than NaCl stress, consistent with depression of nodule growth under drought stress.

Nodule Functioning Under Salinity

There have been several reports that salinity adversely affects nitrogen fixing activity as estimated by the acetylene reduction assay (Yousef and

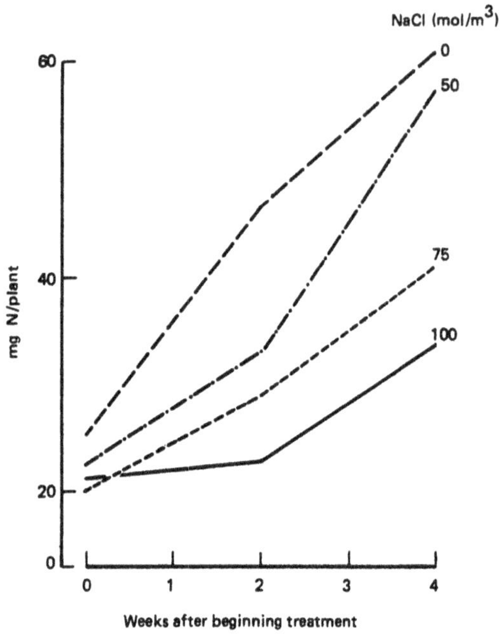

Fig. 2. Effect of NaCl on nitrogen content of *Vicia faba* plants (Yousef and Sprent 1983).

Sprent 1983; Lauter *et al.* 1981). These two papers also indicate that in *V. faba* and *Cicer arietinum*, respectively, nitrogen fixation as estimated by total plant nitrogen (Fig. 2), is depressed and nitrogen fixing plants are more susceptible to salinity than plants given combined nitrogen. However, in the studies of Zahran and Sprent discussed above, total plant nitrogen was little affected in spite of adverse effects on nodule structure and nitrogenase (acetylene reduction) activity. Detailed comparison of these data with those of Yousef and Sprent (1983) for the same host/rhizobial combination suggests that these differences may be related to plant growth rate. Where effects of salinity on plant nitrogen were found, plants were grown under the high photon flux densities and long days of summer. Where plant nitrogen was not affected, plants were grown at the lower photon flux densities of winter, although the daylengths were supplemented to give a 16h photoperiod. Growth in dry weight and nitrogen fixed was much slower in the latter. Thus, although nodule structure was adversely affected by stress, plants were still able to fix sufficient nitrogen for their needs. These data further emphasize the major discrepancies which may arise between estimates of nitrogenase activity based on acetylene reduction and the actual amount of nitrogen fixed. Therefore, it seems likely that the effects of salinity at least are likely to be most acute when plants are growing most quickly.

Studies have been carried out on too few species to assess the variability in response of different legume symbioses to salinity. However, basic physiological studies indicate that this area should be explored. For example, *V. faba* and *C. arietinum* accumulate both Na and Cl (Yousef and Sprent 1983; Lauter *et al.* 1981), whereas *Phaseolus vulgaris* excludes Na (Jacoby 1979), and Cl accumulation in *Glycine max* is controlled by a simple major gene system (Abel and Mackenzie 1964).

Nodule Functioning Under Drought

There are now numerous reports from both laboratory and field based experiments showing that drought has a direct adverse effect on nodule activity (Albrecht *et al.* 1984), which may be aggravated by effects on plant photosynthesis. One of the direct effects of drought is on the oxygen permeability of nodules (Pankhurst and Sprent 1975; Weisz *et al.* 1985). If nodule permeability is reduced and even if high rates of photosynthesis are maintained, nodules will be limited by their ability to carry out oxidative phosphorylation (Sheehy *et al.* 1985). For reasons which have recently been discussed (Sprent 1985; 1986), high rates of nitrogen fixation are only likely when nodules are well supplied with water. Some longer-lived, including perennial, nodules can survive quite severe drought in an inactive state and resume activity when water becomes available. Under really dry conditions most potentially nodulated plants appear not to nodulate, except when they have deep roots which can reach the water table or tap deeper, moist soil horizons. The value of many shrubby legumes in arid areas probably lies in their extensive and deep root systems and not in their potential to fix nitrogen, even in nitrogen-poor soils (Sprent 1986). There is, however, considerable potential for both shrubby and herbaceous (annual and perennial) legumes to fix nitrogen in semi-arid areas, especially where there is a definite rainy season, and much room for improvement of drought tolerance of host material (Aparico-Tejo *et al.* 1980).

References

Abel, G. and Mackenzie, A.J. 1964. Salt tolerance of soybean varieties (*Glycine max* Merrill) during germination and later growth. Crop Science 4: 157-161.

Albrecht, S.L., Bennett, J.M. and Boote, K.J. 1984. Relationship of nitrogenase activity to plant water stress in field-grown soybeans. Field Crops Research 8: 61-71.

Aparico-Tejo, P.M., Sanchez-Diaz, M.F. and Pena, J.L. 1980. Nitrogen fixation and stomatal response and transpiration in *Medicago sativa*, *Trifolium repens* and *T. subterraneum* under water stress and recovery. Physiologia Plantarum 48: 1-4.

Chandler, M.R. 1978. Some observations on infection of *Arachis hypogaea* L. by *Rhizobium*. Journal of Experimental Botany 29: 749-755.

Chandler, M.R., Date, R.A. and Roughley, R.J. 1982. Infection and root nodule development in *Stylosanthes* species by *Rhizobium*. Journal of Experimental Botany 33: 47-57.

Engin, M. and Sprent, J.I. 1973. Effects of water stress on growth and nitrogen-fixing activity of *Trifolium repens*. New Phytologist 72: 117-126.

Faria, S.M.de, Franco, A.A., Jesus, R.M.de, Menandro, M.S.de S., Baitello, J.B., Mucci, S.F., Dobereiner, J. and Sprent, J.I. 1984. New nodulating trees from South-East Brazil. New Phytologist 98: 317-328.
Fyson, A. and Sprent, J.I. 1982. The development of primary root nodules on *Vicia faba* L. grown at two temperatures. Annals of Botany 50: 681-692.
Gallacher, A.E. and Sprent, J.I. 1978. The effect of different water regimes on growth and nodule development of greenhouse grown *V. faba*. Journal of Experimental Botany 29: 413-423.
Griffin, D.M. 1981. Water and microbial stress. Advances in Microbial Ecology 5: 91-136.
Jacoby, B. 1979. Sodium recirculation and loss from *Phaseolus vulgaris* L. Annals of Botany 43: 741-744.
Lauter, D.J., Munns, D.N. and Clarkin, K.L. 1981. Salt response of chickpea as influenced by N supply. Agronomy Journal 73: 961-966.
Sheehy, J.E., Minchin, F.R. and Witty, J.F. 1985. Control of nitrogen fixation in a legume nodule: an analysis of the role of oxygen diffusion in relation to nodule structure. Annals of Botany 55: 549-562.
Pankhurst, C.E. and Sprent, J.I. 1975. Effects of water stress on the respiratory and nitrogen-fixing activity of soybean root nodules. Journal of Experimental Botany 26: 287-304.
Shoushtasi, N.H. and Pepper, I.L. 1985a. Mesquite rhizobia isolated from the Sonoran desert: competitiveness and survival in soil. Soil Biology and Biochemistry 17: 803-806.
Shoushtasi, N.H. and Pepper, I.L. 1985b. Mesquite rhizobia isolated from the Sonoran desert: physiology and effectiveness. Soil Biology and Biochemistry 17: 797-802.
Singleton, P.W., Swaify, S.A. and Bohlool, B.B. 1982. Effect of salinity on *Rhizobium* growth and survival. Applied and Environmental Microbiology 44: 884-890.
Skujins, J. 1984. Microbial ecology of desert soils. Advances in Microbial Ecology 7: 49-91.
Sprent, J.I. 1984. Effects of drought and salinity on heterotrophic nitrogen fixing bacteria and on infection of legumes by rhizobia. Pages 295-302 *in* Advances in Nitrogen Fixation Research (Veeger, C. and Newton, W.E., eds.). Martinus Nijhoff, Wageningen, The Netherlands.
Sprent J.I. 1985. Nitrogen fixation in arid environments. Pages 215-229 *in* Plants for Arid Regions (Wickens, G.E., Gooden, J.R. and Field, D.V., eds.). George Allen and Unwin, London, UK.
Sprent, J.I. 1986. Problems and potentials for nitrogen fixation in deserts. *In* Arid Lands: To-day and Tomorrow (Nabham, G.P. and Elias-Cesnik, A., eds.). (In Press).
Sprent, J.I. and McInroy, S.M. 1984. Effects of salinity on growth and nodulation of *Arachis hypogaea*. Page 546 *in* Advances in Nitrogen Fixation Research (Veeger, C. and Newton, W.E., eds.). Martinus Nijhoff, Wageningen, The Netherlands.
Weisz, P.R., Denison, R.F. and Sinclair, T.R. 1985. Response to drought stress of nitrogen fixation (acetylene reduction) rates by field-grown soybeans. Plant Physiology 78: 525-530.
Yousef, A.N. and Sprent, J.I. 1983. Effects of NaCl on growth, nitrogen incorporation and chemical composition of inoculated and NH_4NO_3 fertilized *Vicia faba* (L.) plants. Journal of Experimental Botany 34: 941-950.
Zahran, H.H. and Sprent, J.I. 1986. Effects of sodium chloride and polyethylene glycol on root-hair infection and nodulation of *Vicia faba* L. plants by *Rhizobium leguminosarum*. Planta 167. (In press)..lm 10;.s 2;·iscussion

Discussion

Moawad

Does lentil have one or two infectible periods?

Sprent

I have no direct evidence, but as it is taxonomically closely related to *Vicia* and other genera which seem to have two. I would predict lentil also has two.

Danso

Casuarina will grow in saline environments. Has anyone taken a good look at its ability to nodulate and/or fix nitrogen in saline soils?

Sprent

Some species of *Casuarina* will nodulate in saline soils and also soils of very high pH. How much nitrogen is fixed in these situations, I do not know.

Materon

Can you elaborate on why selecting legume plants for higher rates of photosynthetic activity is not advantageous in semi-arid environments, and how does this affect BNF?

Sprent

Selection for high rates of photosynthesis often means high rates of water loss (e.g. because stomatal density may be increased). If water rather than light or CO_2 is the major factor limiting growth, then this may not give better growth and nitrogen fixation and may even be a disadvantage.

D.P. Beck and L.A. Materon (editors), Nitrogen Fixation by Legumes in Mediterranean Agriculture
© 1988 ICARDA – Printed in the Netherlands. ISBN 90 247 3624 2

The Role of Mycorrhiza in Improving the Establishment and Function of the *Rhizobium*-Legume System under Field Conditions

J.M. BAREA, C. AZCON-AGUILAR and R. AZCON
Soil Microbiology Department, Estacion Experimental del Zaidin (Granada), C S I C, Prof. Albareda 1, 18008-Granada, Spain

Introduction

The scarcity of plant-available P in soils is a limiting factor in legumes because it affects not only plant establishment and growth but also nodulation and N_2-fixation. Nitrogenase activity needs, in fact, 21 moles of ATP to reduce dinitrogen to ammonia. Thus, nodules often contain two to three times more P per unit of dry matter than the root on which they are formed. Additionally, legumes have a restricted root system compared to graminaceous plants so they are relatively poor foragers for phosphate, a nutrient which diffuses very slowly through the soil solution to the rhizosphere from outside this zone (Munns and Mosse 1980).

Apart from the legume-*Rhizobium* symbiosis, legumes are able to form another root symbiosis with certain soil fungi. The resultant mutualistic relationships are the mycorrhiza (Barea and Azcon-Aguilar 1983). Five types of mycorrhiza can be recognized but the vesicular-arbuscular type (VA mycorrhiza) is by far the most widespread (Harley and Smith 1983), and are the most commonly occurring on nodulated, N_2-fixing plants (Barea and Azcon-Aguilar 1983).

After colonizing the root cortex, the fungi develop a network of hyphae in the surrounding soil beyond the root hair zone. Through this external mycelium they take up nutrients from the soil solution, mainly those of low mobility such as phosphate, zinc, and perhaps molybdate, and translocate these ions to the host plant (Tinker and Gildon 1983; Tinker 1984). Fig. 1 illustrates this process in VAM. There are also indications that VAM have a role in reducing drought stress (Graham and Syvertsen 1984; Huang *et al.* 1985). The formation and physiology of VAM have been recently reviewed (Smith 1980; Hayman 1983; Gianinazzi-Pearson and Gianinazzi 1983; Clarkson 1985).

Legumes are usually mycorrhizal when grown under natural conditions. However, legume species differ in their growth responses to mycorrhizal infection. Many legumes are dependent on mycorrhiza for phosphate uptake and therefore, for nodulation in P-deficient soils. Others need mycorrhiza to reach "maximum yield", and finally, in the case of *Lupinus*, we have an example of one of the few plant species which are mycorrhiza-independent (Barea and Azcon-Aguilar 1983).

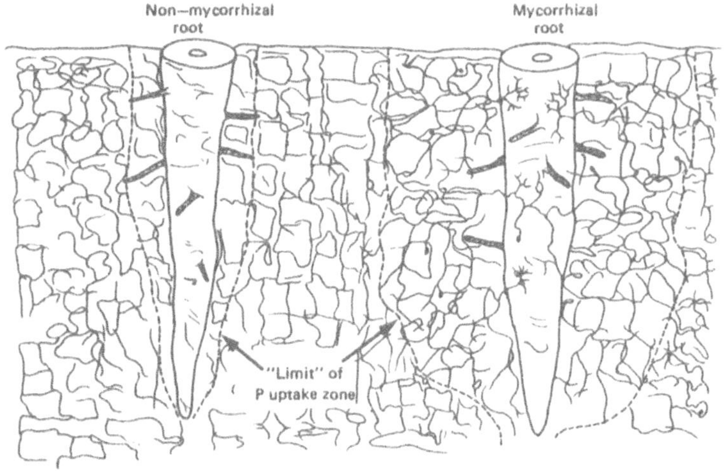

Fig. 1. Schematic representation of the increased volume of soil that a root can use for phosphate uptake due to mycorrhiza.

Unlike the legume-*Rhizobium* symbiosis, the different fungal endophytes do not appear to be host-specific. They do, however, differ in the extent to which they stimulate plant growth and N_2-fixation (Munns and Mosse 1980).

There are several approaches to the elucidation of the physiological aspects of the legume-*Rhizobium*-VA mycorrhiza interactions. One of these tries to ascertain whether mycorrhiza enhance symbiotic N_2-fixation only through the stimulation of host nutrition, or whether they also exert a more direct effect on nodulation and nitrogenase activity (Bowen and Smith 1981). It has been reported that mycorrhizal effects on N_2-fixation preceded any effect on plant growth. This suggests that nodules demand phosphate first. In any case, a steady supply of phosphate to root cells and to adjacent nodules would be stimulating to nodule function. These observations have been reviewed by Barea and Azcon-Aguilar (1983).

Feasibility of Mycorrhizal Inoculation Under Field Conditions

In considering the practical field application of mycorrhizal effects on legume production, there are several points of interest. One of these is concerned with factors determining plant benefits from mycorrhizal inoculation, which include (i) the degree of mycorrhizal dependency of the crop species, (ii) soil fertility level, (iii) size and effectiveness of the native mycorrhizal populations, and (iv) cultivation practices (Hayman 1984). Thus, tests must be developed for the assessment of field situations in which VAM inoculation of legume crops may be feasible (Azcon-Aguilar *et al.* 1986). Another point to be considered is that the production of a high-quality, pathogen-free inoculum is actually a limiting factor for large-scale inoculation with VA mycorrhizal fungi. As these fungi are dependent on an association with living plant roots, the inocula used consist of rhizosphere samples from stock plant cultures or mycorrhizal roots. These inocula can be applied to crops in which the technique of transplanting is part of the normal cultivation practices, as the plants can be preinoculated in containers, nursery beds, or small field plots (Hyman 1984). A technique of seed pelleting, to be applied on a more practical scale, is now being studied for forage legumes.

In spite of the difficulties of inoculation, some field experiments have been carried out in microplots with the inocula presently available. Some of these trials are helping to establish an ecophysiological basis for the interactions between legumes and their two types of microsymbionts. For example, we have developed some of the following assays in the past few years.

Inoculation experiments with *Medicago sativa* and its partners *Rhizobium meliloti* and the VA fungus *Glomus mosseae*, were carried out under field conditions in untreated arable soil to illustrate the interactions in the tripartite symbiosis. The experiments were planted in the same soil but in plots that had received different agronomic practices and which differed from one another in some characteristics. Plot A_2 possessed twice as much available phosphate and three times as many VA propagules as plot A_1 (Fig. 2). In the latter, inoculation with *G. mosseae* was always effective in promoting plant growth,

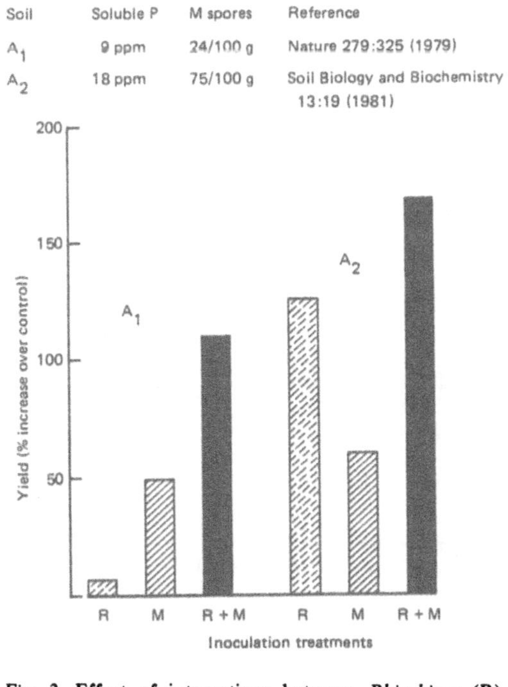

Fig. 2. Effect of interactions between *Rhizobium* (R) and *Mycorrhiza* (M) on the yield of *Medicago sativa* under field conditions.

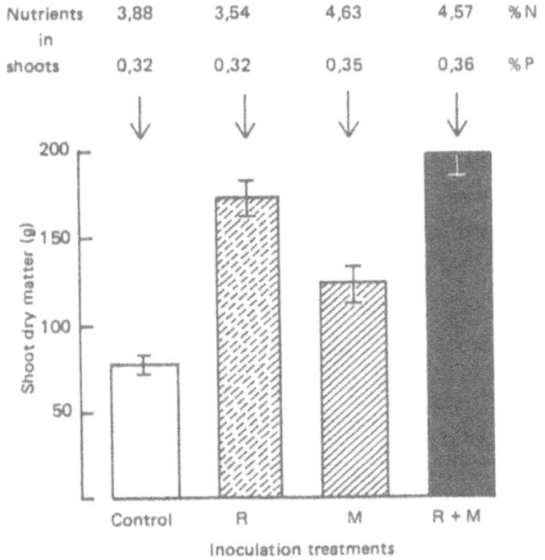

Fig. 3. Shoot dry matter following field inoculation of *Medicago sativa* with *Rhizobium* (R) and *Mycorrhiza* (M), first harvest.

but *R. meliloti* was only able to enhance the growth of *G. mosseae*-inoculated plants. Phosphate was probably the limiting factor for *R. meliloti* activity, because of the scarcity of available nutrient and the small number of spores of the family Endogonaceae in this test soil (A_1). Hence, plants did not respond

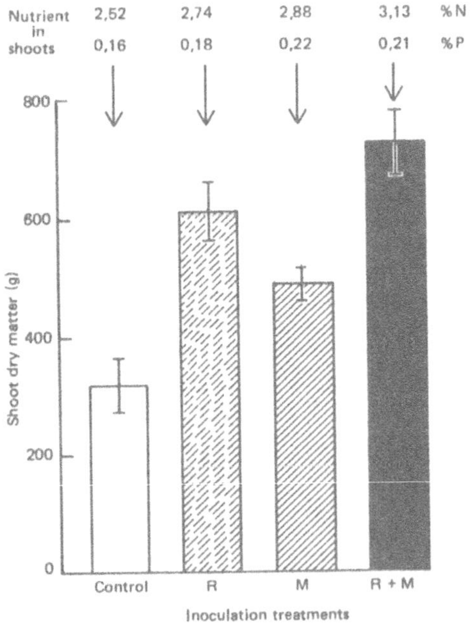

Fig. 4. Shoot dry matter following field inoculation of *Medicago sativa* with *Rhizobium* (R) and *Mycorrhiza* (M), second harvest.

to *R. meliloti* inoculation unless they were also inoculated with VA fungi. In contrast, *R. meliloti* was effective when inoculated alone into plants growing in plot A_2. The efficiency of indigenous VA fungi, together with the higher concentration of available P in the soil, could cause plants to respond to *R. meliloti* inoculated alone. In both cases, however, inoculating with *R. meliloti* plus *G. mosseae* more than doubled the yield compared to the uninoculated controls (Azcon-Aguilar *et al.* 1979; Azcon-Aguilar and Barea 1981).

As alfalfa plants are usually harvested by serial cutting, the persistence of the effect of their microsymbionts was studied by Azcon-Aguilar and Barea (1981). According to their results, it is obvious that the effect of the introduced endophytes persisted through to the second harvest. The dual inoculum (*Rhizobium* plus mycorrhiza) was significantly more effective than any other treatment. Mycorrhizal plants had higher contents (%) of P and N than those inoculated with *Rhizobium* alone, either at the first or the second harvest (Figs. 3 and 4).

The introduction of the forage legume *Hedysarum coronarium* to new habitats was significantly improved by rhizobial and mycorrhizal inoculation (Azcon-Aguilar *et al.* 1982). After several soils had been studied for the feasibility of mycorrhizal inoculation, some field experiments were carried out. In one of these, the effects of chemical (NPK) and biological treatment (*Rhizobium* plus *G. mosseae*) on the growth and nutrition of this legume in the field were compared. The results of the leaf analysis indicate that inoculated plants had the highest N, P, and K concentrations in their leaves. A growth response to microbial inoculation was also evident at harvest. In fact, inoculated plants produced significantly more biomass, and had the highest N and P concentrations in their shoots (Fig. 5).

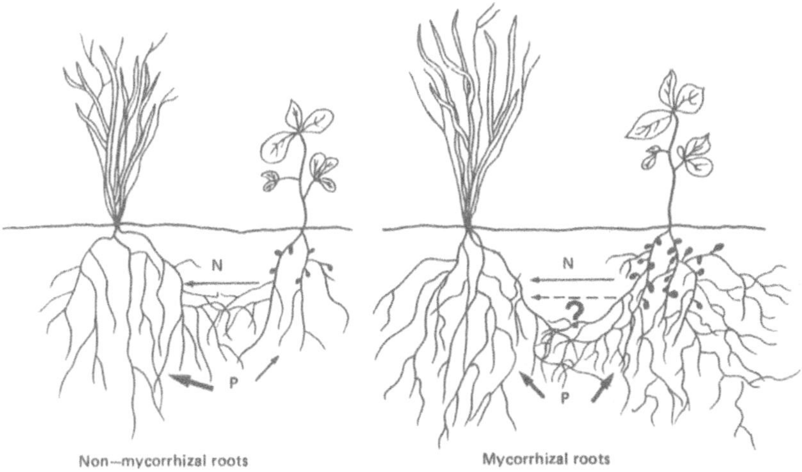

Fig. 5. Effect of chemical (NPK) and biological (*Rhizobium* + *Mycorrhiza*-RM-) treatment on the yield and nutrition of *Hedysarum coronarium* under field conditions.

The Role of Mycorrhiza in Mixed Legume-Grass Swards

It is well known that pasture productivity can be improved by growing grasses and legumes in mixed swards rather than in individual pure stands of either crop. A key factor is the nitrogen input to this pasture system derived from biological nitrogen fixation (BNF) by the legume-*Rhizobium* symbiosis. Part of the fixed nitrogen can be transferred from the legume to the grass (Haynes 1980). It is also commonly assumed that grasses have a competitive advantage over legumes, mainly for phosphate uptake, and tend to dominate the pasture. The introduction of VAM into the system could affect such pastures in several ways. Firstly, because legumes are more mycotrophic than grasses, mycorrhiza would compensate for this imbalance in phosphate uptake ability (Haynes 1980) and legume nodulation and N_2-fixation could be enhanced. Secondly, mycorrhizal hyphae can absorb and translocate $^{15}NH_4$ from soil to root cells (Ames *et al.* 1983; Barea *et al.* 1986) and also facilitate nitrogen transfer between plants (Van Kessel *et al.* 1985; Francis *et al.* 1986). These mycorrhiza-*Rhizobium*-legume-grass interactions are illustrated in Fig. 6.

The implications of VAM in pasture development were studied in field experiments by Barea *et al.* (1986). The main conclusions were: (i) mycorrhizal inoculation was needed for suitable development of clover, (ii) the amount of mycorrhizal roots in the low mycotrophic ryegrass was significantly improved by the highly mycorrhizal clover, and (iii) there was an improvement in the dry matter and total nitrogen yields of grass when mixed with clover, compared with the pure grass stand which could be, at least in part, related to point (ii) above.

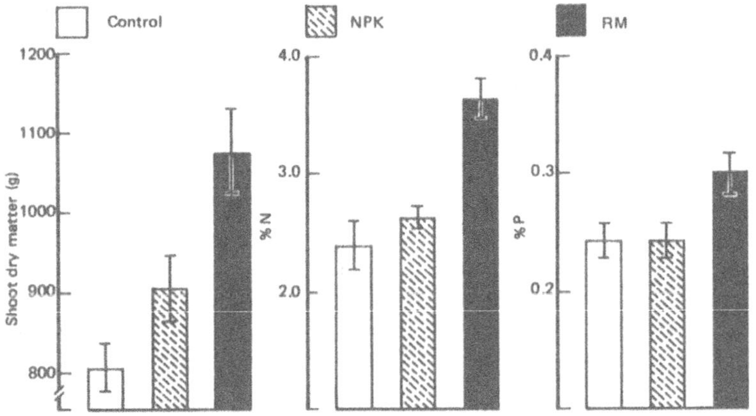

Fig. 6. Schematic representation of the effects of mycorrhiza: (i) improving the competitivity of a legume vs a grass for phosphate (P) uptake: (ii) increasing nodulation and N_2 fixation, and (iii) facilitating N transfer in legume-grass mixed swards.

The Use of ^{15}N Methodology to Evaluate the Mycorrhizal Effect of Improving N_2-Fixation

Methodology using ^{15}N-labelled fertilizers is now widely applied to measure N_2-fixation by *Rhizobium*-legume symbioses in the field. The advantages and difficulties of this technique have been recently reviewed and discussed (Witty 1983; Chalk 1985; Hardarson *et al.* 1984; Danso 1986). It is accepted that the use of ^{15}N-enriched materials offers the only direct method to distinguish between the relative contributions of the three nitrogen sources to "fixing" plants, i.e., soil, fertilizer, and atmosphere. As Danso (1986) stated, a major advantage of using ^{15}N methodologies is the possibility of assessing the influences of some treatments on N_2-fixation distinct from soil or fertilizer nitrogen supply.

A technique using ^{15}N-labelled inorganic fertilizer was applied by Barea *et al.* (1986) to estimate N_2-fixation by the forage legume *H. coronarium* L. and to ascertain the role of vesicular-arbuscular (VA) mycorrhizas on plant nitrogen nutrition throughout a growing season under field conditions. The absence of the specific *Rhizobium* sp. for the forage legume in the test soil (pH 7.5, 8 ppm Olsen P) allowed us to use ^{15}N methodology with the same legume as a reference "non-fixing" crop. The two main treatments were: (i) *Rhizobium* sp. plus 8 kg N/ha as $(NH_4)2SO_4$ with 10% ^{15}N atom excess per cut, and (ii) 40 kg N/ha as $(NH_4)2SO_4$ with 2% ^{15}N atom excess per cut. These treatments were combined with three P treatments: (i) 100 kg P/ha for all the growing season, (ii) a mycorrhizal inoculum (*G. mosseae*), and (iii) the untreated soil as a control.

Mycorrhizal inoculation not only enhanced dry matter yield, nitrogen concentration, and total nitrogen yield, but also the amount of nitrogen derived from soil and from fixation, compared with either control plants or those to which phosphate was added. This indicated that mycorrhiza acted both by a P-mediated mechanism to improve N_2-fixation and by enhancing nitrogen uptake from the soil. The latter agrees with recent findings which indicate that VA mycorrhizal hyphae can translocate and assimilate ammonium, a fact of physiological and ecological interest.

Conclusions

The data reported here support the increasingly recognized assumption that VA mycorrhizas can be used to improve legume productivity by reducing phosphate fertilizer input. However, inoculant production is now limiting the extensive use of this symbiosis, although a system of inoculation using multi-seeded pellets containing *Rhizobium* and mycorrhizal fungi is being investigated.

The use of ^{15}N methodology is suggested for the evaluation of the mycorrhizal role in improving N_2-fixation under field conditions, and to assess the transfer of fixed N_2 from pasture legumes to associated grass as affected by mycorrhiza.

Acknowledgements

This work is included in the Project no. 1759/82 (CAICYT)

References

Ames, R.N., Reid, C.P.P., Porter, L.K. and Cambardella, C. 1983. Hyphal uptake and transport of nitrogen from two ^{15}N-labelled sources by *Glomus mosseae*, a vesicular-arbuscular mycorrhizal fungus. New Phytologist 95: 381-396.

Azcon-Aguilar, C., Azcon, R. and Barea, J.M. 1979. Endo-mycorrhizal fungi and *Rhizobium* as biological fertilizers for *Medicago sativa* in normal cultivation. Nature 279:325-327.

Azcon-Aguilar, C. and Barea, J.M. 1981. Field inoculation of *Medicago* with VA mycorrhiza and *Rhizobium* in phosphate-fixing agricultural soil. Soil Biology and Biochemistry 13: 19-22.

Azcon-Aguilar, C., Barea, J.M., Azcon, R. and Olivares, J. 1982. Effectiveness of *Rhizobium* and VA mycorrhiza in the introduction of *Hedysarum coronarium* in a new habitat. Agriculture and Environment 7: 199-206.

Azcon-Aguilar, C., Barea, J.M., Azcon, R. and Diaz-Rodriguez, R.M. 1986. Assessment of field situations for the feasibility of vesicular-arbuscular mycorrhizal inoculation, using a forage legume as test plant. Agriculture, Ecosystems and Environment 15: 241-252.

Barea, J.M. and Azcon-Aguilar, C. 1983. Mycorrhizas and their significance in nodulating nitrogen-fixing plants. Pages 1-54 *in* Advances in Agronomy (Brady, N.C., ed.). Academic Press, New York, USA.

Barea, J.M., Azcon-Aguilar, C. and Azcon, R. 1986. Use of nuclear techniques to evaluate the mycorrhizal role and its interaction with P fertilizers on symbiotic N_2-fixation by legume-*Rhizobium* systems in improving pasture management. Paper presented at the Second Contractor Meeting on Nuclear Techniques in Pasture Management, March 1986, Vienna, Austria.

Bowen, G.D. and Smith, S.E. 1981. The effects of mycorrhizas on nitrogen uptake by plants. Ecological Bulletin Stockholm 33: 237-247.

Chalk, P.M. 1985. Estimation of N_2 fixation by isotope dilution: An appraisal of techniques involving ^{15}N enrichment and their application. Soil Biology and Biochemistry 17: 389-410.

Clarkson, D.T. 1985. Factors affecting mineral nutrient acquisition by plants. Annual Review of Plant Physiology 36: 77-115.

Danso, S.K.A. 1986. Estimation of N_2-fixation by isotope dilution: An appraisal of techniques involving ^{15}N enrichment and their application-comments. Soil Biology and Biochemistry 18: 243-244.

Francis, R., Finlay, R.D. and Read, D.J. 1986. Vesicular-arbuscular mycorrhiza in natural vegetation systems. IV. Transfer of nutrients in inter- and intra-specific combinations of host plants. New Phytology 102: 103-111.

Gianinazzi-Pearson, V. and Gianinazzi, S. 1983. The physiology of vesicular-arbuscular mycorrhizal roots. Plant and Soil 71: 197-209.

Graham, J.H. and Syvertsen, J.P. 1984. Influence of vesicular-arbuscular mycorrhiza on the hydraulic conductivity of roots of two citrus rootstocks. New Phytology 97: 277-284.

Hardarson, G., Zapata, F. and Danso, S.K.A. 1984. Field evaluation of symbiotic nitrogen fixation by rhizobial strains using ^{15}N methodology. Plant and Soil 82:369-375.

Harley, J.L. and Smith, S.E. 1983. Mycorrhizal Symbiosis. Academic Press, London, UK. 483 pp.

Hayman, D.S. 1983. The physiology of vesicular-arbuscular endomycorrhizal symbiosis. Canadian Journal of Botany 61: 944-963.

Hayman, D.S. 1984. Methods for evaluating and manipulating vesicular-arbuscular mycorrhiza. Pages 95-117 *in* Microbiological Methods for Environmental Biotechnology. Society for Applied Bacteriology (Grainger, J.M. and Lynch, J.M., eds.). Academic Press, New York, USA.

Haynes, R.J. 1980. Competitive aspects of the grass-legume association. Pages 227-261 *in* Advances in Agronomy (Brady, N.C., ed.). Academic Press, New York, USA.

Huang, R.S., Smith, W.K. and Yost, R.S. 1985. Influence of vesicular-arbuscular mycorrhiza on growth, water relations, and leaf orientation in *Leucaena leucocephala* (LAM.) New Phytology 99: 229-243.

Munns, D.N. and Mosse, B. 1980. Mineral nutrition of legume crops. Pages 115-125 *in* Advances in Legume Science (Summerfield, R.J. and Bunting, A.H., eds.). HMSO, London, UK.

Smith, S.E. 1980. Mycorrhizas of autotrophic higher plants. Biological Reviews 55: 275-313.

Tinker, P.B. 1984. The role of microorganisms in mediating and facilitating the uptake of plant nutrients from soil. Plant and Soi! 76: 77-91.

Tinker, P.B. and Gildon, A. 1983. Mycorrhizal fungi and ion uptake. Pages 1-32 *in* Metals and Micronutrients. Uptake and Utilization by Plants (Robb, D.A. and Pierpoint, W.S., eds.). Academic Press, London, UK.

van Kessel, C., Singleton, P.W. and Hoben, H.J. 1985. Enhanced N-transfer from soybean to maize by vesicular-arbuscular mycorrhizal (VAM) fungi. Plant Physiology 79: 562-563.

Witty, J.F. 1983. Estimating N_2-fixation in the field using ^{15}N-labelled fertilizer: Some problems and solutions. Soil Biology and Biochemistry 15: 631-639..lm 10;

Discussion

Sprent

Did you observe any correlation between root hair production and mycorrhizal infection in the woody legumes?

Barea

This correlation is to be expected but we did not study that. We will try to do so shortly because it is interesting.

Eaglesham

Further to your theory that mycorrhizal connection between legume and non-legume may facilitate nitrogen transfer, have you observed mycorrhizal hyphae within nodules?

Barea

The mycorrhizal hyphae do not colonize nodular tissue in herbaceous legumes, but do so in actinorrhizal nodules and, as we observed, in that of the woody legumes.

Ismaili

Do you have any evidence of the effect of mycorrhizas on the resistance of legumes to water stress.

Barea

Yes there is some published information, for example in Barea and Azcon-Aguilar (1983) which is in the reference list of my paper.

Solh

Would there be any technical or handling problems in commercially producing one inoculum containing both *Rhizobium* bacteria and mycorrhizal fungi?

Barea

Yes there are some, but work is in progress at several research centers trying to obtain pellets which include mycorrhizal inoculum, *Rhizobium*, seeds (forage legume), and binding agents. It is hoped that this can be applied by a sowing machine. There is no problem with plantation crops.

Part III

Ecology of Rhizobium and Inoculation

D.P. Beck and L.A. Materon (editors), Nitrogen Fixation by Legumes in Mediterranean Agriculture
 ISBN 90 247 3624 2

Ecological Aspects of the Root-Nodule Bacteria: Competition and Survival

J.M. VINCENT

Emeritus Professor University of Sydney, Sydney, Australia

Summary. A number of environmental factors interact with rhizobia in their symbiosis with legumes. Although greenhouse and light-troom experiments can be useful in studying some of these factors, extrapolation to field conditions can be difficult.

This paper discusses various aspects of rhizobial ecology as well as techniques for defining rhizobial parameters and identifying native and inoculum strains.

The Nature of the Problem

The successful introduction of a strain of *Rhizobium*, its persistence, and relative nodulating success in subsequent seasons depend on many environmental factors. The interaction of these factors with rhizobia is complex and has considerable practical significance. There are comprehensive reviews on the subject by Bushby (1982) and Alexander (1985), but the latter disregards a large part of earlier relevant results. Two other recent contributions merit particular mention (Brockwell *et al.* 1982; van Rensburg and Strijdom 1985) for their direct approach to determining persistence of introduced rhizobial strains.

Relevant environmental factors are those directly affecting the nature of the soil, such as content and nature of the clay fraction (Bushby 1982; Marshall 1975), nutritive or toxic minerals, H^+ concentration, water-aeration balance, temperature, microbiotic antagonists, predators (including *Bdellovibrio*) and bacteriophage, higher plants (particularly legumes susceptible to nodulation by the rhizobia under investigation), and the extent and nature of other rhizobial strains with the same nodulating specificity.

Stages Determining the Outcome

The stages at which strain persistence, growth in soil, colonisation of the root surface, and nodulating success of a competing strain are determined can be visualised for the carry-over of rhizobia from one growing season to another as follows:

$$X_0 \quad \overset{k_p}{\quad} \quad X_1 \quad \overset{k_g}{\quad} \quad X_2 \quad \overset{k_c}{\quad} \quad X_3 \quad \overset{k_n}{\quad} \quad n$$

in which X_0 is the number of viable rhizobia in the soil at the end of the growing season, X_1 is the number persisting in the soil into the next growing season, X_2 is the number in the soil after the initiation of more favourable conditions at the time of fresh growth of specific legume, X_3 is the number colonising the root surface of the new season's host, and n is the proportion of nodules due to the particular rhizobial strain. k_p, k_g, k_c, and k_n are respective constants determining progress at each stage.

In an ideal situation, it would be possible to determine a strain's intrinsic property and reaction to environmental conditions at each of these steps and relate this to its relative survival and nodulating success. In fact, under field conditions, such stepwise determination is more difficult to achieve for those events which take place in a normal, unsterilized, relatively undisturbed soil and on the root surface. Techniques suitable for this purpose and some of the results which have been obtained will be discussed later in this account.

Controlled Experimental Models

Greenhouse and light-room experiments with defined, sterilised substrate (agar or sand) have permitted some factors to be studied with paired or multiply-competing strains (Vincent and Waters 1953; Marques Pinto *et al.* 1974; Labandera and Vincent 1975; Franco 1976; Materon and Vincent 1980). Such experiments have the advantage of ready quantitative recovery and identification of chosen strains, and give the opportunity for determining k_c and k_n (expressed as the competitive index). However, as was recognised when such studies were first undertaken, whilst they are valid for the study of operative factors under specified conditions, they can hardly be extrapolated to the much more complex, ill-defined, undisturbed soil under field conditions. For this, we must continue our search for fully satisfactory techniques which will lend themselves unequivocally to the direct study of rhizobia in the soil and on the root.

Heterogeneity of Naturalised Rhizobial Populations

This aspect, although clearly demonstrated many years ago, has not received enough attention when we assess the capacity of an introduced strain to persist, grow, and nodulate.

There is abundant evidence, when the distinguishing characteristics of naturally-occurring rhizobia found in a restricted area have been determined, that at a given time these rhizobia can be very diverse. Some old and more recent data sufficiently illustrate the point (Table 1).

When based on a single characteristic, occurrence of even greater heterogeneity than these references illustrate is to be expected. For example, we have been unable to find any marked linkage between such identifying features as

Table 1. Heterogeneity of rhizobial "strains" isolated from one locality.

Host	Locality	No of minimal			Reference
		Isolates	Strains	Basis	
Medicago minima	Leyburn, Gunnedah, NSW	20	9	Serology	Hughes and Vincent 1942
Trifolium glomeratum	Woodbine, Tichborne, NSW	20	8	Serology	Hughes and Vincent 1942
T. repens	University, Sydney, NSW	18	6	Serology	Hughes and Vincent 1942
T. glomeratum	Woodbine, Tichborne, NSW	191	13	Effectiveness pattern	Purchase and Vincent 1949
T. glomeratum	Woodbine, Tichborne, NSW	84	10	Serology	Purchase and Vincent 1949
M. polymorpha	Plant Breeding Station, Gunnedah, NSW	63	10	Serology + effect-iveness pattern	Purchase *et al.* 1951
M. laciniata	Plant Breeding Station Gunnedah, NSW	13	5	Serology + effect-iveness pattern	Purchase *et al.* 1951
Phaseolus vulgaris	Norfolk, UK	259	20	Intrinsic antibiotic resistance *M. sativa*	Benyon and Josey 1980 Grove Farm, Norfolk, UK28 3 Electrophoretic
Pisum sativum	Grove Farm, Norfolk, UK	89	10	mobility classes	Young 1985
T. repens	Grove Farm, Norfolk, UK	94	9	with three enzymes	
P. vulgaris	Grove Farm, Norfolk, UK	41	2		

symbiotic effectiveness and serological specificity, so that a combination of the two would provide even more division. At the same time, it should be noted that to label any of these isolates as a strain on the basis of any single shared characteristic is unjustified, although it is reasonable to regard those with demonstrated differences at least as different strains.

Two other points from our earlier work are also relevant. Paired nodules from one plant were about as likely to yield different strains as nodules formed on different plants. Though multiple occupancy of single nodules could not be demonstrated in field-grown plants, paired or multiple strain inocula regularly produce nodules containing more than one strain in plants grown on agar or in sand culture. This could reflect spatial differences in the distribution of rhizobia on the root surface under the different circumstances.

Implications of Strain Heterogeneity

In soil studies, when we express the number of nodules occupied by an added identifiable strain as a proportion, we must remember that we are not dealing with one to one competition but with the success of one or two versus the rest. What do we know about the year by year performance of the different strains that make up this complex competing population? Were we able to follow their performance as individual strains, would we find that success amongst some of them would fluctuate as much as has been found in the relatively short history of added inocula?

Let us consider a hypothetical situation which might reasonably be envisaged on the basis that the relative nodulating success of any strain is likely to be variously affected by soil and other factors, themselves variable from season to season. The kind of result which might emerge is illustrated in Table 2.

With this kind of analysis, were it possible, the inoculum strain might appear no worse in the longer term than other separately identifiable naturalised strains. If, on the other hand, one of the naturally-occurring strains was superior over all or most of the seasons of the test, it could have superior

Table 2. Seasonal fluctuation in strain success (hypothetical).

Strain	Percent representation Season			
	1	2	3	4
Inoculum	*55*	19	10	15
Naturalized				
A	1	*50*	10	5
B	30	15	*60*	5
C	4	10	4	10
Others	10	6	16	*65*

nodulating competence and, if effective in N_2-fixation, it would have potential as an inoculant, at least in regions where its better performance had been demonstrated.

Techniques for Defining Rhizobial Parameters

Complete answers to questions about the performance of rhizobial strains in the complex soil environment are difficult to secure. Even when the studied strain carries a clearly recognisable label, the result will leave a heterogeneous naturalised population largely undefined.

Nodule Occupancy

Nodule occupancy is, however, relatively straightforward permitting the ready recognition of added strains. This may be on the basis of nodule characteristics (such as colour (Cloonan 1963) or clear differences in effectiveness (Ireland and Vincent 1968)), serological typing, or use of strains labelled with antibiotic resistance.

One of the several methods of serological typing may be used, directly with the nodule itself (Means *et al.* 1964) or with cultures derived from nodules. The choice of a serological method must be made carefully with a view to relative simplicity, convenience, and reliability. A method which is more discriminating for some species of rhizobia may not be the best for another. Good examples of the use of the serological label to determine nodulating success and survival of rhizobia are to be found in the studies by Brockwell *et al.* (1982) and Krien and Ham (1985).

Rhizobia Outside the Nodule

Direct recognition and counting of rhizobia in the soil and on the root surface still suffer from some limitations. Two methods are being used fairly extensively: the fluorescent antibody (FA) technique for detecting strains with a distinctive antigen constitution (Schmidt 1974; Reyes and Schmidt 1979), and plate-counting rhizobia with appropriate antibiotic resistance in a medium designed to restrict or distinguish other soil microorganisms (Bushby 1984).

The FA technique is being widely used but suffers from incomplete recovery of a known viable number. However, it is still considered to give valid comparative results (Vidor and Miller 1980).

Direct plating requires a balance between suppression of unwanted general soil microorganisms and avoidance of inhibition of the antibiotic resistant rhizobia being investigated (Bushby 1984). Underestimates in this case required the application of correction factors between 2.7x and 5.3x. If this relatively poor recovery is avoided by simplification of the media, there may be over estimation due to inclusion of non-rhizobial colonies in the count, or under estimation if there is excessive growth of inhibitory soil forms. Certainly, the method is not altogether dependable and appropriate checks must be applied (including confirmatory tests on representative counted colonies). Adjustments are likely to be needed according to the nature of the microflora of the soil being investigated and the labelled *Rhizobium* under study.

Both methods fail to differentiate the negative rhizobial reactors or antibiotic-suppressed forms.

An extension of the plant-infection count coupled with strain identification or other characterisation of rhizobia occupying nodules of the test plants remains a possibility. This method suffers from possible intervention of selective nodulating competitiveness which could cloud a straightforward interpretation. However, the limiting dilutions at which nodules are formed could be restricted to a numerically dominant strain, or to relatively few strains which might be distinguishable without too much difficulty. When the majority of strains are ineffective with a particular host, the greater dilutions will permit quantification and further study at a stage when any effective strain will have been diluted out. The greatest dilution showing effective nodules could also be relevant. A further advantage might be secured by the incorporation of more than one host in the test series, if thereby the plant reaction to majority strains could be modified.

Distinction Between Background Strains

While none of the quick tests suitable for typing a large number of nodules, cultures, or cells provides an analysis within the nonreactors, those tests based on cell cultures from nodules formed in the field or dilution count provide a better opportunity of doing this in a more detailed investigation.

A Possible Ecological Approach

Another approach could be to expose a soil, not previously sown with inoculated seed, separately to different limiting conditions (such as pH, moisture, and temperature) and isolate cultures from the nodules formed at the greatest dilution nodulating a test plant. Cultures developed in this way could be characterised, serologically or by antibiotic label, and used with another recognisable test strain as a counted mixture in soil exposed to the same single factor variables. At least each strain being tested under these conditions would be competing with other identifiable strains shown to have superior survival under particular conditions. Later, planting uninoculated seed of the requisite legume would allow nodule occupancy to be related to the degree of intervening strain survival. Identification of the nodules formed at highest positive nodulating dilution, or the use of FA or antibiotic-plating technique would provide some quantitative evidence as to adaptation to a particular modified environment.

Results reported by Hartel and Alexander (1984) of superior survival of strains of bradyrhizobia from a hot, dry region over those from a cooler, wetter area give some support for this approach, as does progress being made with strain selection designed to suit certain soil characteristics (H.V.B. Bushby, personal communication).

References

Alexander, M. 1985. Ecological constraints on nitrogen fixation in agricultural ecosystems. Pages 163-183 *in* Advances in Microbial Ecology. Volume 8. (Marshall, K.C., ed.). Plenum Press, New York, USA.

Benyon, J.L. and Josey, D.P. 1980. Demonstration of heterogeneity in a natural population of *Rhizobium phaseoli* using variation in intrinsic antibiotic resistances. Journal of General Microbiology 118: 437-442.

Brockwell, J., Gault, R.R., Zorin, M. and Roberts, M.J. 1982. Effects of environmental variables on the competition between inoculum strains and naturalised populations of *Rhizobium trifolii* for nodulation of *Trifolium subterraneum* L. and on rhizobia persistence in the soil. Australian Journal of Agricultural Research 33: 803-815.

Bushby, H.V.A. 1982. Ecology. Pages 35-75 *in* Nitrogen Fixation. Volume 2. *Rhizobium* (Broughton, W.J., ed.). Clarendon Press, Oxford, England.

Bushby, H.V.A. 1984. Colonization of rhizopheres and nodulation of two *Vigna* species by rhizobia inoculated onto seed; influence of soil. Soil Biology and Biochemistry 16: 635-641.

Cloonan, M.J. 1963. Black nodules on *Dolichos*. Australian Journal of Science 26: 121.

Franco, A.A. 1976. Competition amongst rhizobial strains for the colonization and nodulation of two tropical legumes. Plant and Soil 45: 27-48.

Hartel, P.G. and Alexander, M. 1984. Temperature and desiccation tolerance of cowpea rhizobia. Canadian Journal of Microbiology 30: 820-823.

Hughes, D.Q. and Vincent, J.M. 1942. Serological studies of the root-nodule bacteria. III. Tests of neighbouring strains of the same species. Proceedings of the Linnean Society of New South Wales 67: 142-152.

Ireland, J.A. and Vincent, J.M. 1968. A quantitative study of competition for nodule formation. Transactions of the Ninth International Congress on Soil Science 2: 85-93.

Krien, C.S. and Ham, G.E. 1985. Effect of soil temperature and inoculum rate on the recovery of three introduced strains of *Rhizobium japonicum*. Agronomy Journal 77: 484-489.

Labandera, C.A. and Vincent, J.M. 1975. Competition between an introduced strain and native Uruguayan strains of *Rhizobium trifolii*. Plant and Soil 42: 327-347.

Marques Pinto, C., Yao, P.Y. and Vincent, J.M. 1974. Nodulating competitiveness amongst strains of *Rhizobium meliloti* and *R. trifolii*. Australian Journal of Agricultural Research 25: 317-329.

Marshall, K.C. 1975. Clay mineralogy in relation to survival of soil bacteria. Annual Review of Phytopathology 13: 357-373.

Materon, L.A. and Vincent, J.M. 1980. Host specificity and interstrain competition with soybean rhizobia. Field Crops Research 3: 215-224.

Means, U.M., Johnson, H.W. and Date, R.A. 1964. Quick serological method for classifying strains of *Rhizobium japonicum* in nodules. Journal of Bacteriology 87: 547-553.

Purchase, H.F. and Vincent, J.M. 1949. A detailed study of the field distribution of strains of clover nodule bacteria. Proceedings of the Linnean Society of New South Wales 74: 227-236.

Purchase, H.F., Vincent, J.M. and Ward, L.M. 1951. The field distribution of strains of nodule bacteria from species of *Medicago*. Australian Journal of Agricultural Research 2: 261-272.

Reyes, V.G. and Schmidt, E.L. 1979. Population densities of *Rhizobium japonicum* strain 123 estimated directly in soil and rhizospheres. Applied and Environmental Microbiology 37: 854-858.

Schmidt, E.L. 1974. Quantitative autoecological study of microorganisms in soil by immunofluorescence. Soil Science 118: 141-149.

Somasegaran, P. and Hoben, H.J. 1985. Methods in legume-*Rhizobium* technology. University of Hawaii. NifTAL, Project, Paia, Hawaii.

Trinick, M.J. 1969. Identification of legume nodule bacteria by the fluorescent antibody reaction. Journal Applied Bacteriology 32: 181-186.

van Rensburg, H.J. and Strijdom, B.W. 1985. Effectiveness of *Rhizobium* strains used as inoculants after their introduction into soil. Applied and Environmental Microbiology 49: 127-131.

Vidor, C. and Miller, R.H. 1980. Relative saprophytic competence of *Rhizobium japonicum* strains in soils as determined by the quantitative fluorescent antibody technique (FA). Soil Biology and Biochemistry 12: 483-487.

Vincent, J.M. and Waters, L.M. 1953. The influence of host on competition amongst clover root-nodule bacteria. Journal of General Microbiology 9: 357-370.

Young, J.P.W. 1985. *Rhizobium* population genetics: enzyme polymorphism in isolates from peas, clover, beans and lucerne grown at the same site. Journal of General Microbiology 131: 2399-2408.

D.P. Beck and L.A. Materon (editors), Nitrogen Fixation by Legumes in Mediterranean Agriculture
 ISBN 90 247 3624 2

The Microbial Aspects of Faba Bean Culture

N. AMARGER

Laboratoire de Microbiologie des Sols INRA, Dijon, France

Summary. A study of *Rhizobium leguminosarum* and *Vicia faba* was carried out to determine whether or not the microbial partner of the symbiosis is limiting plant potential under French conditions. Isolates of *R. leguminosarum* were obtained from each of 57 soil samples using *V. faba* as trap host. These isolates were used to study symbiotic response in greenhouse experiments. Most isolates formed an effective symbiosis with *V. faba* var Ascott. Five effective isolates were used to study effectiveness, which varied between strains and between soils. In the field, there was no significant difference in yield between inoculated and uninoculated *V. faba*. It was concluded that the populations of *R. leguminosarum* in French soils are abundant, diverse, and effective.

Introduction

Western European countries are dependent on imports for their protein requirements. For the past decade these countries have tried to increase their capacity for plant protein production, and research on crops with the potential for reducing the need for imported protein has been developed. Among grain legumes, *Vicia faba*, *Pisum sativum*, and more recently *Lupinus albus*, were selected as crops with high protein yield potential for most of the EEC countries.

As legumes, these plants produce their own nitrogen through dinitrogen fixation provided there are specific rhizobia in the soil able to form effective root nodules. To show that the microbial partner is not limiting plant potential, the presence of specific rhizobia in French soils and the symbiotic properties of some isolates were studied. The work done with *R. leguminosarum* and *V. faba minor* is reported here.

Presence of *R. leguminosarum* in Soils

To determine whether or not *R. leguminosarum* is present in soils where faba beans can be grown, 57 representative French agricultural soils were sampled. The most probable number of *R. leguminosarum* was determined (Vincent 1970) using *V. sativa* as trap host (Fig. 1). Half of the soils studied

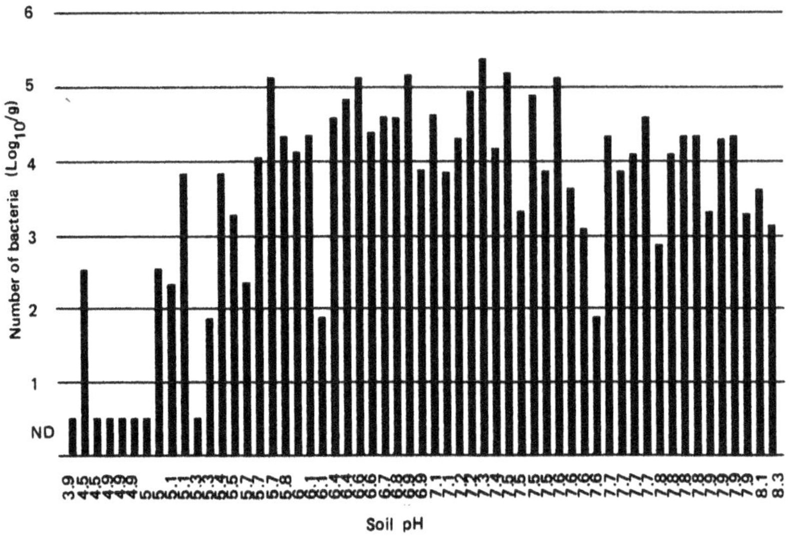

Fig. 1. Numbers of *R. leguminosarum* in idfferent soils.

had more than 10^4 *R. leguminosarum* per gram of soil, one third between 10^2 and 10^4, and 5% between 10 and 10^2. In the last seven soils, the number of *R. leguminosarum*, if any, was less than 10 per gram. The soils with an undetectable number of *R. leguminosarum* were all sandy loam or loam soils with pH less than 5.5.

Most of the soils studied seemed to contain high numbers of *R. leguminosarum*, with the exception of a few acidic soils in the south west of France, so faba beans should be easily nodulated.

Effectiveness of *R. leguminosarum* Strains

R. leguminosarum was isolated from nodules formed on *V. faba* grown in different soils and symbiotic effectiveness was determined in greenhouse experiments. Each isolate was inoculated onto seedlings of *V. faba* var Ascott grown in sand on a nitrogen-free medium and plant dry matter was determined after 2 months of growth. The number of isolates per class of effectiveness is given in Fig. 2. Ten percent of the isolates were totally ineffective and another 10% were poorly effective, giving less than 50% of the dry matter produced by the reference strains. The other isolates were effective with the variety tested.

These results confirmed previous results from a more limited sampling. They show that the majority of *R. leguminosarum* present in the soils studied can form an effective symbiosis with *V. faba* var Ascott.

Interaction between strains of *Rhizobium* and faba bean varieties has already been described (Lawes *et al.* 1978). To determine whether such

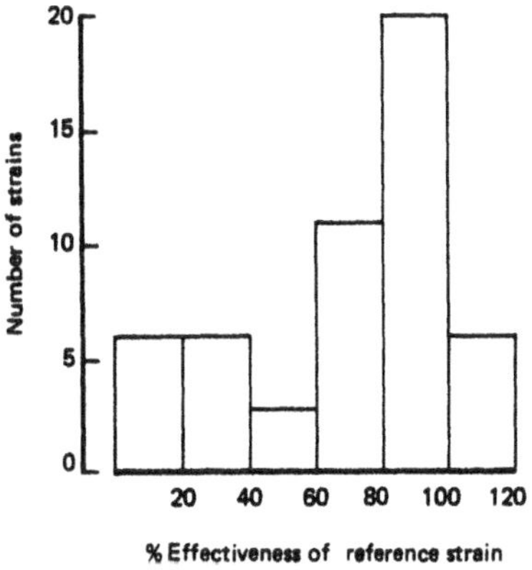

Fig. 2. Number of *R. leguminosarum* isolates in each class of effectiveness.

interactions existed under French conditions, the effectiveness of 12 isolates was estimated on three commercial varieties of *V. faba*, Ascott, Columba, and Strube. The dry matter yields of plants inoculated with the different strains are given in Fig. 3. Although the strain x variety interaction was significant ($P < 0.05$) it was very small compared to the strain effect.

In a second experiment, six strains were inoculated onto four varieties of *V. faba*, G77, G78, G8, and Po ad 74 and to their hybrids, G77 × G8, G77 × Po

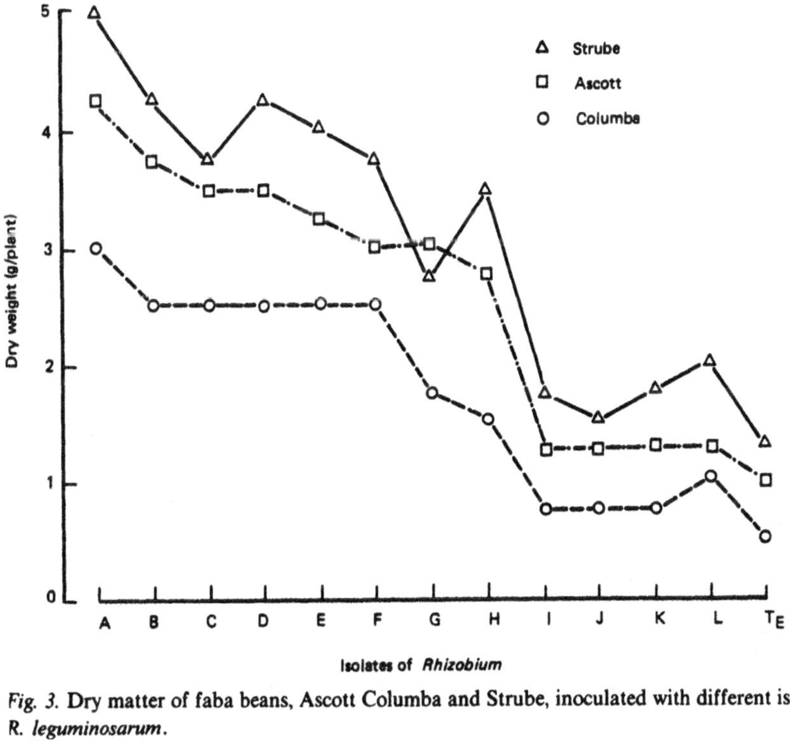

Fig. 3. Dry matter of faba beans, Ascott Columba and Strube, inoculated with different isolates of *R. leguminosarum*.

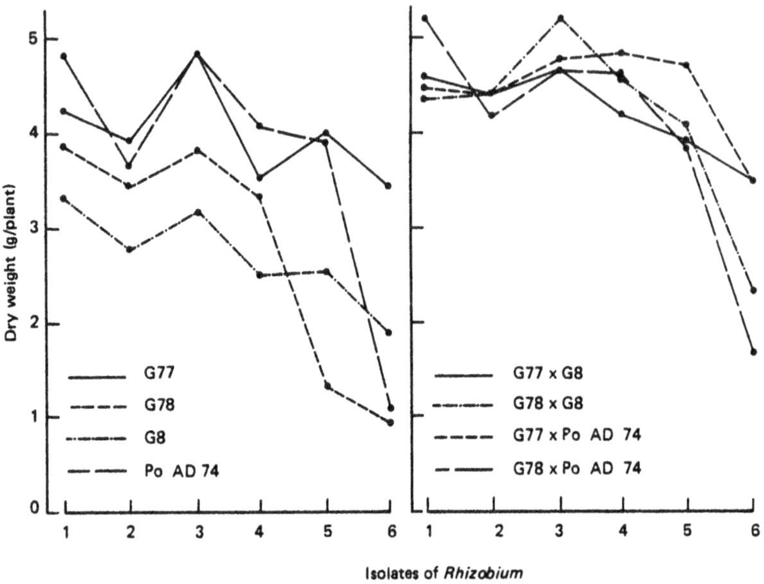

Fig. 4. Dry matter of different varieties of faba beans and their hybrids inoculated with different isolates of *R. leguminosarum*.

ad 74, G78 × G8, and G78 × Po ad 74. Plant dry matter yields are given in Fig. 4. In this case, the strain x variety interaction was highly significant ($P < 0.01$) due to strains 5 and 6 which behave differently with the varieties. Strain 5 was effective with all the varieties but one. Strain 6 was effective with the parent line G77 and the two hybrid lines G77 × Po ad 74 and G77 × G8, but was poorly effective with the other lines. In fact, the low dry matter yields observed with these different combinations were due to a very limited number of nodules on the roots and not to ineffective nodulation. In these cases, the number of nodules per plant was less than 10, while it was more than 200 with the other strain x variety associations. This restricted nodulation character is probably recessive, as hybrids of the two lines with restricted nodulation and a normally nodulating line both nodulated normally.

Interactions of symbiotic properties between faba bean genotypes and rhizobial strains occurred, but they were of very little importance compared to the interactions found by Lawes *et al.* (1978).

Nodulating Competitiveness of *R. leguminosarum* Strains

The competitiveness of five effective isolates was estimated in greenhouse experiments using antibiotic resistant mutants. The five strains were different in competitiveness (Amarger 1984). When mixed in equal quantities with one of the other four strains, the most competitive strain formed 30, 15, 3, and 2 times more nodules.

Table 1. Approximate number of bacteria needed per seed resulting in 50% of the nodules formed by the inoculum strain.

Soil	Inoculum strain	
	A	B
1	8.2×10^3	1.6×10^6
2	3.1×10^5	3.1×10^6
3	5.0×10^5	4.5×10^8
4	1.1×10^6	1.7×10^7

Two of these strains were inoculated onto seeds of faba beans grown in four different soils containing between 7×10^3 and 8×10^4 *R. leguminosarum* per gram. The number of bacteria needed per seed to result in 50% of the nodules being formed by the inoculant strain is given in Table 1. To form 50% of the nodules in the four soils, strain A, which was 10 times more competitive than strain B in the previous experiment, could be inoculated in smaller numbers than strain B. Although the number of indigenous *R. leguminosarum* was the same in soils 1 and 3 and in soils 2 and 4, both inoculant strains formed more nodules in soil 1 than in soil 3 and in soil 2 than in soil 4. The differences between soils could be due to differences in the competitiveness of indigenous strains, differences in the properties of the soil, or both.

Although interaction between strains and soils existed, it was possible to find strains, like strain A, which form the majority of the nodules when inoculated onto plants growing in different soils with an indigenous population of *R. leguminosarum*.

Effectiveness of *R. leguminosarum* in the Field

A field experiment was carried out on a soil from which two isolates out of four were ineffective. One of the antibiotic resistant strains which was highly effective and competitive was inoculated at two levels onto faba bean seeds. The soil contained 7×10^4 *R. leguminosarum* at planting and the number of nodules formed by the inoculant strains was determined after 2 months of growth. There was no significant difference between the grain yield of the inoculated and the uninoculated plots, although the inoculant strain formed 90% or more of the nodules in the inoculated plots. This indicates that if ineffective nodules were formed yield was not affected.

Conclusion

R. leguminosarum is present in high numbers in most French agricultural soils. This confirms findings reported for other European countries (Vincent 1974). Although ineffective strains could be isolated, the majority of strains isolated from nodules of faba beans growing on the different soils were highly effective with *V. faba minor* var Ascott. Some interactions between plant genotypes and rhizobial strains could be found, but their statistical significance in the three varieties studied was much smaller than the rhizobial strain effect.

Nodulation of faba bean with a highly effective inoculant or with a population of indigenous rhizobia containing ineffective strains resulted in similar grain yields, indicating that nitrogen fixation was the same whether it was due to nodules formed by different strains of varying effectiveness or to nodules formed by a single effective strain. A certain proportion of nodules fixing little or no nitrogen may be present on roots without altering the total amount of fixed nitrogen.

From the studies reported here, it can be concluded that the populations of *R. leguminosarum* present in French agricultural soils are abundant, diverse, and effective enough to support the nitrogen demand of the *V. faba* varieties presently cultivated in France. If, in some cases, evidence of inadequate populations was given, strains selected for their effectiveness and their ability to compete successfully with indigenous populations are available.

References

Amarger, N. 1984. Evaluation of competition in *Rhizobium* spp. Pages 300-305 *in* Current Perspectives in Microbial Ecology (Klug, M.J. and Reddy, C.A., eds.). American Society for Microbiology, Washington, USA.

Lawes, D.A., Mytton, L.R., El-Sherbeeny, M.H. and Sorwli, F.K. 1978. Symbiotic variation in field beans (*Vicia faba*). Annals of Applied Biology 88: 466-468.

Vincent, J.M. 1970. A manual for the practical study of root nodule bacteria. IBP Handbook no 15. Blackwell Scientific Publications, Oxford, England.

Vincent, J.M. 1974. Root-nodule symbioses with *Rhizobium*. Pages 266-341 *in* The Biology of Nitrogen Fixation. North Holland Publishing Company, Amsterdam, The Netherlands.

Discussion

Broughton

Do you have any evidence that the *R. leguminosarum* strains you selected for *V. faba* are any different from those isolated using *P. sativum* as host?

Amarger

The effectiveness spectrum of *R. leguminosarum* isolated from *V. faba* was not different from the effectiveness spectrum of *R. leguminosarum* isolated from peas. Eighty percent of isolates either from *V. faba* or *P. sativum* were found to be effective on *V. faba*, while 95% of the same isolates (either from faba bean or *Pisum*) were effective on peas.

D.P. Beck and L.A. Materon (editors), Nitrogen Fixation by Legumes in Mediterranean Agriculture
 ISBN 90 247 3624 2

Numerical Effects of *Rhizobium* Population on Legume Symbiosis

J. BROCKWELL [1], D.F. HERRIDGE [2], LINDA J. MORTHORPE [3], and R.J. ROUGHLEY [4]

[1] *Division of Plant Industry, New South Wales Dept. of Agriculture, CSIRO*
[2] *ARC, TamworthCanberra, A.C.T.*
[3] *ARC, TrangieAustralia*
[4] *HRS, Gosford*

Summary. Some of the ecological aspects of legume inoculation are discussed in this paper, with particular emphasis on numbers of rhizobia. Other aspects discussed include the ecology of rhizobia in soils previously free of rhizobia and where rhizobia are already present, nitrate effects on the symbiosis, the ecology of rhizobia in relation to the environment, and inoculation strategy.

Introduction

Many of the arable soils within ICARDA's mandate region have been used, on and off, for growing crop legumes for centuries, if not millennia. It should follow that these soils contain large, adapted populations of *Rhizobium* spp. In addition, the widespread natural occurrence of species of *Medicago*, *Melilotus*, *Trigonella*, *Trifolium*, and miscellaneous genera (Post 1932; Davis 1970) make it likely that substantial numbers of rhizobia also exist in the non-arable soils. However, there is evidence that populations of *Rhizobium* spp. are not as large or as widespread in the region as might be anticipated (Damirgi 1969; Hamdi *et al.* 1978). Moreover, in recent decades, cultivation has spread to areas that were previously rangeland with no history of growing crop legumes and presumably no population of appropriate rhizobia. This paper deals with some of the ecological considerations in determining whether inoculation of a particular legume intended for a particular location is necessary to optimise nitrogen fixation and, if so, what inoculation strategy should be adopted. Emphasis is placed on numerical considerations.

A series of experiments on the population dynamics of *Rhizobium japonicum** has been conducted in Australian environments that are ecologically distinct: (i) where a soybean inoculant is introduced into a soil previously free of *R. japonicum*, and (ii) where inoculant is introduced into soil containing an established (resident) population of *R. japonicum*. The results of these experiments and others indicate strongly that, in soil previously free of rhizobia, the

* Bergey *et al.* (1984) now use *Bradyrhizobium japonicum* instead of *Rhizobium japonicum*; we prefer *R. japonicum* for reasons of familiarity.

extent of nodulation depends on the application rate of inoculant, its distribution through the soil profile, its survival, and the nitrate content of the soil. These same factors remain influential in the presence of established rhizobia, but the size of the resident population is important in determining the extent of inoculant-induced nodulation.

Some of the findings from this work which relate to the population dynamics of inoculant strains and their rhizobial competitors are relevant to inoculation strategy, not only for soybeans but for legumes generally.

The work has also been a catalyst in formulating a research programme to develop a non-microbiological procedure to estimate the abundance, effectiveness, and diversity of *Rhizobium* populations occurring naturally in soil. The development of such a procedure would allow agriculturalists who have no access to microbiological facilities to determine if and where legume inoculation is necessary and to plan appropriate inoculation strategies for legume establishment. This in turn would assist in planned development of legume productivity. For instance, it would be an important step in the establishment of medic-based pastures to increase the quantity and quality of sheep meat produced in marginal cropping areas and rangelands of West Asia, or in obtaining increased yields of pulses grown for human consumption. Collaborating organizations are the University of Jordan, ICARDA, the New South Wales Department of Agriculture, and CSIRO Division of Plant Industry. The programme is funded by the Australian Centre for International Agricultural Research (ACIAR).

Autecology of Inoculant in Soil Previously Free of Rhizobia

Rate of Inoculation

Experiments with soybeans conducted at three locations in New South Wales on soils initially free of *R. japonicum* showed that inoculation rate considerably influenced the development of the symbiosis.

In an experiment at Breeza (Table 1), increasing rates of inoculant sprayed into the seedbed led to greater rhizosphere colonization, improved nodulation and nitrogen fixation, and larger post-harvest populations of *R. japonicum* in the soil. The absence of any treatment differences in grain nitrogen illustrates that soil nitrate and nodulation are interactive and complementary in meeting the nitrogen requirements of a soybean crop grown on a soil with high nitrate levels.

In an experiment at Trangie (Table 2), the seed coat inoculation treatments were distinguished by the strength of bonding between seed and inoculant: gum slurry > water slurry > dry dusting. Much of the inoculant was lost from the dry dusting treatment as the seed passed through the seeder, whereas the gum slurry treatment retained almost all its inoculant. As a consequence the various treatments represented different rates of inoculant introduced into the

Table 1. Effect of form and rate of inoculation on the development of soybean symbiosis on a vertisol initially containing 30 ug NO_3/g and free of *R. japonicum* at Breeza, NSW (Herridge *et al.* 1984).

Form and rate of inoculation (n = normal)	Rhizosphere colonization-day 23 (log_{10} rhizobia/plant)	Nodule dry weight-day (g/plant)	Ureides in xylem sap * day 112 (% of total nitrogenous solutes)	N in grain day 154		Rhizobia in soil day 184 (log_{10} rhizobia/g)
				(kg/ha)	(%)	
Nil	nil	0.001	17	138	4.64	0.45
Spray - n	3.00	0.12	44	172	5.46	3.15
Spray - 10n	3.02	0.23	not done	182	5.47	4.52
Spray - 100n	4.16	0.26	54	175	5.64	4.62
Seed coat - n	4.99	0.27	not done	184	5.49	4.63

* Index of N_2 fixation (Herridge 1984).

Table 2. Effect of form of inoculation on rhizosphere colonization, nodule development, and yield of soybeans grown on a grey clay soil at Trangie, NSW (Morthorpe *et al.* 1985).

Form of inoculation	Rhizosphere colonization-day 28 (log_{10} rhizobia/plant)	Nodule score (0–5)		Grain yield (t/ha)
		day 43	day 113	
Dry dusting of seed	0.30	0.3	2.3	2.96
Water slurry on seed	2.70	1.4	4.1	3.50
Gum slurry on seed	3.78	1.6	4.1	3.32
Sprayed into seedbed	2.10	0.6	4.7	3.33
Uninoculated	nil	0.04	0.2	1.99

seedbed. These results also illustrate a relationship between increasing inoculation rates and improved symbiosis.

In an experiment at Leeton (Table 3), there were five inoculation rates and two levels of soil nitrate induced by land pre-treatment. The data show

Table 3. Correlations between five rates of inoculation (1/25 to 25 x normal) and nine features of soybean symbiosis on a well-structured, grey-black clay soil initially free of *R. japonicum* at Leeton, NSW (Brockwell *et al.* 1985).

Symbiotic feature	Correlation (r)	Significance of r
Nodule score (0–5) day 23	0.990	< 0.001
Nodule score (0–5) day 114	0.914	< 0.01
Crop N (kg/ha) day 78	0.821	< 0.05
Crop N (%) day 78	0.936	< 0.01
Crop N (kg/ha) day 114	0.925	< 0.01
Seed N (%) final harvest–low nitrate soil	0.898	< 0.01
Seed N (%) final harvest–high nitrate soil	0.964	< 0.001
Seed N (kg/ha) final harvestlow nitrate soil	0.978	< 0.001
Seed N (kg/ha) final harvesthigh nitrate soil	0.841	< 0.02

Table 4. Recovery of *R. japonicum* (log_{10} rhizobia/g soil) from four zones of the profile of soil * inoculated by spraying rhizobia directly into the seedbed (Herridge *et al.* 1984).

Rate of inoculation (n = normal)	Depth of sample (cm)			
	3–6	6–9	9–12	12–15
n	2.27	1.27	nil	nil
100n	4.05	2.78	0.83	0.98

* Soil samples taken within 2h of inoculation, stored at 4°C overnight, and enumerated (Brockwell *et al.* 1975).

consistently high correlations between inoculation rate and the various features of the symbiosis measured throughout crop growth.

In similar work by Smith *et al.* (1981), soybean nodulation responded to increasing rates of inoculation of up to 3.9×10^9 rhizobia per cm of row (10000 X normal), and at least 1.0×10^7 organisms per cm of row were required to maximize grain yield.

Distribution of Inoculant Through the Soil Profile

There is a fundamental difference between seed coat inoculation and spray inoculation in which rhizobia are delivered directly into the seedbed. With seed coat inoculation, centres of rhizobial populations are concentrated in the vicinity of every seed, whereas with spray inoculation, the inoculant population is evenly distributed along the length of the seedbed and therefore the numbers in the vicinity of each seed are smaller. Examples of how *R. japonicum* was distributed through the soil profile shortly after spray inoculation are presented in Table 4. The implications of this are shown in Table 2; the spray inoculation treatment resulted in plants with fewer nodules than in seed coat inoculation (gum slurry, water slurry) during early growth (day 43). Nodulation was improved as the crop approached maturity (day 113).

Survival of Inoculant

The drastic effect of high temperature on inoculant viability has long been recognized (Bowen and Kennedy 1959) but the data in Table 5 illustrate a substantial loss of viability of *R. japonicum* even at a moderate temperature (28°C). The number of rhizobia applied as inoculant (normal rate) in experiments in 1981 was equivalent to 1.1×10^{12} per hectare. Next day, less than 5% of these organisms was recoverable and it must be assumed that further mortality would have occurred before the rhizosphere effect of young soybean seedlings induced multiplication of the survivors. In experiments conducted in

Table 5. Effect of ambient temperature at time of sowing on survival of *R. japonicum* inoculant * (Brockwell *et al.* 1987).

	Number of rhizobia ($\log_{10}$ rhizobia/ha)	
	at 28° C (1981)	at 38° C (1982)
Inoculant application	12.04	11.95
Inoculant recovery from soil soon after sowing	10.71	9.26
Inoculant recovery (%)	4.64	0.20

* Soil samples taken from seedbed within 2h of inoculation, stored at 4° C overnight, and enumerated (Brockwell *et al.* 1975).

1982, when ambient temperature on the day of sowing was 38° C, inoculant mortality in the first 24 hours was 20 times greater.

Population Fluctuations

The figures in Table 6 trace the development of *R. japonicum* populations during the first year after inoculation of soybean sown in a vertisol at Breeza, NSW. Features of the data include the degree of inoculant mortality between time of inoculation and the formation of a rhizosphere, and the substantial increase in rhizobial populations in the soil after the crop had been harvested.

Populations of *R. japonicum* shortly after harvest no doubt comprised rhizobia of soil, rhizosphere, and nodule origin, and it is a matter for conjecture how many rhizobia each of those three components contributed to the total population. Between 6 and 10 months after sowing, the soil populations increased 200-fold. We submit that this increase was due mainly to release of rhizobia from nodules as they disintegrated (Paau *et al.* 1980), which is consistent with the conclusion of Kuykendall *et al.* (1982) that best establishment of rhizobia occurred via the nodule.

Table 6. Fluctuations in populations of *R. japonicum* during and after a soybean crop at Breeza, NSW (Brockwell *et al.* 1987).

Source of population	Size of population ($\log_{10}$ rhizobia/ha)
In soil before sowing	nil
At inoculation of seedbed	12.04
In seedbed 24h after sowing	10.71
In rhizosphere (day 14)	8.00
In rhizosphere (day 28)	8.55
In rhizosphere (day 42)	10.33
In soil 6 months after sowing	12.09
In soil 10 months after sowing	14.36
In soil 12 months after sowing	14.39

Nitrate Effects on the Symbiosis

The inhibitory effects of combined nitrogen on nodulation and nitrogen fixation are well known (Raggio and Raggio 1962; Vincent 1965; Dixon 1969). An example of increased nodulation associated with depletion of soil nitrate is shown in Table 7. However, the inhibitory effects of soil nitrate on nodulation can be offset by raising inoculation rates (Table 8). The influence of soil nitrate on the net input of nitrogen to the ecosystem from well-nodulated soybeans is shown in Table 9. On the high nitrate site, the grain removed 125 kg N/ha from the soil and the residues remaining after harvest contained 81 kg/ha of fixed nitrogen, a net loss from the system of 44 kg N/ha. On the low nitrate site, the grain removed only 65 kg N/ha from the soil whereas the residues contained 104 kg/ha of fixed nitrogen, a net gain of 39 kg N/ha.

The mode of action of nitrate in reducing nodulation is not clearly understood. Munns (1968a; 1968b), for instance, regarded it as a complex of

Table 7. Interaction between level of soil nitrate and nodulation of soybeans (Herridge *et al.* 1984).

	42/35 days (nod.) (NO_3)	62/63 days (nod.) (NO_3)
Nodule dry weight (g/plant)	0.03	0.26
Soil nitrate (0–15 cm) (μg/g)	30	12 (μg/g)

Table 8. Interactive effects of nitrate * and inoculation rate on number of nodules/plant in soybeans 70 days after sowing at Breeza, NSW.

Rate of inoculation (n = normal)	Nitrogen application (kg/ha)			
	0	100	200	300
1000 n	98	54	40	37
100 n	76	21	6	9
n	16	2	0	3
nil	0	0	0	0

* Nitrate in plots without additional N, approximately 13 kg/ha, 0-30 cm, at sowing.

Table 9. Residual values of fixed nitrogen from crops of soybeans grown on soils containing high and low levels of nitrate (Bergersen *et al.* 1985).

		Source of N (kg/ha)		
		Fixation	Soil	Total
High soil nitrate	Crop N	143	274	417
	Grain N	62	125	187
	Residual N	81	149	230
Low soil nitrate	Crop N	244	120	364
	Grain N	140	65	205
	Residual N	104	55	159

Table 10. The influence of nitrate and resident rhizobia on the rhizosphere populations ($\log_{10}$ rhizobia/plant) and nodulation (nodules/plant in parentheses) of Bragg (nitrate-sensitive) and of nitrate-tolerant lines of soybean grown on a vertisol at Breeza, NSW.

Resident rhizobia	Soil nitrate nitrogen (kg/ha, 0–120 cm)	Inoculant strain				
		Bragg	Mutant 1007	Mutant 1116	Korean 466	Korean 468
Site 4	260	3.20	6.19	4.38	6.62	5.94
R. japonicum: nil		(0.9)	(19.7)	(14.2)	(20.7)	(17.2)
Site 3	68	5.32	6.24	6.79	5.70	5.56
R. japonicum: nil-low		(12.5)	(45.8)	(14.1)	(32.3)	(20.3)
Site 2	140	6.62	7.00	6.42	6.34	6.62
R. japonicum: high		(10.9)	(122.9)	(42.2)	(47.3)	(41.1)

effects on root-hair curling, root-hair infection, and the rate of development of infected root hairs. Dazzo *et al.* (1981), on the other hand, suggested that the nitrate ion interfered with the expression of recognition phenomena that are precursors of the infection process. More recently, Sherwood *et al.* (1984) found that the capacity of the white clover lectin, trifoliin, to bind to *R. trifolii* was reduced by 98% in the presence of 15 mM KNO_3. We have evidence which is consistent with these hypotheses. The soybean cultivar Bragg, whose nodulation is sensitive to nitrate, was compared with two nitrate tolerant mutants of Bragg and two Korean lines also tolerant to nitrate in an experiment on five soils at Breeza, with different nitrate contents and resident *R. japonicum*. Some of the data are presented in Table 10. The normal rate of inoculation was used for all lines, the dimensions of the rhizosphere populations of *R. japonicum* were estimated when the seedlings were 14 days old, and nodule formation was observed 50 days later. At a low level of nitrate (site 3), there were no major differences between lines. At a high level of nitrate (site 4), the rhizosphere populations of nitrate sensitive Bragg were depressed between 10 and 1000 times compared with the nitrate tolerant lines and this was reflected in nodule number. At site 2, the plants were independent of inoculation because of the large resident population of *R. japonicum*; there were no differences in rhizosphere populations but the nodulation of Bragg, although adequate, was significantly less than that of the nitrate tolerant lines. Thus, the capacity of the tolerant lines to generate large rhizosphere populations of rhizobia is one factor in their ability to produce nodules in the presence of high levels of nitrate.

Autecology of Inoculant in Soil Already Occupied by Rhizobia

It is not generally appreciated to what extent populations of rhizobia already established in soil outnumber and form a barrier against inoculant

Table 11. Relationships between the sizes of populations of rhizobia applied as seed inoculant and those already resident in the soil.

Resident soil rhizobia (0–15 cm depth)		Inoculated soybean seed	
no./g soil	$\log_{10}$ rhizobia/ha	$\log_{10}$ rhizobia/ha	approx. ratio (resident: inoculant)
10	10.35	10.78	1:1
1000	12.35	10.78	100:1
100000	14.35	10.78	10000:1

* 15000 rhizobia/seed, 80 kg seed/ha. Figures for inoculated lucerne seed (20000 rhizobia/seed, 5 kg seed/ha) are identical.

strains (Table 11). Populations of as few as 1000 resident rhizobia per gram of soil reduce the chances of successful legume inoculation to a very low level (Brockwell 1981).

Antibiotic resistant mutants of *R. japonicum* CB1809 were used to inoculate soybeans in a vertisol at Breeza containing resident populations of strain CB1809. Subsequently, the populations of rhizobia that developed in rhizospheres were counted and the strains responsible for nodule formation were identified.

The data in Table 12 show that rhizosphere populations of the resident CB1809 increased between day 14 and day 28, but the numbers of inoculant CB1809str decreased slightly. The inoculant had little or no impact on nodulation in the presence of large populations of resident rhizobia but, in the presence of small populations, formed the majority of nodules. The proportion of nodules formed by inoculant was greater than its representation in the rhizosphere relative to resident rhizobia (Table 13). This apparent imbalance might be attributed to a concentration of inoculant in that zone of the seedbed where infectable foci on the seedling roots first form, compared with the random distribution of soil rhizobia throughout the profile. If so, it may be

Table 12. Effects of size of populations of resident[1] (CB1809) and inoculant (CB1809str) *R. japonicum* on the rhizosphere colonization and nodule occupancy in soybeans grown at Breeza, NSW (Brockwell *et al.* 1987).

R. japonicum populations in soil at sowings (log rhiz/ha)		Rhizosphere colonization (log rhiz/pl)				Nodules (%, day 41) formed by	
Resident	Inoculant	Day 14		Day 28			
1809	1809str	1809	1809str	1809	1809str	1809	1809str
9.80	11.95	nd[2]	2.50	2.10	2.10	0	100
9.80	14.11	nd	4.00	3.90	3.30	11	89
14.20	11.95	6.33	2.15	7.10	2.40	100	0
14.20	14.11	6.20	2.70	6.60	2.40	97	3

1. Resident rhizobia present for 1 year, high rate of inoculant mortality.
2. Not detected.

Table 13. Effects of a small resident [1] population of *R. japonicum* CB1809 on the rhizosphere colonization and nodule occupancy by inoculant CB1809strspc applied at increasing rates to soybeans grown at Breeza, NSW (Brockwell *et al.* 1987).

R. japonicum populations in soil at sowing (log rhiz/ha)		Rhizosphere colonization-day 42 (log rhiz/pl)		Nodules (%, day 100) formed by	
Resident 1809	Inoculant 1809strspc	1809	1809str	1809	1809strspc
10.73	nil	2.70	nd [2]	100	0
10.73	12.45	5.60	4.10	51	49
10.73	14.64	7.35	4.30	24	76

1. Resident rhizobia present for 2 years, low rate of inoculant mortality.
2. Not detected.

that nodule occupancy is largely determined by the component of a mixed population of competing rhizobia which first colonizes that region of the root where nodules subsequently form (Kosslak *et al.* 1983). As the plants aged, as their roots extended beyond the zone of inoculant placement, and as rhizosphere populations of resident rhizobia increased substantially, there was also an increase in the proportion of nodules occupied by those residents (Brockwell *et al.* 1987).

Implications for Inoculation Strategy

A number of conclusions emerge:

(i) Rate of inoculation (Table 1) and the form of inoculation (where that form has an effect on the rate, Table 2), enhance rhizosphere colonization and nodule formation. This improvement in the early symbiosis is often (Table 3), but not always, associated with higher yield.

(ii) The form of inoculation influences the distribution of inoculant through the soil profile (Table 4). While inoculant applied to the seed concentrates at a number of points along the row, liquid inoculant tends to diffuse through the seedbed.

(iii) Considerable loss of viability occurs immediately after the inoculant is introduced into the seedbed and is exacerbated by hot weather (Table 5).

(iv) Populations of rhizobia increase once the host has formed a rhizosphere. Very large increases occur after harvest, probably due to the release of rhizobia from disintegrating nodules (Table 6).

(v) The inhibitory effects of soil nitrate (Table 7) can be offset with increased rates of inoculation (Table 8).

(vi) Soil nitrate and nodulation are interactive and complementary in meeting the nitrogen requirements of a well-nodulated legume. In high nitrate soil, the majority of seed nitrogen is derived from the soil but in low nitrate soil most of the seed nitrogen is fixed from the atmosphere (Table 9). This accounts for the common observation that nodulation responses in the early symbiosis are not always reflected in yield.

(vii) The capacity of nitrate tolerant legumes to generate large rhizosphere populations of rhizobia is one factor in their ability to produce nodules in the presence of high levels of nitrate (Table 10).

(viii) Although resident rhizobia soon dominate inoculant strains in colonization of the rhizosphere (Table 12), the proportion of nodules formed by the inoculant was greater than its rhizosphere representation (Table 13). This may be due to an advantage of strategic placement.

Ultimate domination of inoculant strains by resident rhizobia is a common occurrence (Dudman and Brockwell 1968; Roughly *et al.* 1976; Ellis *et al.* 1984). In our experiments, the superiority of CB1809 was apparently a consequence of it being the first strain of *R. japonicum* introduced into the soil where, in the absence of competition from other rhizobia, it soon developed a very large, well-dispersed population. This represented a more potent source of inoculum for soybeans than the smaller numbers of inoculant rhizobia introduced subsequently, which were further diminished by environmental hazards at and immediately following sowing. Thus, the competitive superiority of background rhizobia over subsequent inocula is explicable in purely numerical terms. We submit that, for an inoculant strain to be a successful competitor against a background population, it must possess particular competitive characteristics, e.g. bacteriocin production (Schwinghamer 1975), and it must be mobile. Without an ability to migrate from the site of inoculation to the close proximity of centres of populations of background rhizobia, it is unlikely that an inoculant strain could express any competitive capability it might have. Hamdi (1971) demonstrated that movement of rhizobia through soil was restricted in the absence of free moisture. Likewise, Schwinghamer and Brockwell (1978) showed that the ability of bacteriocinogenic strains to suppress sensitive rhizobia was stronger in "wet" peat than in "damp" peat. These observations suggest that mobility of rhizobia is often dependent on environmental factors. However, Howieson and Ewing (1986) found some evidence of differential mobility among strains of rhizobia.

Some of the findings relating to the population dynamics of inoculant strains and their rhizobial competitors already established in soil are germane to inoculation strategy (Brockwell *et al.* 1987; Herridge *et al.* 1987). Seeding legumes in hot weather is wasteful of inoculant (at 38°C there is > 99.8% loss of viability by the next day) (Bowen and Kennedy 1959). However, since that practice is often unavoidable, it is clearly desirable that inoculant strains be tolerant of high temperature (Munevar and Wollum 1982) or better protected from its deleterious effects (Kremer and Peterson 1982). Obvious benefits accrue from using the highest feasible rates of inoculation to increase nodulation and nitrogen fixation (Herridge *et al.* 1984; Bergersen *et al.* 1985). Our results indicated that soil rhizobia were most susceptible to competition from inoculant rhizobia shortly after germination of the host and that this period should be the special focus of inoculation strategy. Early nodulation of the plant by the inoculant strain can be induced, even in the presence of large populations of competing soil bacteria, by placement of inoculant in that zone of the seedbed where infection foci on the roots will first appear. Accurate placement and concentration of rhizobia might be achieved by using seeding equipment with an attachment that permits granular inoculant to be deposited into the seedbed (Scudder 1975; Muldoon *et al.* 1980). There may be some scope for enhancing nodulation away from the immediate site of inoculation with inoculants containing strains of rhizobia that are vigorously motile (Ames and Bergman 1981). The quality (numerical standard) of inoculant might be improved and the quantities used can be increased several-fold. Seed and inoculant should be sown into moist and, if feasible, cool seedbeds. Populations of background rhizobia might be established by inoculating a prior crop. *Rhizobium* is essentially a rhizosphere organism and previous rhizospheres (perhaps of other species, not necessarily legumes) could provide a vehicle for its multiplication (Diatloff 1969; Robert and Schmidt 1983; Hiltbold *et al.* 1985). It may be feasible to select more freely-nodulating soybeans (Carroll *et al.* 1985; Herridge and Betts 1985).

Sophisticated improvements in inoculation strategy will not by themselves negate the competitive advantages of naturalized rhizobia adapted to the environment and randomly distributed through the soil. The susceptibility of legume roots to infection by rhizobia is a transient phenomenon (Bhuvaneswari *et al.* 1981) so the further the infectable zone extends beyond the site of inoculation, the greater the opportunity for nodule formation by soil rhizobia. Later, when those nodules decay and release their contents, they are a potent source of further soil colonization by the infecting rhizobia (Kuykendall *et al.* 1982). Even assuming that the potential for soil colonization by inoculant and soil rhizobia from nodule disintegration is a function of relative nodule number, when the two sorts of bacteria are mixed by cultivation the resident rhizobia are likely to become the dominant component of the mixed population by weight of numbers and then to dominate in nodulating subsequent crops.

Ecology of Rhizobia as Related to Vegetation, Soil, Climate and Land Management

Demographic pressures that have developed in West Asia since World War II have led to changing patterns of agriculture. Cropping has extended to land previously used solely for grazing, thereby forcing flocks more and more into low rainfall areas (< 150 mm).

The provision of productive leguminous pastures would help to alleviate this situation. Good legume pastures would reduce grazing pressures and the rate of land degradation in pastoral areas and marginal cropping zones, and would increase the quantity and quality of livestock in the region (McWilliam 1982). The areas most suited to the introduction of legumes are elevated and/or non-arable lands and where fallow is a component in the rotation. The Mediterranean climate in West Asia and the alkalinity of many soils suggest that annual medics (*Medicago* spp.) have the greatest potential for this purpose (Radwan *et al.* 1978). However, medics cannot express this potential unless the soils contain compatible populations of the effective root-nodule bacterium, *Rhizobium meliloti*.

Although *R. meliloti* is endemic to West Asian soils, practically nothing is known about the size or nitrogen-fixing diversity of the populations occurring in various soils, although there are exceptions (Damirgi 1969; Hamdi *et al.* 1978). Such information is basic to the determination of which medics can be established and what ameliorative measures may be necessary for establishment to be productive. Some evidence indicates that the absence of effective, naturalised strains of *R. meliloti* from some soils is a major constraint on the establishment and productivity of annual medics.

Therefore, information about the size and symbiotic characteristics of rhizobial populations in soil is a prerequisite for successful legume establishment. Microbiological procedures for obtaining this information already exist, but they are of no real practical value because they are too time-consuming and require facilities that are rarely available, particularly in developing countries. The object of the research programme alluded to in the introduction is to provide non-microbiological means of acquiring information about the size and diversity of populations of rhizobia in soil. The hypothesis behind the proposal is that the nature of soil-borne populations of rhizobia (number, nitrogen-fixing capacity) is determined by factors closely related to physical features of the landscape (such as vegetation, precipitation, elevation, soil type, soil pH, exchange capacity, soil structure, and land management), and that therefore, the rhizobial status of a soil can be predicted by reference to these physical features and without recourse to microbiological facilities (Brockwell and Robinson 1970). Complementary investigations using the *Medicago* spp./*R. meliloti* symbiosis as a model system will be conducted in Jordan, Syria and Australia (New South Wales, South Australia). It may be feasible to extend the work to include other legume/*Rhizobium* associations.

References

Ames, P. and Bergman, K. 1981. Competitive advantage provided by bacterial motility in the formation of nodules by *Rhizobium meliloti*. Journal of Bacteriology 148: 728-729.

Bergersen, F.J., Turner, G.L., Chase, D.L., Gault, R.R. and Brockwell, J. 1985. The natural abundance of ^{15}N in an irrigated soybean crop and its use for the calculation of nitrogen fixation. Australian Journal of Agricultural Research 36: 411-423.

Bergey, D.H., Holt, J.G. and Krieg, N.R. (eds). 1984. Bergey's Manual of Systematic Bacteriology, Vol. 1. Williams and Wilkins, Baltimore, USA.

Bhuvaneswari, T.V., Bhagwat, A.A. and Bauer, W.D. 1981. Transient susceptibility of root cells in four common legumes to nodulation by rhizobia. Plant Physiology 68: 1144-1149.

Bowen, G.D. and Kennedy, M. 1959. Effect of high soil temperatures on *Rhizobium* spp. Queensland Journal of Agricultural Science 16: 177-197.

Brockwell, J. 1981. A strategy for legume nodulation research in developing regions of the Old World. Plant and Soil 58: 367-382.

Brockwell, J., Diatloff, A., Grassia, A. and Robinson, A.C. 1975. Use of wild soybean (*Glycine ussuriensis* Regel and Maack) as a test plant in dilution-nodulation frequency tests for counting *Rhizobium japonicum*. Soil Biology and Biochemistry 7: 305-311.

Brockwell, J., Gault, R.R., Chase, D.L., Turner, G.L. and Bergersen, F.J. 1985. Establishment and expression of soybean symbiosis in a soil previously free of *Rhizobium japonicum*. Australian Journal of Agricultural Research 36: 397-409.

Brockwell, J., Roughley, R.J. and Herridge, D.F. 1987. Population dynamics of *Rhizobium japonicum* strains used to inoculate three successive crops of soybeans. Australian Journal of Agricultural Research 38: 61-74.

Brockwell, J. and Robinson, A.C. 1970. Observations on the natural distribution of *Rhizobium* spp. relative to physical features of the landscape. Pages 438-441 *in* Proceedings of the Eleventh International Grassland Congress.

Carroll, B.J., McNeil, D.L. and Gresshoff, P.M. 1985. A supernodulation and nitrate-tolerant (*nts*) soybean mutant. Plant Physiology 78: 34-40.

Damirgi, S.M. 1969. Populations of the naturalized alfalfa nodule bacteria in central Iraqi soils. Iraqi Journal of Agricultural Science 4: 37-44.

Davis, P.H. 1970. Flora of Turkey and the East Aegean Islands, Vol. 3. University Press, Edinburgh, UK.

Dazzo, F.B., Hrabek, E.M., Urbano, M.R., Sherwood, J.E. and Truchet, G. 1981. Regulation of recognition in the *Rhizobium*-clover symbiosis. Pages 292-295 *in* Current Perspectives in Nitrogen Fixation Proceedings of the Fourth International Symposium on Nitrogen Fixation (Gibson, A.H. and Newton W.E., eds.). Australian Academy of Science, Canberra, Australia.

Diatloff, A. 1969. The introduction of *Rhizobium japonicum* to soil by seed inoculation of non-host legumes and cereals. Australian Journal of Experimental Agriculture and Animal Husbandry 9: 357-360.

Dixon, R.O.D. 1969. Rhizobia (with particular reference to relationships with host plants). Annual Review of Microbiology 23: 137-158.

Dudman, W.F. and Brockwell, J. 1968. Ecological studies of root-nodule bacteria introduced into field environments. I. A survey of field performance of clover inoculants by gel immune diffusion serology. Australian Journal of Agricultural Research 19: 739-747.

Ellis, W.R., Ham, G.E. and Schmidt, E.L. 1984. Persistence and recovery of *Rhizobium japonicum* inoculum in a field soil. Agronomy Journal 76: 573-576.

Hamdi, Y.A. 1971. Soil-water tension and the movement of rhizobia. Soil Boilogy and Biochemistry 3: 121-126.

Hamdi, Y.A., Yousef, A.N., Amira Al-Azawi, S.K. and Al-Baquari, M. 1978. Distribution of *Rhizobium meliloti* and *R. trifolii* in Iraqi soils. Soil Biology and Biochemistry 10: 149-150.

Herridge, D.F. 1984. Effects of nitrate and plant development on the abundance of nitrogenous solutes in root-bleeding and vacuum-extracted exudates of soybean. Crop Science 24: 173-179.

Herridge, D.F. and Betts, J.H. 1985. Nitrate tolerance in soybean: variation between genotypes. Page 32 *in* Nitrogen Fixation Research Progress (Evans, H.J., Bottomley, P.J. and Newton, W.E., eds.). Martinus Nijhoff, Dordrecht, The Netherlands.

Herridge, D.F., Roughley, R.J. and Brockwell, J. 1984. Effect of rhizobia and soil nitrate on the establishment and functioning of the soybean symbiosis in the field. Australian Journal of Agricultural Research 35: 149-161.

Herridge, D.F., Roughley, R.J. and Brockwell, J. 1987. Low survival of *Rhizobium japonicum* inoculant leads to reduced nodulation, nitrogen fixation and yield of soybean in the current crop but not in the subsequent crop. Australian Journal of Agricultural Research 38: 75-82.

Hiltbold, A.E., Patterson, R.N. and Reed, R.B. 1985. Soil populations of *Rhizobium* in a cotton-corn-soybean rotation. Soil Science Society of America Journal 49: 343-348.

Howieson, J.G. and Ewing, M.A. 1986. Acid tolerance in the *Rhizobium meliloti-Medicago* symbiosis. Australian Journal of Agricultural Research 37: 55-64.

Kosslak, R.M., Bohlool, B.B., Dowdle, S. and Sadowsky, M.J. 1983. Influence of environmental factors on inter-strain competition in *Rhizobium japonicum*. Applied and Environmental Microbiology 46: 870-873.

Kremer, R.J. and Peterson, H.L. 1982. Effect of inoculant carrier on survival of *Rhizobium* on inoculated seed. Soil Science 134: 117-125.

Kuykendall, L.D., Devine, T.E. and Cregan, P.B. 1982. Positive role of nodulation on the establishment of *Rhizobium* in subsequent crops of soybean. Current Microbiology 7: 79-81.

McWilliam, J.R. 1982. Pasture and forage systems. A proposal for research in North Africa and West Asia. Report prepared for the Director General, ICARDA, Aleppo, Syria.

Morthorpe, L.J., Gault, R.R. and Brockwell, J. 1985. Soybean inoculation trial at Trangie, 1984/85. Macquarie Irrigator, New Ser. 1: 1-2. New South Wales Department of Agriculture, Dubbo, NSW, Australia.

Muldoon, J.F., Hume, D.J. and Beversdorf, W.D. 1980. Effects of seed- and soil-applied *Rhizobium japonicum* inoculants on soybeans in Ontario. Canadian Journal of Plant Science 60: 399-409.

Munevar, F. and Wollum, A.G., II. 1982. Response of soybean plants to high root temperature as affected by plant cultivar and *Rhizobium* strain. Agronomy Journal 74: 138-142.

Munns, D.N. 1968a. Nodulation of *Medicago sativa* in solution culture. II. Compensating effects of nitrate and of prior nodulation. Plant and Soil 28: 246-257.

Munns, D.N. 1968b. Nodulation of *Medicago sativa* in solution culture. III. Effects of nitrate on root hairs and infection. Plant and Soil 29: 33-47.

Paau, A.S., Bloch, C.B. and Brill, W.J. 1980. Developmental fate of *Rhizobium meliloti* bacteroids in alfalfa nodules. Journal of Bacteriology 147: 1480-1490.

Post, C.E. 1932. Flora of Syria, Palestine and Sinai, Vol. 1. American University Press, Beirut, Lebanon.

Radwan, M.S., Al-Fakhry, A.K. and Al-Hasan, A.M. 1978. Some observations on the performance of annual medics in northern Iraq. Mesopotamia Journal of Agriculture 13: 55-67.

Raggio, M. and Raggio, N. 1962. Root nodules. Annual Review of Plant Physiology 13: 109-128.

Robert, F.M. and Schmidt, E.L. 1983. Population changes and persistence of *Rhizobium phaseoli* in soil and rhizospheres. Applied and Environmental Microbiology 45: 550-556.

Roughley, R.J., Blowes, W.M. and Herridge, D.F. 1976. Nodulation of *Trifolium subterraneum* by introduced rhizobia in competition with naturalized strains. Soil Biology and Biochemistry 8: 403-407.

Schwinghamer, E.A. 1975. Properties of some bacteriocins produced by *Rhizobium trifolii*. Journal of General Microbiology 91: 403-413.

Schwinghamer, E.A. and Brockwell, J. 1978. Competitive advantage of bacteriocin- and phage-producing strains of *Rhizobium trifolii* grown in mixed culture in nutrient broth and in peat. Soil Biology and Biochemistry 10: 383-387.

Scudder, W.T. 1975. *Rhizobium* inoculation of soybeans for sub-tropical and tropical soils. I. Initial field trials. Soil and Crop Science Society of Florida Proceedings 34: 79-82.

Sherwood, J.E., Truchet, G.L. and Dazzo, F.B. 1984. Effect of nitrate supply on the *in vivo* synthesis and distribution of trifoliin A, a *Rhizobium trifolii*-binding lectin, in *Trifolium repens* seedlings. Planta 162: 540-547.

Smith, R.S., Ellis, M.A. and Smith, R.E. 1981. Effect of *Rhizobium japonicum* inoculant rates on soybean nodulation in a tropical soil. Agronomy Journal 73: 505-508.

Vincent, J.M. 1965. Environmental factors in the fixation of nitrogen by the legume. Pages 384-435 *in* Soil Nitrogen (Bartholomew, W.V. and Clark, F.C., eds.). American Society of Agronomy, Madison, Wisconsin, USA.

Discussion

Danso

Dr Peter Greshoff has found that with his supernodulating soybean, ability to nodulate in the presence of high soil nitrogen is dictated by the plant, while in your case, this is being attributed to *Rhizobium* numbers. In this case, will I be right to suggest that by simply increasing rhizobial numbers in the rhizosphere through e.g., inoculation, nitrate inhibition of nodulation could be overcome?

Brockwell

I see no conflict between Dr Greshoff's results and our own. In addition, our results show clearly that nitrate inhibition of nodulation can be partially overcome by increasing rates of inoculation which in turn leads to increased rhizosphere populations.

Materon

I would like to comment that Materon and Vincent, in a paper entitled 'Host specificity and interstrain competition with soybean rhizobia' (Field Crops Research 3: 215-224), utilized the strain CB1809 to inoculate soybean lines Lee, Hardee, and several stable crosses. In some instances, it was found that nodulation success was not influenced by the superior representation of CB1809 in the root population in the presence of one or more competitors. Nodulation success, or lack of it, was attributed to host incompatibility with CB1809 and other inoculant strains not including CC709. Soybean genotypes bearing the Rj2 gene are symbiotically incompatible with strains belonging to serogroups 122 and C1.

D.P. Beck and L.A. Materon (editors), Nitrogen Fixation by Legumes in Mediterranean Agriculture
© 1988 ICARDA – Printed in the Netherlands. ISBN 90 247 3624 2

Identification and Characterization of Rhizobia Using the ELISA Technique

A.R.J. EAGLESHAM and M.J. SINCLAIR
Boyce Thompson Institute, New York, USA

Summary. Methods available for the identification of rhizobia in ecology work are summarized. Since 1978 the enzyme-linked immunosorbent assay (ELISA) has become a popular serological tool. A brief review of the ELISA-rhizobia literature is presented, followed by a comparison of the serology of cowpea isolates from semi-arid and humid environments in West Africa.

Identification of Rhizobial Strains

Studies on the ecology of *Bradyrhizobium* require the ability to recognize strains. For model-system investigation of rhizobial competitiveness, the ability to distinguish between strains is necessary and the interpretation of data from field legume inoculation trials is often assisted by the capacity to determine the relative proportions of root nodules occupied by the applied and indigenous types. Where the diversity of native soil populations of rhizobia are to be characterized, simple, rapid tools are desirable for the delineation of types.

To assist in rhizobial identification, the legume microbiologist has a broad range of methods from which to choose. The nodule occupant may affect nodule appearance; ineffective strains often form smaller, white- or green-pigmented nodules, and some bradyrhizobia produce nodules of unusual color due to synthesis of an ancillary pigment (Cloonan 1963; Eaglesham 1985; Eaglesham *et al.* 1982; Kuykendall and Elkan 1976). Distinctive colony appearance on agar media is another convenient and potentially useful characteristic (Kuykendall and Elkan 1976; Marques Pinto *et al.* 1974; Meyer and Peuppke 1980; Stowers and Eaglesham 1984). The degree of absorption of congo-red dye incorporated into agar media can vary among strains (Kneen and LaRue 1983) and has been used as an easy marker (Bromfield and Jones 1979). Mutants resistant to (an) antibiotic(s) have been extensively used (Schwinghamer and Dudman 1980), because strains so marked can be easily and definitively recognized in cultures from soil and nodules, when grown in the presence of the appropriate antibiotic(s). Auxotrophic mutants are similarly useful (Johnston and Beringer 1975). However, checks are necessary with genetic markers for variation from wild type in terms of effectiveness and/or competitiveness (Bromfield and Jones 1979; Hagedorn 1979; Johnston and

Beringer 1975; Pankhurst 1977; Schwinghamer 1964; 1967; Schwinghamer and Dudman 1973). Problems of altered symbiotic potential can be circumvented by using strains with characteristic patterns of intrinsic resistance to antibiotics (Bromfield *et al.* 1982; Eaglesham 1987; Jenkins and Bottomley 1985; Josey *et al.* 1979; Kingsley and Bohlool 1983; Kremer and Peterson 1982; La Favre and Focht 1983; Rupela *et al.* 1982; Sinclair and Eaglesham 1984). Serology has been used probably more than any other tool for the identification, classification, and characterization of rhizobia, and is second only to gel-electrophoresis of proteins in discriminatory power (Dughri and Bottomley 1984; Kamicker and Brill 1986). The agglutination test has been used for many years in the study of antigenic relationships of rhizobia (Bushnell and Sarles 1939; Damirgi *et al.* 1967; Dudman 1977; Graham 1963; Means *et al.* 1964; Parker and Grove 1970; Schwinghamer and Dudman 1980; Somasegaran and Hoben 1985; Vincent 1970) and the related techniques, gel-immune diffusion (Damirgi *et al.* 1967; Dudman 1964; 1977; Fuquay *et al.* 1984; Schwinghamer and Dudman 1980; Sinha and Peterson 1980; Somasegaran and Hoben 1985; Vincent 1970; Vincent and Humphrey 1970; Vincent *et al.* 1973), and immunofluorescence (Robert and Schmidt 1985; Schmidt 1973; Schmidt *et al.* 1968; Schwinghamer and Dudman 1980; Somasegaran and Hoben 1985; Trinick 1969; Vincent 1970) have been widely used as powerful aids for rhizobial ecology work.

Kishinevsky and Bar-Joseph (1978) reported the adaptation of an enzyme-linked immunosorbent assay (ELISA) for the identification of rhizobia in culture and in nodules. Enzyme-labelled antibodies had been used for some years for the detection of viruses in mammalian tissue (Nakane and Pierce 1966) and had then been developed as a quantitative assay (Clark and Adams 1977; Engvall and Perlmann 1972). The advantages of the method have resulted in ELISA gaining prominence as a tool for rhizobial research (Table 1). These advantages are rapidity, sensitivity (less antiserum and antigen required), and relative simplicity without the necessity of expensive equipment. The ELISA method combines serological specificity with enzyme-catalyzed sensitivity.

ELISA

The process fundamental to the ELISA technique is the specific adsorption of rhizobial antigens to antibodies conjugated to an enzyme (usually alkaline phosphatase), followed by application of a chromogenic substrate (usually p-nitrophenyl phosphate). Color development is proportional to the amount of enzyme present, as determined by the degree of antigen/antibody interaction. The technique has obvious quantitative potential, (and has been adapted for purposes of rhizobial cell enumeration, Table 1) and because its efficiency is largely independent of the antigen:antibody ratio, detection of rhizobia is possible over a wide range of concentrations.

Table 1. Citations of rhizobial work using ELISA.

Type of study	Source
Identification of cultured rhizobia	Asanuma *et al.* 1985; Berger *et al.* 1979; Fuhrmann and Wollum II 1985; Kishinevsky and Bar-Joseph 1978; Kishinevsky and Gurfel 1980; Kishinevsky and Maoz 1983; Morley and Jones 1980; Olsen and Rice 1984; Olsen *et al.* 1983.
Identification of nodule occupants	Asanuma *et al.* 1985; Berger *et al.* 1979; Eklund 1984; Fuhrmann and Wollum 1985; Hodgson and Waid 1981; IITA 1981; Jones and Morley 1981a; 1981b; Kishinevsky and Bar-Joseph 1978; Kishinevsky and Gurfel 1980; Kishinevsky and Maoz 1983; Nambiar et al. 1984; Ranga Rao *et al.* 1985; Rice *et al.* 1984.
Indigenous populations	Ahmad *et al.* 1981; Eagleshamet al. 1984.
Enumeration	Martennsson and Gustafsson 1985; Nambiar and Anjaiah 1985; Olsen et al. 1983; Renwick and Jones 1985.
Methodological adaptations	Asanuma *et al.* 1985; Fuhrmann and Wollum 1985; Kishinevsky and Maoz 1983; Kishinevsky *et al.* 1984; Martennsson *et al.* 1984. Morley and Jones 1980; Nambiar and Anjaiah 1985; Olsen and Rice 1984; Renwick and Jones 1985.

Much of the rhizobial ELISA work has involved a solid support for antigens or antisera. Microplates have been widely used (Ahmad *et al.* 1981; Asanuma *et al.* 1985; Fuhrmann and Wollum II 1985; Kishinevsky and Bar-Joseph 1978; Morley and Jones 1980; Nambiar *et al.* 1984; Olsen *et al.* 1983). These polystyrene 96-well plates are treated to adsorb protein. Glass agglutination slides provide a re-usable, hence less expensive, alternative (Berger *et al.* 1979; Hodgson and Waid 1981). Two forms of ELISA have been employed, direct and indirect. The direct double-antibody sandwich (DAS) method requires initial surface-adsorption of the purified immunogamma globulin (YG) followed by the antigen and then by enzyme-conjugated YG (YG-conjugate) (Asanuma *et al.* 1985; Kishinevsky and Bar-Joseph 1978; Morley and Jones 1980; Nambiar *et al.* 1984; Olsen and Rice 1984; Olsen *et al.* 1983; Renwick and Jones 1985). The major disadvantage of the direct method is that its scope of applicability is limited by the number of antigen-specific YG-conjugates available. This problem is obviated by the indirect method with which the antigens are initially applied to the solid phase, followed by YG and then goat (or swine or sheep) anti-rabbit YG-conjugate (Ahmad *et al.* 1981; Asanuma *et al.* 1985; Berger *et al.* 1979; Fuhrmann and Wollum II 1985; Olsen and Rice 1984); thus, a single YG-conjugate (which is available commercially) can be used in conjunction with a number of separately strain-specific YG preparations. An indirect sandwich procedure utilizing pepsin-digested F(ab')$_2$-specific YG (Turner 1981) has been reported (Nambier and Anjaiah 1985), but appears to have little advantage over the standard indirect method. The indirect approach eliminates problems caused by variation in reactivity of home-lab-produced conjugates. On the other hand the greater sensitivity of the DAS method (Asanuma *et al.* 1985; Barbara and Clark 1982; Olsen and Rice 1984), would make it the technique of choice if the supply of antigen were limited, e.g. from very small nodules. Significant non-specific adsorption of goat anti-rabbit YG to nodule antigens has been observed (Kishinevsky and Maoz 1983; Kishinevsky *et al.* 1984), although other work has demonstrated excellent agreement between culture and nodule antigens (Fuhrmann and Wollum II 1985). This non-specificity was eliminated by using staphylococcal protein A conjugated to alkaline phosphatase in place of goat anti-rabbit YG-conjugate (protein A binds specifically to the Fc (Turner 1981) region of immunoglobulins (Kronvall *et al.* 1970). Interestingly, the indirect protein A method was more sensitive than the direct sandwich method (Kishinevsky and Moaz 1983; Kishinevsky *et al.* 1984). In addition to problems of false positives resulting from non-specific reactions, it is important to be aware of possibilities of false negatives. Insignificant ELISA readings have been obtained with YG preparations which gave high titres in agglutination tests (Kishinevsky and Gurfel 1980; Olsen *et al.* 1983). Clearly, adequate preliminary investigations are required to ensure acceptable limits of specificity and sensitivity according to the nature of the study. Fuhrmann and Wollum (1985) recently showed that YG purification was not necessary for their routine identification of *Bradyrhizobium japonicum*, and Asanuma *et al.* (1985) examined conditions of nodule storage and adaptations of microplate methods to facilitate ELISA use at field sites.

Occasional poor reproducibility of data from microplates has been ascribed to within- or between-plate variability causing uneven passive adsorption of antibody or antigen preparations (Clark and Adams 1977; Fuhrmann and Wollum II 1985; Olsen and Rice 1984). Plates which are manufacturer-certified for well uniformity are now available, but the problem can be circumvented by evaporative drying of antigens onto the well surface, albeit with some loss of sensitivity (Olsen and Rice 1984), or by heat fixation onto glass (Berger *et al.* 1979), or by dispensing with the solid phase with a tube/centrifugation technique (Martennsson and Gustafsson 1985; Martennsson *et al.* 1984).

"Checker-board" assays are required to determine the appropriate concentrations of antigen, YG, and YG-conjugate. A number of these have been reported and they provide useful guidelines (Asanuma *et al.* 1985; Berger *et al.* 1979; Kishinevsky and Maoz 1983; Martennsson *et al.* 1984; Morley and Jones 1980). Morley and Jones (1980) demonstrated an improvement of four orders of magnitude in the sensitivity of the ELISA technique with a fluorescent rather than a chromogenic substrate; however, the sensitivity of their color assay was relatively poor (Kishinevsky and Bar-Joseph 1978). Whatever form of the ELISA technique is used, sensitivity and economy of use of YG are ensured. Detection of as few as $2x10^2$ cells is possible (Martennsson *et al.* 1984) but $5x10^4$ is closer to the norm (Kishinevsky and Bar-Joseph 1978).

Clearly, detection limit is more important for work with nodules than with cultured cells. Positive detection of nodule antigens is possible with less than 80 μg of nodule tissue (Kishinevsky and Bar-Joseph 1978). Tube-grown clover nodules of only 0.1-0.3 mg fresh weight were of sufficient size to assay with three antisera (Hodgson and Waid 1981). Heat treatment of antigens improved specificity (Kishinevsky and Bar-Joseph 1978), and generally (Furhmann and Wollum II 1985; Kishinevsky and Bar-Joseph 1978; Kishinevsky and Gurfel 1980; Kishinevsky and Maoz 1983) but not always (Berger *et al.* 1979; Martennsson *et al.* 1984) improved sensitivity. There is disagreement also on the desirability of heat treatment of cells used for antiserum production. Heat treatment (Fuhrmann and Wollum II 1985; Olsen *et al.* 1983) eliminates flagellar antigens, denatures somatic and internal protein antigens, and in producing LPS-specific YG, should improve assay specificity. However, the applicability of this rationale appears to depend on the *Rhizobium* under study; comparison of anti-whole-cell YG and anti-LPS YG did not reveal a greater specificity with the latter (Martennsson *et al.* 1984).

Because readings can be made by visual scoring (Berger *et al.* 1979; IITA 1981) the ELISA technique can be used without special equipment and away from the laboratory. Where large numbers of assays are involved, the use of microplates is almost essential, and multi-pipettes are available to facilitate loading. Precision and accuracy of color density readings are, obviously, improved by the availability of a spectrophotometer, and the least expensive visible-range models are quite adequate. If funds allow, an automatic plate-reader will not only speed data acquisition but, by making several readings, it also allows determination of reaction slopes with further improvement in data precision. Its versatility and economy of use of antisera are likely to make ELISA the method of choice for many future rhizobial studies.

Serological (ELISA) Comparison of Rhizobia from Semi-Arid and Humid Tropical Soils of West Africa

Introduction

Within the genus *Bradyrhizobium*, only one species (*japonicum*) is designated (Jordan 1984), reflecting the fact that the soybean microsymbiont has been more thoroughly studied than other slow-growing types. Although there is a great deal of data in the literature on strains of *Bradyrhizobium* spp., relatively few strains have been studied, e.g. strains 32H1 and CB 756 are often assumed to be typical of the group as a whole. In fact, even with colony morphology (Jordan 1984), we have found that $>50\%$ of isolates from cowpeas (all slow-growers) in West African soils are atypical in their production of copious slime.

We have been engaged in a program of characterization of cowpea rhizobia in soils at three locations in West Africa (Ahmad, Eaglesham and Hassouna

1981; Ahmad *et al.* 1981; Eaglesham and Ayanaba 1984; Eaglesham *et al.* 1982; 1984; Sinclair and Eaglesham 1984). A preliminary serology study of these rhizobia indicated differences which correlated with colony type (Ahmad, Eaglesham and Hassouna 1981). However, strains of the atypical, "wet" colony morphology were generally poorly reactive with the five YG preparations used and could not be positively evaluated. We therefore produced additional antisera using strains representative of the three soils and repeated the study, part of which is reported here.

Materials and methods

A total of 54 strains of West African cowpea rhizobia were examined which originated from Maradi in Niger (13° N, annual rainfall 600mm, soil pH 6.1), Ibadan in Nigeria (7° N, 1260mm, pH 5.4), and Onne in Nigeria (5° N, 2500mm, pH 4.6) (Ahmad *et al.* 1981).

Production of antisera, preparation of YG, and the indirect ELISA procedure (using goat anti-rabbit YG conjugated to alkaline phosphatase) were as previously described (Ahmad, Eaglesham and Hassouna 1981).

The ELISA absorbance values (405nm) were normalized against the appropriate positive control (e.g. YG_a x $antigen_a$) so that results from different experiments could be compiled into a single data set (Table 2). To ensure accurancy of the positive control data, five replicate determinations were made and any outlying values were eliminated using the Q test (Dean and Dixon 1951).

Antisera were produced against the strains, IRc 252 and IRc 334a (from Onne, both atypical "wet" colony types), IRc 489b and IRc 500a (Ibadan, "wet"), IRc 256 and IRc 299 (from Onne, typical slow-growing types which we designate as "dry"), and IRc 389d and IRc 409a (Maradi, "dry"). The

Table 2. ELISA cross-reaction of eight West African rhizobial strains and their antisera.

	Antiserum							
	IRc 389d	IRc 299	IRc 409a	IRc 256	IRc 252	IRc 334a	IRc 489b	IRc 500a
IRc 389d	1.0 [a]	0.2	0.6	0.1	0.1	b		
IRc 299	0.4	1.0	0.3	0.2	0.9			
IRc 409a	0.9	0.3	1.0	0.7	0.3			
IRc 256	0.4	0.3	1.1	1.0				
IRc 252					1.0	1.1	0.9	0.6
IRc 334a					1.0	1.0	0.9	0.6
IRc 489b					1.0	1.2	1.0	0.6
IRc 500a					1.0	1.1	0.9	1.0

[a] Normalized absorbance values (NAV)

$$NAV_X = \frac{A_{405} \text{ for } (\text{antiserum}_a \times \text{antigen}_x)}{A_{405} \text{ for } (\text{antiserum}_a \times \text{antigen}_a)}$$

[b] blank indicates < 0.1.

strains were chosen because they were judged to be representative on the basis of several criteria, most importantly nodulation host-promiscuity (Eaglesham *et al.* 1984); 97% of the Maradi strains are "dry", 97% of the Ibadan strains are "wet", and the Onne strains are 60:40 "wet":"dry" (Sinclair and Eaglesham 1984). A fully reciprocal YG x strain data set was determined, and the 54 strains were assayed for serological relatedness to the eight YG preparations.

Results

The normalized ELISA values for the eight representative strains assayed with the YG-preparations are given in Table 2. Two distinct groups emerged, one composed only of "wet" strains, the other only of "dry" strains. Some significant cross-reaction occurred with anti-IRc 252 (Onne "wet'), but it was particularly strong only with IRc 299, also from Onne.

The complete 54 strain x 8 YG data set will be presented in detail elsewhere.

Discussion

The complete data set confirmed the conclusion drawn from Table 2, i.e., that the "wet" and "dry" strains are antigenically distinct. There were indications of this in a previous study (Ahmad, Eaglesham and Hassouna 1981), but antisera to "wet" strains were required for substantiation. The "wet" strains were more homogeneous antigenically than the "dry", as has been observed for other characteristics (Sinclair and Eaglesham 1984). Sub-groupings among the "dry" strains contained strains from both Maradi and Onne, showing that no clear distinction could be drawn between them, despite differences in origin. This indicates that the semi-arid and humid environments had relatively little influence on strain serology, or on cowpea strain preference, in contrast to reports with soybean (Damirgi *et al.*1967) and clover (Dughri and Bottomley 1984). Clearly, caution is necessary when choosing strains for soil ecology studies when serological techniques will be used for detection, particularly where cowpea rhizobia are concerned; exotic and indigenous strains must be compared for cross-reaction. For example, Nambiar and Anjaiah (1985) in India found that a peanut field-study with a strain from the Middle East was compromised by a high frequency of positive reactions with nodules on uninoculated plants.

Clearly, official recognition of the existence of "wet" bradyrhizobia is overdue. SDS-polyacrylamide gel protein patterns of "wet" and "dry" cowpea isolates indicate that they are fundamentally different types (M.L. Maina, M.D. Stowers and A.R.J. Eaglesham, unpublished), and serological data support this. Only studies at the genetic level will determine whether they represent different biovars or, perhaps, different species.

Acknowledgement

We acknowledge the support of the United Nations Development Programme, Grant No. GLO/77/013.

References

Ahmad, M.H., Eaglesham, A.R.J. and Hassouna, S. 1981. Examining serological diversity of "cowpea" rhizobia by the ELISA technique. Archives of Microbiology 130:281-287.

Ahmad, M.H., Eaglesham, A.R.J., Hassouna, S., Seaman, B., Ayanaba, A., Mulongoy, K. and Pulver, P. 1981. Examining the potential for inoculant use with cowpeas in West African soils. Tropical Agriculture (Trinidad) 58: 325-335.

Asanuma, S., Thottappilly, G., Ayanaba, A. and Ranga Rao, V. 1985. Use of the enzyme-linked immunosorbent assay (ELISA) in the detection of *Rhizobium* both in culture and from root nodules of soybeans and cowpeas. Canadian Journal of Agriculture 31:524-528.

Barbara, D.J. and Clark, M.F. 1982. A simple indirect ELISA using $F(ab)_2$ fragments of immunoglobulin . Journal of General Virology 58:315-322.

Berger, J.A., May, S.N., Berger, L.R. and Bohlool, B.B. 1979. Colorimetric enzyme-linked immunosorbent assay for the identification of strains of *Rhizobium* in culture and in the nodules of lentils. Applied and Environmental Microbiology 37:642-646.

Bromfield, E.S.P. and Jones, D.G. 1979. The competitive ability and symbiotic effectiveness of doubly-labelled antibiotic resistant mutants of *Rhizobium trifolii*. Annals of Applied Biology 91:211-219.

Bromfield, E.S.P., Stein, M. and White, R.P. 1982. Identification of *Rhizobium* strains on antibiotic concentration gradients. Annals of Applied Biology 101:269-277.

Bushnell, O.A. and Sarles, W.B. 1939. Investigations upon the antigenic relationships among the root-nodule bacteria of the soybean, cowpea, and lupine cross-inoculation groups. Journal of Bacteriology 38:401-410.

Clark, M.F. and Adams, A.N. 1977. Characteristics of the microplate method of enzyme-linked immunosorbent assay for the detection of plant viruses. Journal of General Virology 34:475-483.

Cloonan, M.J. 1963. Black nodules in *Dolichos*. Australian Journal of Science 26:121.

Damirgi, S.M., Frederick, L.R. and Anderson, I.C. 1967. Serogroups of *Rhizobium japonicum* in soybean nodules as affected by soil types. Agronomy Journal 59:10-12.

Dean, R.B. and Dixon, W.J. 1951. Simplified statistics for small numbers of observations. Analytical Chemistry 23:636-638.

Dudman, W.F. 1964. Immune-diffusion analysis of the extracellular soluble antigens of two strains of *Rhizobium meliloti*. Journal of Bacteriology 88:783-794.

Dudman, W.F. 1977. Serological methods and their application to dinitrogen-fixing organisms. Pages 487-508 *in* A Treatise on Dinitrogen Fixation (Hardy, R.W.F. and Gibson, A.H., eds.). John Wiley and Sons, New York, USA.

Dughri, M.H. and Bottomley, P.J. 1984. Soil acidity and the composition of an indigenous population of *Rhizobium Trifolium* in nodules of different cultivars of *Trifolium subterraneum* L. Soil Biology and Biochemistry 16:405-411.

Eaglesham, A.R.J. 1985. Comparison of nodulation promiscuity of US- and Asian-type soya beans. Tropical Agriculture (Trinidad) 62:105-109.

Eaglesham, A.R.J. 1987. The use of intrinsic antibiotic resistance (IAR) for rhizobium study. Pages 185-204 *in* Practical Symbiotic Nitrogen Fixation Methodology (Elkan, G.H., ed.). Marcel Dekker, Inc.

Eaglesham, A.R.J. and Ayanaba, A. 1984. Tropical stress ecology of rhizobia, root nodulation and legume fixation. Pages 1-35 *in* Current Developments in Biological Nitrogen Fixation (Subba Rao, N.S., (ed.). Oxford and IBH, New Delhi, India.

Eaglesham, A.R.J., Ahmad, M.H., Hassouna, S. and Goldman, B.J. 1982. Cowpea rhizobia producing dark nodules: use in competition studies. Applied and Environmental Microbiology 44:611-618.

Eaglesham, A.R.J., Stowers, M.D., Sinclair, M.J., Ayanaba, A. and Goldman, B.J. 1984. Patterns of diversity of cowpea-miscellany rhizobia from three West African locations. Pages 119-125 *in* Symbiotic Nitrogen Fixation, Vol. 1. (Ghai, B.S., ed.). USG, Ludhiana.

Eklund, E. 1984. Behaviour of selected Finnish red clover *Rhizobium* inoculants under field conditions. Page 336 *in* Advances in Nitrogen Fixation Research (Veeger, C. and Newton, W.E., eds.). Martinus Nijhoff, The Hague, The Netherlands.

Engvall, E. and Perlmann, P. 1972. Enzyme-linked immunosorbent assay, ELISA. III. Quantitation of specific antibodies by enzyme-labelled anti-immunoglobulin in antigen-coated tubes. Journal of Immunology 109:129-135.

Fuhrmann, J. and Wollum II, A.G. 1985. Simplified enzyme-linked immunosorbent assay for routine identification of *Rhizobium japonicum* antigens. Applied and Environmental Microbiology 49:1010-1013.

Fuquay, J.I., Bottomley, P.J., and Jenkins, M.B. 1984. Complementary methods for the differentiation of *Rhizobium meliloti* isolates. Applied and Environmental Microbiology 47:663-669.

Graham, P.H. 1963. Antigenic affinities of the root nodule bacteria of legumes. Antonie van Leeuwenhoek 29:281-291.

Hagedorn, C. 1979. Relationship of antibiotic resistance to effectiveness in *Rhizobium trifolii* populations. Soil Science Society of America Journal 43:921-925.

Hodgson, A.L.M. and Waid, J.S. 1981. Use of enzyme-linked immunosorbent assay (ELISA) to identify bacteriocinogenic strains of *Rhizobium trifolii* in nodules following a mixed strain inoculation. Page 430 *in* Current Perspectives in Nitrogen Fixation (Gibson, A.H. and Newton, W.E., eds.). Australian Academy of Science, Canberra, Australia.

IITA (International Institute of Tropical Agriculture) 1981. Inoculants for soybeans in tropical soils. Pages 60-62 *in* Research Highlights 1980. IITA, Ibadan, Nigeria.

Jenkins, M.B. and Bottomley, P.J. 1985. Composition and field distribution of the population of *Rhizobium meliloti* in root nodules of uninoculated field-grown alfalfa. Soil Biology and Biochemistry 17:173-179.

Johnston, A.W.B. and Beringer, J.E. 1975. Identification of the *Rhizobium* strains in pea root nodules using genetic markers. Journal of General Microbiology 87:343-350.

Jones, D.G. and Morley, S.J. 1981a. Marker techniques, competitive ability and host:strain preference in the clover: *Rhizobium* symbiosis. Page 429 *in* Current Perspectives in Nitrogen Fixation (Gibson, A.H. and Newton, W.E., eds.). Australian Academy of Science, Canberra, Australia.

Jones, D.G. and Morley, S.J. 1981b. The effect of pH on host plant 'preference' for strains of *Rhizobium trifolii* using fluorescent ELISA for strain identification. Annals of Applied Biology 97:183-190.

Jordan, D.C. 1984. Family III. Rhizobiaceae. Pages 235-251 *in* Bergey's Manual of Systematic Bacteriology, Vol. 1. 9th edition. (Kreig, N.R. and Holt, J.G., eds.). Williams and Wilkins Co., Baltimore, USA.

Josey, D.P., Beynon, J.L., Johnston, A.W.B. and Beringer, J.E. 1979. Strain identification in *Rhizobium* using intrinsic antibiotic resistance. Journal of Applied Bacteriology 46:343-350.

Kamicker, B.J. and Brill, W.J. 1986. Identification of *Bradyrhizobium japonicum* nodule isolates from Wisconsin soybean farms. Applied and Environmental Microbiology 51:487-492.

Kingsley, M.T. and Bohlool, B.B. 1983. Characterization of *Rhizobium* sp. (*Cicer arietinum* L.) by immunofluorescence, immunodiffusion, and intrinsic antibiotic resistance. Canadian Journal of Microbiology 29:518-526.

Kishinevsky, B. and Bar-Joseph, M. 1978. *Rhizobium* strain identification in *Arachis hypogaea* nodules by enzyme-linked immunosorbent assay (ELISA). Canadian Journal of Microbiology 24:1537-1543.

Kishinevsky, B. and Gurfel, D. 1980. Evaluation of enzyme-linked immunosorbent assay (ELISA) for serological identification of different *Rhizobium* strains. Journal of Applied Bacteriology 49:517-526.

Kishinevsky, B. and Maoz, A. 1983. ELISA identification of *Rhizobium* strains by use of enzyme-labelled protein A. Current Microbiology 9:45-49.

Kishinevsky, B., Maoz, A., Gurfel, D. and Nemas, C. 1984. A serological study of *Rhizobium* strains using direct and indirect enzyme-linked immunosorbent assays (ELISA). Page 346 *in* Advances in Nitrogen Fixation Research (Veeger, C. and Newton, W.E., eds.). Martinus Nijhoff/Junk, The Hague, The Netherlands.

Kneen, B.E. and LaRue, T.A. 1983. Congo red absorption by *Rhizobium leguminosarum*. Applied and Environmental Microbiology 45:340-342.

Kremer, R.J. and Peterson, H.L. 1982. Nodulation efficiency of legume inoculation as determined by intrinsic antibiotic resistance. Applied and Environmental Microbiology 43:636-642.

Kronvall, G., Seal, U.S., Finstad, Y. and Williams, R.C. 1970. Phylogenetic insight into evolution of mammalian FD fragment of YG globulin using a staphylococcal protein A. Journal of Immunology 104:140-147.

Kuykendall, L.D. and Elkan, G.H. 1976. *Rhizobium japonicum* derivatives differing in nitrogen-fixing efficiency and carbohydrate utilization. Applied and Environmental Microbiology 32:511-519.

La Favre, J.S. and Focht, D.D. 1983. Conservation in soil of H_2 liberated from N_2 fixation by Hup^- nodules. Applied and Environmental Microbiology 46:304-311.

Marques Pinto, C., Yao, P.Y. and Vincent, J.M. 1974. Nodulating competitiveness amongst strains of *Rhizobium meliloti* and *R. trifolii*. Australian Journal of Agricultural Research 25:317-329.

Martennsson, A.M. and Gustafsson, J.G. 1985. Competition between *Rhizobium trifolii* strains for nodulation during growth in a fermenter, and in soil-based inoculants, studied by ELISA. Journal of General Microbiology 131:3077-3082.

Martennsson, A.M., Gustafsson, J.G. and Ljunggren, H.D. 1984. A modified, highly sensitive enzyme-linked immunosorbent assay (ELISA) for *Rhizobium meliloti* strain identification. Journal of General Microbiology 130:247-253.

Means, U.M., Johnson, H.W. and Date, R.A. 1964. Quick serological method of classifying strains of *Rhizobium japonicum* in nodules. Journal of Bacteriology 87:547-553.

Meyer, M.C. and Pueppke, S.G. 1980. Differentiation of *Rhizobium japonicum* strain derivatives by antibiotic sensitivity patterns, lectin binding, and utilization of biochemicals. Canadian Journal of Microbiology 26:606-612.

Morley, S.J. and Jones, D.G. 1980. A note on a highly sensitive modified ELISA technique for *Rhizobium* strain identification. Journal of Applied Bacteriology 49:103-109.

Nakane, P.K. and Pierce, G.B. 1966. Enzyme-labelled antibodies: preparation and application for the location of antigens. Journal of Histochemistry and Cytochemistry 14:929-931.

Nambiar, P.T.C. and Anjaiah, V. 1985. Enumeration of rhizobia by enzyme-linked immunosorbent assay (ELISA). Journal of Applied Bacteriology 58:187-193.

Nambiar, P.T.C., Srinivasa Rao, B. and Anjaiah, V. 1984. Studies on competition, persistence, and methods of application of a peanut *Rhizobium* strain NC 92. Peanut Science 11:83-87.

Olsen, P.E. and Rice, W.A. 1984. Minimal antigenic characterization of eight *Rhizobium meliloti* strains by indirect enzyme-linked immunosorbent assay (ELISA). Canadian Journal of Microbiology 30:1093-1099.

Olsen, P.E., Rice, W.A., Stemke, G.W. and Page, W.J. 1983. Strain-specific serological techniques for the identification of *Rhizobium meliloti* in commercial alfalfa inoculants. Canadian Journal of Microbiology 29:225-230.

Pankhurst, C.E. 1977. Symbiotic effectiveness of antibiotic-resistant mutants of fast- and slow-growing strains of *Rhizobium* nodulating *Lotus* species. Canadian Journal of Microbiology 23:1026-1033.

Parker, C.A. and Grove, P.L. 1970. The rapid serological identification of rhizobia in small nodules. Journal of Applied Bacteriology 33:248-252.

Ranga Rao, V., Ayanaba, A., Eaglesham, A.R.J. and Thottappilly, G. 1985. Effects of *Rhizobium* inoculation of field-grown soybeans in Western Nigeria and assessment of inoculum persistence during a two year fallow. Tropical Agriculture 62:125-130.

Renwick, A. and Jones, D.G. 1985. A comparison of the fluorescent ELISA and antibiotic resistance identification techniques for use in ecological experiments with *Rhizobium trifolii*. Journal of Applied Bacteriology 58:199-206.

Rice, W.A., Olsen, P.E. and Page, W.J. 1984. ELISA evaluation of the competitive abilities of two *Rhizobium meliloti* strains. Canadian Journal of Microbiology 30:1187-1190.

Robert, F.M. and Schmidt, E.L. 1985. Somatic serogroups among 55 strains of *Rhizobium phaseoli*. Canadian Journal of Microbiology 31:519-523.

Rupela, O.P., Josey, D.P., Toomsan, B., Mittal, S., Dart, P.J. and Thompson, J.A. 1982. Application of inherent antibiotic resistance to ecological studies of rhizobia. Pages 625-629 *in* Biological Nitrogen Fixation Technology for Tropical Agriculture (Graham, P.H. and Harris, S.C., eds.). CIAT, Cali, Colombia.

Schmidt, E.L. 1973. Fluorescent-antibody techniques for the study of rhizobial ecology. Bulletins from the Ecological Research Committee (Stokholm) 17:67-76.

Schmidt, E.L., Bankole, R.O. and Bohlool, B.B. 1968. Fluorescent antibody approach to the study of rhizobia in soil. Journal of Bacteriology 95:1987-1992.

Schwinghamer, E.A. 1964. Association between antibiotic resistance and ineffectiveness in mutant strains of *Rhizobium* spp. Canadian Journal of Microbiology 10:221-233.

Schwinghamer, E.A. 1967. Effectiveness of *Rhizobium* as modified by mutation for resistance to antibiotics. Antonie van Leeuwenhoek 33:121-136.

Schwinghamer, E.A. and Dudman, W.F. 1973. Evaluation of spectinomycin resistance as a marker for ecological studies with *Rhizobium* spp. Journal of Applied Bacteriology 36:263-273.

Schwinghamer, E.A. and Dudman, W.F. 1980. Methods of identifying strains of diazotrophs. Pages 337-365 *in* Methods for Evaluating Biological Nitrogen Fixation (Bergersen, F.J., ed.). John Wiley and Sons, New York, USA.

Sinclair, M.J. and Eaglesham, A.R.J. 1984. Intrinsic antibiotic resistance in relation to colony morphology in three populations of West African cowpea rhizobia. Soil Biology and Biochemistry 16:247-251.

Sinha, R.C. and Peterson, E.A. 1980. Homologous serological analysis of *Rhizobium meliloti* strains by immunodiffusion. Canadian Journal of Microbiology 26:1157-1161.

Somasegaran, P. and Hoben, H. 1985. Methods in Legume-*Rhizobium* Technology. NifTAL Project, University of Hawaii, Hawaii, USA.

Stowers, M.D. and Eaglesham, A.R.J. 1984. Physiological and symbiotic characteristics of fast-growing *Rhizobium japonicum*. Plant and Soil 77:3-14.

Trinick, M.J. 1969. Identification of legume nodule bacteroids by the fluorescent antibody reaction. Journal of Applied Bacteriology 32:181-186.

Turner, M.W. 1981. Structure and function of immunoglobulins. Pages 1-57 *in* Structure and Function of Antibodies (Glynn, L.E. and Steward, M.W., eds.). John Wiley and Sons, New York, USA.

Vincent, J.M. 1970. A Manual for the Practical Study of Root-Nodule, Bacteria. IBP Handbook No. 15. Blackwell Scientific Publications, Oxford, England, UK.

Vincent, J.M. and Humphrey, B. 1970. Taxonomically significant group antigens in *Rhizobium*. Journal of General Microbiology 63:379-382.

Vincent, J.M., Humphrey, B. and Skrdleta, V. 1973. Group antigens in slow-growing rhizobia. Archives of Microbiology 89:79-82.

Discussion

Sylvester-Bradley

Some tropical forage legume strains which are "dry" on YMA pH 6.8 are "wet" on YMA pH 5.5. i.e., it seems that the strains are less "stressed" on YMA pH 5.5. They also grow better on medium of this pH.

Eaglesham

We have not examined the effect of low pH on the "wet" / "dry" phenotype. However, exposure of "wet" strains to high temperature or 0.5% NaCl often results in a less slimy colony morphology. We should examine "dry" Onne strains at ph 5.5, or lower.

Ruiz-Argueso

Do you find a clear-cut separation of "dry" and "wet" strains or do you also find strains with intermediate slime production capacity?

Eaglesham

Only about 3% of the 300 strains examined cannot easily be categorized.

Broughton

Are your "wet" and "dry" *Bradyrhizobium* types equivalent to the two types of *B. japonicum* J. Stanley *et al.* (J. Bact. 1985) described which differ in arrangement at *nif* genes?

Eaglesham

I don't know. Dr. A. Szolay and co-workers at Boyce Thomson have examined the *nif* genes of IRc 78, a "dry" strain from Marendi. "Wet" strains have not been genetically studied so far.

D.P. Beck and L.A. Materon (editors), Nitrogen Fixation by Legumes in Mediterranean Agriculture
 – Printed in the Netherlands. ISBN 90 247 3624 2

Selection of *Rhizobium* Strains

J.A. THOMPSON

Agricultural Research Centre, Tamworth, NSW, Australia

Summary. This paper discusses the need for legume inoculation and the symbiotic, ecological, and cultural characteristics which can be used for *Rhizobium* strain selection. Other aspects relating to inoculation, such as sources of strains for testing, genetic manipulation, and the use of single or multistrain inocula are also discussed.

Is Strain Selection Necessary?

Implicit in the decision to become involved in *Rhizobium* strain selection is the fact that a particular legume of interest is expected to benefit from inoculation in the chosen environment. Too often, strain selection is initiated without the knowledge that the legume will respond or that the inoculum will produce nodules, and not uncommonly sparse nodulation on a commercial crop is cited as a reason for using inoculum. Whatever the situation, it is essential that the need for inoculation is established.

Date (1977) classified hosts into three groups on the basis of their symbiotic responses: (a) PE (= promiscuous and effective) for hosts which nodulate effectively with a wide range of strains and include *Arachis*, *Cajanus*, and *Vigna*; (b) PI (= promiscuous and ineffective) for hosts which nodulate with a wide range of strains, often ineffectively, and include the *Vicia-Lens-Pisum-Lathyrus* group (Date, personal communication); and (c) S (= specific), a group which is particularly relevant to the Middle East region and contains *Cicer*, *Medicago*, and *Trifolium*. Progressing through these groups, the likelihood of local strains of *Rhizobium* being available becomes less and the need for inoculation greater.

The early experimental designs to determine the need for inoculant proposed by Nutman (1976) and Vincent (1970) have been simplified to those advocated by Brockwell (1971), Date (1977), and NifTAL (Harris 1979; Singleton *et al.* 1985). The basic treatments are (a) uninoculated control to check the presence and effectiveness of local strains, (b) inoculated treatment with a *Rhizobium* strain known, or at least reported, to be effective with the host, and (c) an inoculated treatment plus nitrogen to ensure a well grown plant and to determine whether nitrogen fixation by the inoculated treatment is adequate.

If it is possible that other nutrients are limiting, all three treatments should be examined with and without non-nitrogenous fertilizer. Date (1977) has presented a wide range of interpretable results in detail while pointing out that the design may readily be made more sophisticated with additional treatments or more critical evaluation made of particular aspects, using such tools as serology, to determine the success of the inoculant.

While lack of response to inoculation or nitrogen fertilizer does not necessarily remove the requirement for inoculation, there is little point in proceeding with strain selection for inocula unless other information justifies further investigation. The INLIT trials of NifTAL (Harris 1979) were designed in three stages: (a) a field evaluation using the best known available strains; (b) pot evaluation of local isolates; and (c) field evaluation using the best (b)-strains against the (a)-experiment inoculum. Presumably, however, stage (b) would only follow a response in stage (a).

There are certainly many situations where inoculation is not considered to be necessary, e.g. cowpeas in West African soils (Asanuma and Ayanaba 1985), peanut in South East Queensland (Diatloff 1975), and faba bean and pea in Northern Europe (Lawes 1980; Newbould 1983; Nutman 1976). It seems probable that vast areas of traditional sowing of the world's well established legume crops (e.g. chickpea and pigeonpea in India, and faba bean in China) are unlikely to respond, provided the genetic base of the crop remains unchanged.

Reliance on natural populations has been followed by a further step at IITA in Nigeria. Promiscuously nodulating soybean lines have been included in the breeding programme, and lines have now been developed with desirable agronomic characteristics as well as being able to nodulate with most of the local rhizobia (Dashiell *et al.* 1985). This enterprising approach has its critics and has recently been described as "selection of mediocrity" (Rennie 1985). It is also probable that promiscuous nodulation by the selected host germplasm in the Nigerian environment may not be so successful in the presence of a different soil population. Nevertheless, the approach provides a practical alternative to inoculant production and its attendant distribution problems in developing countries.

Criteria of Strain Selection

The number of criteria used for strain selection has expanded steadily so that recent lists now cover the following.

Symbiotic characteristics

Ability to infect and fix nitrogen with the host under favourable, controlled conditions.

Ecological characteristics

a) Competitive ability with other strains for infection sites on the roots.
b) Survival and persistence in soil in the presence and absence of the host.
c) Special characteristics such as resistance to antibiotics in soil, low soil pH, fungicides, etc.

Cultural characteristics

a) Ability to multiply in broth.
b) Ability to survive in inoculant carrier and on seed.
c) Stability during storage and growth.

Methods of Strain Selection

Symbiotic characteristics

The symbiotic success of a strain with a host is the most important measurable characteristic, and is most readily measured by growing the plant in nitrogen-free medium. Plant dry matter (or nitrogen) production in excess of an unnodulated control is a direct measure of the success of the symbiosis to the time of harvest. The procedures, under bacteriologically controlled conditions, have been well documented and fully described (Vincent 1970; Gibson 1980) and in fact have changed little over decades. The test assemblies range from a tube containing seedling agar or vermiculite, a semi-enclosed agar tube, or a modified Leonard ar assembly, to open pots of sand, vermiculite, or perlite which require elaborate precautions to prevent contamination. The most recent addition to the list is the growth pouch proposed by Porter *et al.* in 1966.

For small-seeded legumes, all assemblies can be used to measure the effectiveness of fixation, while large-seeded legumes require the larger assemblies. Crop legumes, for which the principal yield parameter is grain yield, are not readily assessed under bacteriologically controlled conditions, and Gibson (1980) concluded that only field studies provide a reliable evaluation of crop legumes. However, provided nitrogen levels of soil or growth medium are low, and no native infective *Rhizobium* is present, yields of plant material or nitrogen and observation (and perhaps measurement) of nodules is all that is necessary to evaluate symbiotic performance.

Under such circumstances, Thompson *et al.* (1974) found that strain responses with five *Trifolium* species could be equally well evaluated in terms of yield per unit area when grown as swards, rows, or single spaced plants. In the competitive situations, mean plant dry weight was not a reliable measure but visual rating was most successful with rows, as all treatments can be

Table 1. The place for evaluation techniques in field testing of strain effectiveness.

	Rhizobium	absent	*Rhizobium*		present	
	Low N	High N	Low N		High N	
			E[1]	I	E	I
Yield of dry matter (N or grain)	+		+		?	
Nodulation Quantity	+	+	?	?		
Identification			+	+	+	+
Nitrogen Fixation						
Acetylene reduction [2]	(+)	+		(+)		+
^{15}N	(+)	+		(+)		+
Ureides	(+)	+		(+)		+

= Of value.

[1] Procedure now essentially discredited for this purpose (Minchin *et al.* 1983; Witty and Minchin 1986).

[2] E = effective, I = ineffective.

readily scanned. It is important to bear in mind that we are generally attempting to rank strains and absolute measurements may be secondary.

Evaluation of symbiotic performance is, at best, time consuming and costly, and it is important to avoid unnecessary use of sophisticated but often costly measurement techniques such as acetylene reduction, ^{15}N isotope dilution, or ureide determinations (Table 1).

Even if the acetylene reduction technique were considered to be a reliable procedure, and this is now clearly not so for a variety of reasons (Minchin *et al.* 1983; 1985), neither it nor the other methods can provide any information other than dry matter or nitrogen yield under low nitrogen, *Rhizobium*-free conditions. Even when acetylene reduction was well correlated with nodulation parameters and grain yield of chickpea (Rupela and Dart 1979) or groundnut (Arunachalam *et al.* 1984), its use was unnecessary. In the extreme situation posed by the long delays in field sampling of pigeonpea nodules, correlations are very poor (Thompson *et al.* 1981) and the effort is completely unjustified.

A study by Rennie (1984) illustrates some of the problems in using nitrogen balance (NB) and ^{15}N isotope dilution (ID) estimates of fixation by strains of *R. phaseoli*. As would be expected, under low nitrogen, *Rhizobium*-free conditions both estimates of nitrogen fixation were well correlated and, while ID commonly exceeded NB, the ranking of strains was similar. Under such conditions, there seems little reason to use the more expensive technique.

With increased soil nitrogen, nitrogen uptake also generally increased and, not only did the ranking of strains differ from the other site in terms of amounts of nitrogen fixed, but this was not related in any clear fashion to the changes which occurred in soil nitrogen uptake. It was clearly more meaningful to rank strains in terms of nitrogen fixed rather than percentage of nitrogen fixed (= % Ndfa, Rennie 1984). While Rennie concluded that it was not possible to predict when to use NB with confidence, the same comment may reasonably be applied to the ^{15}N dilution method, and cost clearly favours NB. Vasilas and Ham (1984), using different levels of nitrogen on soybean, concluded that the difference method gave as accurate a measure as the isotope dilution procedure.

In the presence of local ineffective rhizobia in low-nitrogen soil, successful nodulation by effective inoculant strains should result in a dry matter and plant nitrogen yield response comparable to the *Rhizobium*-free situation, and sophisticated methodology is unnecessary.

However, the presence of local effective rhizobia in soils of high or low nitrogen can mask all plant dry matter and nitrogen differences, and if there is a reason to pursue strain evaluation under these circumstances, nodulation success must be determined in terms of success of nodule occupancy by the inoculum. Alternative measurements of nitrogen derived from fixation by ^{15}N dilution are unlikely to provide any further information.

With those species on which the uriede index technique can be used, it has proved to be a useful indicator of the proportion of the plant's nitrogen being fixed (Herridge 1984), and is particularly useful in examining the responses of different hosts to soil nitrate-nitrogen levels (Herridge and Betts 1985). It would seem to have a place similar to ^{15}N isotope dilution procedures for evaluation of strains under high soil nitrogen conditions, but cannot be expected to differentiate between inputs of effective inoculant and local strains.

In so far as symbiotic characteristics are generally considered separately from ecological characteristics, there is considerable merit in restricting such tests to the low nitrogen, *Rhizobium*-free situation where simple yield and nodulation measurements suffice to rank strains.

The time-consuming nature of evaluating strain effectiveness favours the development of simple preliminary screening procedures. The relationship of seed isozyme groups to effectiveness has been reported by Robinson *et al.* (1976) and Date (1984) and shows promise.

Ecological characteristics

The ecological criteria should be tested under selected or, where practicable, contrived situations which allow the evaluation of a particular character which should preferably not be confounded with measurement of symbiotic characteristics. Competitive ability clearly requires a competitive situation and is most directly tested in the field. Alternatively, field soil can be tested under controlled conditions in pots or soil cores (Vincent 1970). One of the simple criteria for evaluation is percentage nodule occupancy, while suitable serological or marking techniques must be used to establish the success of infection of particular strains.

Survival and persistence in soil in the absence of the host is fairly easily studied, albeit often with laborious MPN counts, but the use of FA serology by Bohlool and Schmidt (1970) provides an alternative.

Criteria of significance are commonly regionally significant, and not necessarily of general application. This particularly applies to antibiotic resistance which has been dramatically illustrated by Scotti *et al.* (1982) in the cerrado soils of Brazil. However, screening for resistance is a relatively simple procedure and resistant strains have proved to be successful as inocula (Dobereiner *et al.* 1981).

Caution must be exercised in the decision to select for tolerance to low pH in rhizobia, not least because of the difficulties of the laboratory techniques, which frequently seem to be difficult to transfer between laboratories. The complex nature of the low pH soil, with differing importance of associated aluminium and manganese levels, adds confusion to the possibility that the rhizobia are more tolerant than the host. In fact, Munns *et al.* (1981) suggest that "efforts to improve on acid tolerance should centre on soybean not rhizobia" although he concedes there is a case for selecting rhizobia for mung and cowpea (Munns 1984). Thornton and Davey (1984) showed that *R. trifollii* strains selected for acid tolerance resulted in larger populations in soils stored in the laboratory. Howieson and Ewing (1986) found that several strains of *R. meliloti*, originating from acid soils in Italy, were markedly superior to Australian commercial strains when used on moderately acid soils. However, the success was only evident with host species which themselves were acid tolerant, and the authors make no claim that the range of the host could be extended by using an acid tolerant *Rhizobium* strain.

While salinity is undoubtedly a major problem in plant growth, little is known of the salt tolerance of rhizobia, and it is commonly believed that the salinity tolerance of the host is lower than that of the rhizobia themselves (Parker *et al.* 1977; Sprent 1984).

The studies of Vincent and his co-workers on both temperate and tropical pasture species (Vincent and Waters 1953; Marques Pinto *et al.* 1974; Labandera and Vincent 1975; Franco and Vincent 1976; Materon and Vincent 1980) illustrated the difficulties of generalizations about the competitive ability of a strain and its relationship to any single feature of the host-*Rhizobium* symbiosis. It is safe to state that selection on the basis of particular ecological criteria does not necessarily result in superior strains in the field. Strains of *R. meliloti* from various sources, known to be effective on the basis of field trials in *Rhizobium*-free sites, were examined by van Rensberg and Strijdom (1982) for competitive ability in paired comparison in soil in pots. The results were not consistent with field success as inocula in competitive situations and "placed doubt on the relevance of singling out competitive ability for special attention when selecting a strain for inoculant production".

These results are a timely warning that our selection criteria should have field significance to justify the chosen screening procedure, but until we separate the components of the competitive ability of a strain, our selections will always run the risk of encountering a new aspect not previously met. While strains with superior tolerance to temperature and desiccation can be selected from hotter, drier environments (Hartel and Alexander 1984), soils in such environments have many other variable attributes which may apply other selection pressures to the selected strains. As pointed out by Bushby (1982), the significance of soil physical reactions is little studied and a much better understanding of these interactions is necessary before we can claim to design suitable selective ecological criteria, and to extrapolate these to the field situation.

Meanwhile, strain comparisons are often made under field conditions where soil nitrogen levels are sufficiently high to, at least partly, support the legume and/or native rhizobia which are present. Such testing clearly involves other ecological as well as symbiotic criteria but often the only data presented are those of yield. An example is the decision of the All India Pulse Improvement Program to accept the chickpea strain IC76 for use in Indian inoculants solely on the basis of increases in grain yield measured on nine research stations (ICRISAT 1984).

Cultural characteristics

The ability to multiply in broth and survive in carrier and on seed can be readily examined using normal microbiological procedures.

The question of a strain's stability is not one which necessarily lends itself to evaluation, and generally, strains selected by other criteria are used until there is evidence of mutation. Herridge and Roughley (1975) not only showed that cowpea *Rhizobium* CB756 contained a number of colony variants, in turn related to symbiotic effectiveness, but some of these remained unstable and did not breed true.

The Australian inoculant quality authorities (UDALS and subsequently AIRCS) have periodically encountered instability involving loss of effectiveness or infectivity (Date 1969; Roughley 1976; Roughley and Pulsford 1982). Other cases have also become evident during commercial inoculant production. For example, Corbin *et al.* (1977) selected CB1189 and CC1192 as suitable strains for chickpea for which no native rhizobia are present in Australia. But CB1189 was difficult to grow to high count broths in commercial fermenters, so CC1192, for which there were no such problems, has since been the only strain used commercially. A competitive, effective strain, WU290, was used for some years in Australian inocula before becoming non-infective. The high incidence of changes in effectiveness occurring in the field may also ultimately govern the continued use of some strains (Brockwell *et al.* 1981).

In South Africa, Van Rensberg and Strijdom (1982) in South Africa found that the gumminess of *R. meliloti* strain RF10 was a problem in inoculant manufacture and it was to be replaced, although it had already proven to be a successful inoculant.

Greater use of chemicals, such as fungicides and insecticides, as seed dressings necessitates screening for resistance separately from the evaluation of symbiotic capacity, especially with mutants which may lose effectiveness or infectivity (Rai 1985). Further, there is a risk that the laboratory evaluation of resistance does not correlate with the reaction by plants from treated seed (Staphorst and Strijdom 1976). Therefore it is important that the screening procedure is carefully evaluated.

The medium on which the cultures are stored is also important. Media more closely resembling soil conditions from which the strain was obtained are recommended to ensure the retention of stress-resistant isolates (Date 1982). It is suggested that the standard practice of searching for stress tolerant rhizobia in existing collections which were originally isolated onto standard non-stress media, may not be a fruitful approach (Date and Halliday 1979). Parker *et al.* (1977) also queried whether nutritionally luxurious media may favour the selection of strains with reduced saprophytic competence.

The risk of mutation is minimised by minimal subculturing and long-term storage by lyophilisation. While stress and frequent subculturing favour instability, it is unlikely that the logistics of the operation would favour their use in selecting for stability.

Source of Strains for Testing

The choice of strains for evaluation in a strain selection program can range from local isolates, through strains already evaluated in other parts of the region or country, to cultures from overseas collections, or isolates from a "Center of Diversity" of the host of interest.

The current acceptance of the need for germplasm collection and conservation, so evident with particular crops in the International Agricultural Research Centers, has as a basic tenet the concept of Centers of Diversity and thus the existence of diverse material which may be useful in crop improvement (Table 2). While many plant collections also of necessity focus on secondary centres of diversity, or even modern areas of adaptation, it is rare that legume collections are accompanied by collections of nodule material. Apart from the problem that the need to collect seed generally makes it impractical to collect nodule material, there is a lack of appreciation amongst plant collectors of the significance of *Rhizobium*. The likelihood that plant genetic variability is accompanied by rhizobial variability is not even addressed in relevant texts on genetic conservation of plant germplasm (Witcombe and Erskine 1984; Frankel and Bennett 1970), the global survey of leguminous root nodulation (Allen and Allen 1981), or the majority of papers addressing particular legumes in the International Legume Conference of 1978 (Summerfield and Bunting 1980).

The need for inocula for the wide range of forage legumes examined in northern Australia by the CSIRO Cunningham Laboratory, Brisbane has at times made it essential to return at least to countries of origin. The apparent specificity of local Australian rhizobia with *S. capitata* (t'Mannetje 1977) was not borne out when isolates from the country of origin were included (Date

Table 2. World centers of diversity of crop legumes (based on Zohary 1970).

China	*Glycine max*
	Vigna angularis
	Phaseolus vulgaris, (recessive form; secondary center)
India	*Cicer arietinum*
	Phaseolus aconitifolia
	Phaseolus calcaratus
	Dolichos biflorus
	Vigna unguiculata, subsp. *cylindrica* * (primary center)
	subsp. *unguiculata* * (secondary center)
	Cajanus cajan **
Central Asia	*Pisum sativum*
	Lens esculenta
	Cicer arietinum
Near East	*Cicer arietinum* (secondary center)
	Lens esculenta (a large endemic group of varieties)
	Pisum sativum (a large endemic group, secondary center)
Mediterranean	*Vicia faba*
East Africa	*Cicer arietinum* (a center)
	Lens esculenta (a center)
	Pisum sativum (one of the centers)
West Africa	*Vigna unguiculata* subsp. *unguiculata* * (secondary center)
Central America	*Phaseolus vulgaris*
South America (Peru-Ecuador-Bolivia)	*Phaseolus lunatus*
South America (Brazil-Paraguay)	*Arachis hypogaea*

* Steele and Mehra (1980)
** van der Maesen (1981)

1984), and the inference (Date and Norris 1979) is clearly that a wider range of genetic material was then encountered in the rhizobia.

Lie and Goktan (1984) have pointed out that spread of plants in the new environment is accompanied by narrowing of both host and *Rhizobium* germplasm.

Recent studies of Keyser *et al.* (1985), focussed not only on the known narrow genetic base of US soybeans (Delanney *et al.* 1983), but also suggest that the rhizobia in the US soils are themselves from a very narrow genetic base. From the geographic pattern in Asia of the incidence of incompatibility in host germplasm with rhizobia isolated from US soils, Devine and Breithaupt (1980) conclude that the host and microsymbionts in use in US evolved in different areas of Asia.

Diverse populations do develop in new environments. In Australia, Brockwell *et al.* (1968) found a wide range of effectiveness of local strains infecting *Trifolium subterraneum*, which has only been naturalised in Australia for approximately 100 years. A similar range of effectiveness was evident in collections from the Middle East where the genus occurs naturally. The latter collection certainly yielded some highly effective isolates but it was not possible to compare the collections on all criteria to establish which showed the greater diversity. Certainly it remains possible that the Australian populations may have encountered particular ecological factors which favour the selection of local strains. However, despite claims of superiority of local Western Australian *R. trifolii* strains over an introduced Tasmanian strain (Chatel and Parker 1973), the numbers of strains involved are too low to support claims of local superiority.

Keya (1985) has adopted a very pragmatic approach to strain selection. Not only is he concerned about the effort and facilities directed towards local strain collection and evaluation, but he points out that lack of effective strains is not a limiting factor and that these are readily obtained from laboratories in developed countries. Alexander (1982) has expressed similar views to Keya (1985) and recently Frederick (1985) listed a number of strains claimed to be highly effective and/or competitive and adaptable to changing environments with the major grain legumes. While no distinction is made between host cultivars, such a list provides a useful, readily available set of controls against which local isolates can be tested. The World Catalogue of *Rhizobium* collections (Skinner *et al.* 1983) provides a wider selection of strains and meanwhile collections continue to be developed at MIRCENS and International Agricultural Research Centers. It is arguable that, unless these collections are preserving diverse germplasm from centres of diversity, much effort may be misdirected in their development.

A good example of the universality of an effective strain is *Rhizobium* strain NC92 (of Brazilian origin) and groundnut cultivar Robut 33 (Nambiar *et al.* 1984). Strain NC92 has generally been superior to local strains in many areas of India, especially with cultivar Robut 33. Although the possibility that the superiority of the strain is more symbiotic than ecological is secondary to its widespread success in this case, the importance of the host cannot be ignored. A good example of the importance of the host is that of the recognition that native *R. trifolii* strains, which were ineffective with introduced white clover, were effective with the indigenous clover in Ireland (Sherwood and Masterson 1974).

The contrary view to that of Alexander and Keya is presented by Williams (1984) who claims that "*Rhizobium* strains — for particular environmental conditions — are rarely transferable". However, he subsequently states that there were no consistent differences between local and introduced isolates in an experimental situation. Freire (1982) has also expressed concern that one of the problems of inoculant production in Latin America is dependence on national or imported inoculants rather than locally selected strains.

While there is no doubt that some local conditions necessitate special strain attributes, the majority of sites are generally chosen to be suitable for a potential host, and hence its rhizobia, and Keya's (1985) claim deserves serious evaluation before continuing the common commitment to local strain selection.

A choice between a highly effective, competitive strain and a moderately effective, strongly competitive strain may be best resolved in favour of the latter on the basis that effectiveness is often not fully expressed because of limitations of other plant factors on plant growth (Brockwell 1985).

Genetic Manipulation of Strains

There is no doubt that the production and use of deficient *Rhizobium* mutants will extend our knowledge of the symbiotic system (Rolfe and Shine 1984), but genetic alteration of *Rhizobium* to provide strains superior in field performance has yet to be achieved.

As in selection of available strains, there may well be more to be gained by construction of a strongly competitive strain than a super-effective strain (Brockwell 1985).

Considerable effort is being directed towards understanding and manipulating the symbiotic system but the factors involved in ecological success remain ill-understood, so that it may not even be possible at this stage to adequately define the characters which lead to field success. The challenge is there for the genetic manipulators. Of 500 *R. trifolii* field isolates examined for a wide range of desirable attributes, not one gave high levels of performance in all symbiotic, ecological, and cultural criteria tested (Brockwell *et al.* 1968).

In discussing nitrogen fixation studies in developing countries, Parker (1985) cautioned against too much optimism in the application of genetic engineering to making better legumes or better rhizobia and advocates concentration "on the wonderful range of legumes that exists and not run after rainbows".

Single or Multistrain Inocula?

The theoretical advantages and disadvantages of single strain inocula have been well documented, but it is significant that those favouring single strains are generally associated with inoculant quality control: in Australia, Vincent (1970), Date and Roughley (1977), and Thompson (1980); in Uruguay and Brazil (Freire 1982); and in New Zealand (Anon. 1979). Clearly, the checking of strain identity in the final product is simplest with fewest strains.

Critical comparisons of single versus multistrain inocula are rare and generally dependent on so few strains that generalizations are not possible. For example, Rennie and Dubetz (1984) found that one mixture of strains was as good as the individuals, while the other mixture was detrimental to soybean yield.

Gault (1981) pointed out that at least some US inoculant manufacturers include up to five or six strains for one or a range of appropriate hosts "in the hope that at least one strain will adapt to the environment". Such an approach would seem to be in conflict with Burton's (1981) admirable sentiment that "a multiple strain inoculum cannot be used as a substitute for knowledge of strain performance".

Nevertheless, there is no doubt that many multistrain inocula are produced without adequate justification. In the present state of knowledge, the use of multistrain inocula is not justified until the use of a particular combination is shown to be successful in a particular environment with particular hosts (Rennie and Dubetz 1984).

Conclusions

1. The cost in time and money of *Rhizobium* selection is only justified if
   (a) a need for inoculation of the host of interest is demonstrated, and
   (b) it is known that the selected strains can be utilized successfully by an efficient inoculant industry.

If inoculation is necessary but an efficient inoculant industry is not possible, the availability of host germplasm compatible with local rhizobia should be investigated.

2. The chances of finding symbiotically suitable strains are likely to be greatest from collections made at the Center(s) of Diversity of the host, rather than areas of recent naturalization.

3. The current state of knowledge is not sufficient to provide critical tests of ecological characteristics to predict field performance in the presence of naturalized rhizobia.

4. The case for ecological superiority of local isolates is not generally established and initial ecological screening should always include known effective strains from other collections.

In this context, the justification for local *Rhizobium* collections must be critically examined.

References

Anonymous 1979. The Inoculant and Coated Seed Testing Service. New Zealand Ministry of Agriculture and Fisheries, Seed Testing Station, Palmerston North, New Zealand.

Allen, O.N. and Allen, E.K. 1981. The Leguminosae. University of Wisconsin Press, Madison, USA. 812 pp.

Alexander, M. 1982. Research to enhance nitrogen fixation: misplaced emphasis? Pages 208-229 *in* Priorities in Biotechnology Research for International Development. National Academy Press, Washington, USA.

Arunachalam, V., Pungle, G.D., Dutta, M., Nambiar, P.T.C. and Dart, P.J. 1984. Efficiency of nitrogenase activity and nodule mass in predicting the relative performance of genotypes assessed by a number of characters in groundnut (*Arachis hypogaea*). Experimental Agriculture 20: 303–309.

Asanuma, S. and Ayanaba, A. 1985. Evaluation of selected *Rhizobium* strains for symbiotic effectiveness on cowpeas under tropical conditions. Pages 504-511 *in* Biological Nitrogen Fixation in Africa (Ssali, H. and Keya, S.O., eds.). University of Nairobi, Kenya.

Bohlool, B.B. and Schmidt, E.L. 1970. Immunofluorescence detection of *Rhizobium japonicum* in soils. Soil Science 110: 229–236.

Brockwell, J. 1971. An appraisal of an IBP experiment on nitrogen fixation by nodulated legumes. Plant and Soil, Special Volume: 265–272.

Brockwell, J. 1985. Environmental interactions influencing inovative practices in legume inoculation. Pages 943–950 *in* Proceedings of the World Soybean Research Conference III (Shibles, R., ed.). Westview Press, Boulder, USA.

Brockwell, J., Dudman, W.F., Gibson, A.H., Hely, F.W. and Robinson, A.C. 1968. An integrated programme for the improvement of legume inoculant strains. Transactions of the Ninth International Congress Soil Science 2: 103–114.

Brockwell, J., Gault, R.R., Mytton, L.R., Schwinghamer, E.A. and Gibson, A.H. 1981. Intrastrain stability in symbiotic effectiveness in *Rhizobium* spp. Page 433 *in* Current Perspectives in Nitrogen Fixation (Gibson, A.H. and Newton, W.E., eds.). Australian Academy of Science, Canberra, Australia.

Burton, J.C. 1981. Selection of rhizobia for use in commercial inoculants. Pages 338-339 *in* Current Perspectives in Nitrogen Fixation (Gibson, A.H. and Newton, W.E., eds.). Australian Academy of Science, Canberra, Australia.

Bushby, H.V.A. 1982. Ecology. Pages 35–75 *in* Nitrogen Fixation, Volume 2, *Rhizobium* (Broughton, W.J., ed.). Clarendon Press, Oxford, UK.

Chatel, D.L. and Parker, C.A. 1973. Survival of field-grown rhizobia over the dry summer period in Western Australia. Soil Biology and Biochemistry 5: 415–423.

Corbin, E.J., Brockwell, J., and Gault, R.R. 1977. Nodulation studies on chickpea (*Cicer arietinum*). Australian Journal of Experimental Agriculture and Animal Husbandry 17: 126–134.

Dashiell, K.E., Kueneman, E.A., Root, W.R., and Singh, S.R. 1985. Breeding tropical soybean for superior seed longevity and for nodulation with indigenous rhizobia. Pages 133–139 *in* Soybean in Tropical and Subtropical Cropping Systems (Shanmugasundaram, S. and Sulzberger, E.W., eds.). Asian Vegetable Research and Development Center, Taiwan.

Date, R.A. 1969. A decade of legume inoculant quality control in Australia. Journal of the Australian Institute of Agricultural Science 35: 27–37.

Date, R.A. 1977. The development and use of legume inoculants. *In* Biological Nitrogen Fixation in Farming Systems in the Tropics (Ayanaba, A. and Dart, P.J., eds.). John Wiley and Sons, New York, USA.

Date, R.A. 1982. Collection, isolation, characterization and conservation of *Rhizobium*. Pages 95–109 *in* Nitrogen Fixation in Legumes (Vincent, J.M., ed.). Academic Press, Sydney, Australia.

Date, R.A. 1984. *Rhizobium* for *Stylosanthes*. Pages 243–256 *in* The Biology and Agronomy of *Stylosanthes* (Stace, H.M. and Edye, L.A., eds.). Academic Press, Sydney, Australia.

Date, R.A. and Halliday, J. 1979. Selecting *Rhizobium* for acid infertile soils of the tropics. Nature 277: 62–64.

Date, R.A. and Norris, D.O. 1979. *Rhizobium* screening of *Stylosanthes* species for effectiveness in nitrogen fixation. Australian Journal of Agricultural Research 30: 85–104.

Date, R.A. and Roughley, R.J. 1977. Preparation of legume seed inoculants. Pages 243–275 *in* A Treatise on Dinitrogen Fixation. Section IV Agronomy and Ecology (Hardy, R.W.F. and Gibson, A.H., eds.). John Wiley, New York, USA.

Delannay, S., Rodgers, D.M., and Palmer, R.G. 1983. Relative genetic contributions among ancestral lines to North American soybean cultivars. Crop Science 23: 944–949.

Devine, T.E. and Breithaupt, B.H. 1981. Frequencies of nodulation response alleles Rj2 and Rj4 in soybean plant introduction and breeding lines. USDA Technical Bulletin 1628. US Government Printing Office, Washington, USA.

Diatloff, A. 1975. Effective natural nodulation of peanuts in Queensland. Queensland Journal of Agricultural and Animal Sciences 32: 96–100.

Döbereiner, J., Scotti, N.M.H. Sa and Vargas, M.A.T. 1981. Resistance to streptomycin of *Rhizobium* isolates from cerrado and Amazon soils. Page 434 *in* Current Perspectives in

Nitrogen Fixation (Gibson, A.H. and Newton, W.E., eds.). Australian Academy of Science, Canberra, Australia.
Franco, A.A. and Vincent, J.M. 1976. Competition amongst rhizobial strains for the colonization and nodulation of two tropical legumes. Plant and Soil 45: 27–48.
Frankel, O.H. and Bennett, E. (eds.). 1970. Genetic Resources in Plants-Their Exploration and Conservation. Blackwell, Oxford, England.
Frederick, L.R. 1985. Ecological manipulation to improve legume nodulation. Pages 197–212 *in* Biological Nitrogen Fixation in Africa (Ssali, H. and Keya, S.O., eds.). University of Nairobi, Kenya.
Freire, J.R.J. 1982. Research into the *Rhizobium*-Leguminosae symbiosis in Latin America. Plant and Soil 67: 227–239.
Gault, R.R. 1981. A study of developments and trends associated with legume inoculants for use with new legume crops. *Rhizobium* Newsletter 26: 16–44.
Gibson, A.H. 1980. Methods for legumes in glasshouses and controlled environment cabinets. Pages 139–184 *in* Methods for Evaluating Biological Nitrogen Fixation (Bergersen, F.J., ed.). John Wiley, Chichester, England.
Harris, S. (ed.). 1979. Planning an International Network of Legume Inoculation Trials. NifTAL Project, University of Hawaii, Hawaii, USA.
Hartel, P.G. and Alexander, M. 1984. Temperature and desiccation tolerance of cowpea rhizobia. Canadian Journal of Microbiology 30: 820–828.
Herridge, D.F. 1984. Effects of nitrate and plant development on the abundance of nitrogenous solutes in root-bleeding and vacuum-extracted exudates of soybean. Crop Science 25: 173–179.
Herridge, D.F. and Betts, J.H. 1985. Nitrate tolerance in soybean: variation between genotypes. *In* Nitrogen Fixation Research Progress. Martinus Nijhoff, Boston, USA.
Herridge, D.H. and Roughley, R.J. 1975. Variation in colony characteristics and symbiotic effectiveness of *Rhizobium*. Journal of Applied Bacteriology 38: 19–27.
Howieson, J.G. and Ewing, M.A. 1986. Acid tolerance in the *Rhizobium meliloti-Medicago* symbiosis. Australian Journal of Agricultural Research 37: 57–64.
ICRISAT (International Crops Research Institute for the Semi-Arid Tropics). 1984. Pages 129–130–*in* Annual Report 1983. Patancheru, A.P., India.
Keya, S.O. 1985. State of the art on biological nitrogen fixation in Africa. Pages 30-46 *in* Biological Nitrogen Fixation in Africa (Ssali, H. and Keya, S.O., eds.). University of Nairobi, Kenya.
Keyser, H.H., Sadowsky, M.J. and Bohlool, B.B. 1985. Fast-growing soybean rhizobia. Pages 926–934 *in* Proceedings of the World Soybean Research Conference III (Shibels, R., ed.). Westview Press, Boulder, USA.
Labandera, C.A. and Vincent, J.M. 1975. Competition between an introduced strain and native Uruguayan strains of *Rhizobium trifolii*. Plant and Soil 42: 327–347.
Lawes, D.A. 1980. Recent developments in understanding, improvement and use of *Vicia faba*. Pages 625–636 *in* Advances in Legume Science (Summerfield, R.J. and Buhnting, A.H., eds.). Royal Botanic Gardens, Kew, England.
Lie, T.A. and Goktan, D. 1984. Gene centres, a source of genetic variants in symbiotic nitrogen fixation: the symbiotic response of the cultivated pea to *Rhizobium leguminosarum* strains from Europe and the Middle East. Plant and Soil 82: 359–367.
Marques Pinto, C., Yao, Y.P. and Vincent, J.M. 1974. Nodulating competitiveness amongst strains of *Rhizobium meliloti* and *R. trifolii*. Australian Journal of Agricultural Research 25: 317–329.
Materon, L.A., and Vincent, J.M. 1980. Host specificity and interstrain competition with soybean rhizobia. Field Crops Research 3: 215–224.
van der Maesen, L.J.G. 1981. Taxonomy of *Cajanus*. Pages 9–14 *in* Proceedings of the International Workshop on Pigeonpeas. Volume 2. ICRISAT, Patancheru, India.
Minchin, F.R., Sheehy, J.E. and Witty, J.F. 1985. Factors limiting N_2 fixation by the legume *Rhizobium* symbiosis. Pages 285–291 *in* Nitrogen Fixation Research Progress (Evans, H.J., Bottomley, P.J. and Newton, W.E., eds.). Martinus Nijhoff, Dordrecht, The Netherlands.

Minchin, F.R., Witty, J.F., Sheehy, J.E. and Muller, M. 1983. A major error in the acetylene reduction assay: decreases in nodular nitrogenase activity under assay conditions. Journal of Experimental Botany 34: 641–649.

t'Mannetje, L. 1977. A revision of varieties of *Stylosanthes guianensis* (Aubl). South West Australian Journal of Botany 25: 347–362.

Munns, D.N. 1984. Coping with soil acidity in legume symbioses. Pages 137–140 *in* Proceedings of the Seventh Australian Legume Nodulation Conference, 14–17 February 1984, Sydney, Australia.

Munns, D.N., Hohenberg, J.S., Richetti, T.L. and Lauter, D.J. 1981. Soil acidity tolerance of symbiotic and nitrogen-fertilized soybeans. Agronomy Journal 73: 407–410.

Nambiar, P.T.C., Dart, P.J., Srinivasa Rao, B. and Ravishankar, H.N. 1984. Response of groundnut (*Arachis hypogaea* L.). Oleagineux 39: 150–155.

Newbould, P. 1983. The application of inoculation in agriculture. Pages 417–442 *in* Temperate Legumes: Physiology, Genetics and Nodulation (Jones, D.G. and Davies, D.R., eds.). Pitman, Boston, USA.

Nutman, P.S. 1976. IBP field experiments on nitrogen fixation by nodulated legumes. Pages 211–237 *in* Symbiotic Nitrogen Fixation (Nutman, P.S., ed.). Cambridge University Press, London, England.

Parker, C.A. 1985. Nitrogen fixation for developing countries with special reference to Africa. Pages 10–29 *in* Biological Nitrogen Fixation in Africa (Ssali, H. and Keya, S.O., eds.). University of Nairobi, Kenya.

Parker, C.A., Trinick, M.J. and Chatel, D.L. 1977. Rhizobia as soil and rhizosphere inhabitants. Pages 311-352 *in* A Treatise on Dinitrogen Fixation. Section IV. Agronomy and Ecology (Hardy, R.W.F. and Gibson, A.H., eds.). John Wiley, New York, USA.

Porter, F.E., Nelson, I.S. and Wold, E.K. 1966. Plastic pouches. Crops and Soils 18(8): 10–11.

Rai, R. 1985. Competitiveness and effectiveness of carbendazin-resistant mutant and indigenous *Rhizobium* strains of chickpea (*Cicer arietinum* Linn.) in calcereous soil. Journal of Agricultural Science, Cambridge 104: 341–344.

Rennie, R.J. 1984. Comparison of N balance and ^{15}N isotope dilution to quantify N_2 fixation in field-grown legumes. Agronomy Journal 76: 785–790.

Rennie, R.J. 1985. Discussion, Section II. Page 169 *in* Proceedings of a Symposium on Soybean in Tropical and Subtropical Cropping Systems on 26 Sept-1 Oct 1983, Tsukaba, Japan (Shanmugasundaram, S. and Sulzberger, E.W., eds.). AVRDC, Taiwan.

Rennie, R.J. and Dubetz, S. 1984. Multistrain vs single strain *Rhizobium japonicum* inoculants for early maturing (00 and 000) soybean cultivars: N_2 fixation quantified by ^{15}N isotope dilution. Agronomy Journal 76: 498–502.

van Rensburg, H.J. and Strijdom, B.W. 1982. Competitive abilities of *Rhizobium meliloti* strains considered to have potential as inoculants. Applied and Environmental Microbiology 44: 98–106.

Robinson, P.J., Date, R.A., and Megarrity, R.G. 1976. Nodulation of *Stylosanthes guyanensis*: prediction of *Rhizobium* effectiveness response with seed isozyme patterns. Australian Journal of Agricultural Research 27: 381–389.

Rolfe, B.G. and Shine, J. 1984. *Rhizobium*-Leguminosae symbiosis: the bacterial point of view. Pages 95–128 *in* Genes Involved in Microbe-Plant Interactions (Varma, D.P.S. and Hohn, Th., eds.). Springer Verlag, Wein.

Roughley, R.J. and Pulsford, D.J. 1982. Production and control of legume inoculants. Pages 193-209 *in* Nitrogen Fixation in Legumes (Vincent, J.M., ed.). Academic Press, Sydney, Australia.

Roughley, R.J. 1976. The production of high quality inoculants and their contribution to legume yield. Pages 125–136 *in* Symbiotic Nitrogen Fixation in Plants (Nutman, P.S., ed.). University Press, Cambridge, England.

Rupela, O.P. and Dart, P.J. 1979. Research on symbiotic nitrogen fixation by chickpea at ICRISAT. Pages 161–167 *in* Proceedings of the International Workshop on Chickpea Improvement, 28 Feb – 2 Mar 1979, Hyderabad, India. ICRISAT, Patancheru, India.

Scotti, M.R., Sa, N.M.H., Vargas, M.A.T. and Dobereiner, J. 1982. Susceptibility of *Rhizobium* strains to antibiotics: a possible reason for inoculation failure in cerrado soils. Pages 195–205 *in* Biological Nitrogen Fixation Technology for Tropical Agriculture (Graham, P.H. and Harris, S.C., eds.). CIAT, Cali, Colombia.

Sherwood, M.T. and Masterson, C.L. 1974. Importance of using the correct test host in assessing the effectiveness of indigenous populations of *Rhizobium trifolii*. Irish Journal of Agricultural Research 13: 101–108.

Skinner, F.A., Hamatova, E. and McGowan, V.F. 1983. World Catalogue of *Rhizobium* Collections. World Data Center for Microorganisms, Department of Microbiology, University of Queensland, Australia.

Singleton, P., Burton, J., Cady, F., Davis, R. and Halliday, J. 1985. Developing cost-effective *Rhizobium* technology for the tropics and subtropics. Pages 275–283 *in* Soybean in Tropical and Subtropical Cropping Systems (Shanmugasundaram, S.S. and Sulzberger, E.W., eds.). AVRDC, Taiwan.

Sprent, J.I. 1984. Effects of drought and salinity in heterotrophic nitrogen fixing bacteria. Pages 295–302 *in* Advances in Nitrogen Fixation Research (Veeger, C. and Newton, W.E., eds.). Martinus Nijhoff, The Hague, The Netherlands.

Staphorst, J.L. and Strijdom, B.W. 1976. Effects on rhizobia of fungicides applied to legume seed. Phytophylactica 8: 47–54.

Steele, W.M. and Mehra, K.L. 1980. Structure, evolution and adaptation to farming systems and environments in *Vigna*. Pages 393–404 *in* Advances in Legume Science (Summerfield, R.J. and Bunting, A.H., eds.). Royal Botanic Gardens, Kew, England.

Summerfield, R.J. and Bunting, A.H. (eds.). 1980. Advances in Legume Science. Royal Botanic Gardens, Kew, England.

Thompson, J.A. 1980. Production and quality control of carrier based legume inoculants. *In* Methods for Evaluating Biological Nitrogen Fixation (Bergersen, F.J., ed.). John Wiley and Sons, Chichester, England.

Thompson, J.A., Kumar Rao, J.V.D.K., and Dart, P.J. 1981. Measurement of inoculation response in pigeonpea. Pages 249–253 *in* Volume 1, Proceedings of the International Workshop on Pigeonpea, 15–19 Dec 1980, Patancheru, India. ICRISAT, Patancheru, India.

Thompson, J.A., Roughley, R.J. and Herridge, D.F. 1974. Criteria and methods for comparing the effectiveness of *Rhizobium* strains for pasture legumes under field conditions. Plant and Soil 40: 511–524.

Thornton, F.C. and Davey, C.B. 1984. Saprophytic competence and acid-tolerant strains of *Rhizobium trifolii* in acid soils. Plant and Soil 80: 337–344.

Vasilas, B.L. and Ham, G. 1984. Nitrogen fixation in legumes: an evaluation of measurement techniques. Agronomy Journal 76: 759-764.

Vincent, J.M. 1970. A Manual for the Practical Study of Root-Nodule Bacteria. IBP Handbook No. 15. Blackwell, Oxford, England.

Vincent, J.M. and Waters, L.M. 1953. The influence of the host and competition amongst clover root-nodule bacteria. Journal of General Microbiology 9: 357–370.

Williams, P.M. 1984. Current use of legume inoculant technology. Pages 173-200 *in* Biological Nitrogen Fixation, Ecology, Technology and Physiology (Alexander, M., ed.). Plenum Press, New York, USA.

Witcombe, J.R. and Erskine, W. (eds.). 1984. Genetic Resources and Their Exploitation – Chickpeas, Faba Beans and Lentils. Martinus Nijhoff, The Hague, The Netherlands.

Witty, J.F. and Minchin, F.R. 1986. The acetylene reduction assay; myths and mysteries. Workshop on Biological Nitrogen Fixation, 13-18 April 1986. ICARDA, Aleppo, Syria.

Zohary, D. 1970. Centres of diversity and centers of origin. Pages 33-42 *in* Genetic Resources in Plants – Their Exploration and Conservation (Frankel, O.H. and Bennett, E., eds.). Blackwell, Oxford, England.

Discussion

Broughton
You mentioned the possibility of introducing "engineered" strains. Which characteristics do you believe to be important to engineer?

Thompson
Strains with extremely narrow host-range (different rhizobia for different varieties) and competitiveness.

Brockwell
Two characters I would like to see in engineered strains are:
(i) Mobility – the ability to move away from the site of inoculation, and
(ii) the capacity to quickly generate large populations in the host rhizosphere.

Sylvester-Bradley
I was interested in your comments on the importance of using local strains. In general we have found that the proportion of effective local strains is no higher than in strains from the center of origin. However, in some cases, we have found that a higher proportion of strains from where the legume has been growing for some time (more than 3 years) in introduced pastures is effective. We have found this in Colombia for both *Pueraria phaseoloides* (group of strains isolated with CIAT 3652) and *Desmodium heterophyllum* (group of strains isolated with CIAT 2469). These two legumes originated in SE Asia.

Solh
What would be your advice on the use of introduced rhizobial strains to be breeders like those at ICARDA working in an area where endogenous rhizobia are available and they are screening germplasm for an international or regional program?

Thompson
It is essential to evaluate legumes with effective nodulation, and the potential for improvement by inoculation needs to be established by comparison of growth with and without nitrogen fertilizer. For logistic reasons, such testing could be confined initially to parental material.

Roughley
Would you please comment on the role of acetylene reduction for selection of strains for effectiveness or the comparison of the relative effectiveness of a group of strains.

Thompson
Any doubts we may have had about the limitations of the acetylene reduction technique for strain evaluation have been dispelled by Dr. Witty's paper at this workshop. There seems to be no place for a technique which suffers from such major errors, and only provides us with acceptable data when it is correlated with other more readily measured variables.

D.P. Beck and L.A. Materon (editors), Nitrogen Fixation by Legumes in Mediterranean Agriculture
© 1988 ICARDA – Printed in the Netherlands. ISBN 90 247 3624 2

Selection of Rhizobia for Inoculation of Forage Legumes in Savanna and Rainforest Soils of Tropical America

ROSEMARY SYLVESTER-BRADLEY, DACIER MOSQUERA, and JESUS EDUARDO MENDEZ

Centro International de Agricultura Tropical (CIAT), Colombia, South America

Summary. Legumes which are adapted for use as forage plants in association with grasses on the acid problem soils of tropical America are being selected by a network of national agricultural programmes coordinated by the Tropical Pastures Programme at CIAT. Inoculants containing rhizobia are used for crop legumes in many of the countries involved, and many legumes yield better when inoculated. Evaluation of rhizobial inoculants for tropical forage legumes is therefore considered to be justified. A large collection of rhizobial isolates has been built up, and methods for screening them and evaluating inoculation responses in the field avoiding the use of complex or expensive equipment, have been developed, with a view to their use in isolated places. The same control strains are tested at different sites, nitrogen mineralization is minimized by using reduced tillage establishment, and there is one treatment where high nitrogen availability is ensured by making frequent applications.

An international collaborative training and research programme has been initiated to assess responses to rhizobial inoculation of locally selected legumes. An agronomist and a microbiologist are responsible for each trial, and both are involved at all stages of planning and execution. This approach has been adopted in view of the recognition that such trials are simple to execute but require the experience of both disciplines, and can fail if this is not available. Failure may be misinterpreted to mean that the legume does not need to be inoculated, or that the technology is too complex to use in the region.

This paper gives examples of the methods which have been developed. Two experiments for screening rhizobia in the greenhouse and the field with the promising forage legume *Arachis pintoi* are described, including details of the methods used. Marked yield increases due to inoculation of this legume with some strains were observed. The most effective strains were isolates from *A. pintoi* and other legumes. Abundant, internally red nodules were formed in the uninoculated controls. This demonstrates that a promiscuous legume may respond to inoculation if suitably selected strains are used.

Introduction

Legumes which are adapted for use as forage plants in association with grasses on the acid problem soils of tropical America are being selected by a network of national agricultural programmes coordinated by the Tropical Pastures Program at CIAT. The objective is to develop technologies which would make beef and milk production in marginal areas into a more economic and less ecologically damaging proposition. The legumes which have shown most promise in improved pastures in savanna areas are *Pueraria phaseoloides*, *Arachis pintoi*, *Centrosema macrocarpum*, *C. brasilianum*, *Centrosema* sp., *Stylosanthes guianensis* (var *pauciflora*), *S. capitata*, and *S. macrocephala*. In rainforest areas, *Desmodium heterophyllum*, *Stylosanthes guianensis* (var *vulgaris*), and *Centrosema pubescens* have also shown promise, and *S. capitata* grows somewhat less vigorously than in savanna areas. *Leucaena leucocephala* also grows well in some rainforest areas, whereas in savanna its development is stunted possibly due to a higher level of Al^{3+} saturation in the deeper soil layers. So far, relatively few shrubby legumes have been tested in either ecosystem.

Most of these legumes are native to Central or South America, except *Desmodium ovalifolium*, *D. heterophyllum*, and *Pueraria phaseoloides* which are from SE Asia.

The soil pH in the areas under study ranges from about 4.0 to 5.5, mostly around 4.5, and aluminium saturation between 70 and 90% of the cation exchange capacity in the surface soil. In most areas, liming is not considered to be economically feasible for pasture establishment, and unless agriculture in the area is sufficiently intensive to include an integrated crop-pasture rotation, the pasture plants must be adapted to grow in unlimed soils. Small applications of Ca, P, Mg, K, S, and micronutrients may be made depending on the socioeconomic status of a given area.

Although it is known that many tropical legumes are promiscuous (nodulate with a wide range of rhizobia) and therefore may not require inoculation in soils which contain abundant and effective rhizobial populations, in acid soils populations may be small or only semi-effective.

Yield increases of many legumes in response to inoculation may therefore be observed even if they are native to the area where they are being sown. Date (1977) recommended a simple three- treatment field experiment (uninoculated, inoculated, and fertilized with nitrogen) for this purpose.

In many of the countries where legumes are being selected as part of an integrated screening programme coordinated by CIAT, (Red Internacional para la Evaluacion de Pastos Tropicales, RIEPT) rhizobial inoculants are used for crops such as soybeans, dry beans, lupins, and also for alfalfa and clover in some cases. However, although many farmers are aware that inoculants can increase yields, in many cases they cannot obtain the inoculants and receive no assistance in using them. Setting up trials to demonstrate the effects of inoculation and improving inoculant distribution and technical assistance should be part of the adoption procedure for newly selected legumes, which in itself requires a concerted effort on the part of extension agencies. In view of these considerations, a collection of about 3000 rhizobial isolates from tropical forage legumes has been built up. Methods have been developed for evaluating the effect of inoculation on growth rate and nitrogen yield of selected legumes in the greenhouse and the field, without the need for sophisticated or expensive equipment.

In this paper, an example of the methods which have been developed is presented, using *Arachis pintoi*, a legume native to central and coastal regions of Brazil. Their use in a collaborative training and research programme for agronomists and microbiologists working at national institutes is discussed.

Materials and Methods

Preselection of rhizobial strains on Arachis pintoi using undisturbed soil cores from Carimagua

An evaluation of this method has been described by Sylvester-Bradley *et al.* (1983). Further details are given in Spanish and English in the manual available from CIAT (CIAT 1986a), and are also presented below.

PVC tubing with a diameter of 4 inches, commonly used for drainpipes in Colombia, was sawn into 25 cm lengths, each of which was sharpened with a rotary sander on the inner edge of one end. Large wooden mallets were made from heart wood and used to introduce the sharpened end of the tubes into the soil to within about 2 cm of their upper edge. The cores were taken from undisturbed savanna at Carimagua (Meta, Colombia, 150m above sea level, 4.5° N, 71.5° W, 2200 mm annual rainfall between April and October, mean temperature 28°C). The tubes were dug out with a spade and transported in wooden boxes (containing no more than 20 cores each to facilitate handling) to CIAT. The soil in the cores was relatively undisturbed, and only a small proportion of the tubes were broken or twisted on introducing them into the soil, if when using the mallets care was taken to introduce the cores slowly. Soil was not lost from the bottom of the cores as this only occurs with very sandy soil. Clearly, it is only possible to introduce the tubes during the wet season.

Roots were removed from the soil surface so that about 2 cm of soil were disturbed, and the cores were numbered and weighed. The soil was taken from five cores at the same time to determine its humidity.

Shortly before the day set for planting the cores, which weighed 3-4 kg, were divided into five blocks, each block consisting of cores of similar weight. The cores were placed on plastic plates to prevent contamination with water on the table top. The weight range within a block was not more than 100 g. The required core weight, with 18% water on a wet weight basis, was calculated for the mean weight/core for each block, using the weight of dry soil/core and dividing by the proportional dry weight required (0.82 in this case). Deionized water was added to maintain the calculated weight for each block throughout the experiment.

The cores were fertilized with (mg/core) 50 P, 128 Ca, 30 K, 60 S, 40 Mg, 5

Zn, 1 Cu, 0.5 B, and 0.4 Mo using 180 Ca $(H_2PO_4).2H_2O$, 141 $CaCO_3$, 59 K_2SO_4, 34 flowers of sulphur, 59 MgO, 19.4 $ZnSO_4.7H_2O$, 2.22 $CuSO_4$, 3.8 $Na_2B_4O_7.10H_2O$, and 0.89 $Na_2MoO_4.2H_2O$, the calculations being made on the basis of surface area/core (88.3 cm^2). No Cl^- sources were used since this had previously been found to be toxic. The insoluble compounds ($CaCO_3$, flowers of sulphur, and MgO) were added by diluting them to 5g/core with sieved soil which was sprinkled on the surface. The other nutrients were added in solution.

Thirty rhizobial isolates were grown on YMA 5.5 plates (10 g mannitol, 0.5 g K_2HPO_4. 0.1 g $MgSO_4.7H_2O$, 0.2 g NaCl, 0.01 g $FeCl_3.6H_2O$, 0.2 g $CaCl_2$, 100 ml yeast water (supernatant from 100 g yeast/l water, steamed and centrifuged), 20 g Difco Bacto agar, and 5 ml bromocresol purple solution at 0.5% in 0.016 N NaOH, all in 1000 ml water; pH adjusted to 5.5 after autoclaving). Peat-based inoculants were made by suspending the rhizobia in 10 ml YM broth and adding 20 g peat from La Selva farm near Rio Negro, Medellin which had been dried in the sun, ground with a hammer mill and sieved through 100 mesh/inch. Inoculants were kept in small thin plastic bags in the refrigerator until the day of planting.

Seeds of *Arachis pintoi* were pre-germinated by soaking in water and incubating on water agar in petri dishes. Only seeds without contaminants were sown. 160 cores were labelled with treatment and block numbers (30 strains, with uninoculated and N fertilized controls) and four small holes about 0.5 cm deep were made in the surface of each core. A small quantity (0.2–0.5 g) of the appropriate inoculant was placed in each hole, then a pregerminated seed was placed on top and covered with a little soil. Cross contamination was avoided by using aseptic techniques.

The soil surface was watered frequently (twice a day) in the first week after sowing to avoid drying out. When the plants were about 2 cm high, they were thinned to 2/core and an open-ended plastic tube, 15 cm long and 5 cm diameter with an aluminium foil cap, was placed in the centre of the soil surface. The rest of the surface was covered with 65 g waxed sand * (Vincent 1970) which prevents cross contamination, growth of algae, and reduces soil temperatures.

The weight of the cores was adjusted with deionized water through the tube, taking into account the weight of soil used for fertilization, sand, and irrigation tube.

After 2 weeks and every following 2 weeks, 58 mg urea/core were added through the irrigation tube to the nitrogen fertilized control treatment, adding an equivalent amount of water to the unfertilized treatments. After 6 weeks, the cores were fertilized with an additional 180 mg/core $Ca(H_2PO_4).2H_2O$ and 59 mg/core K_2SO_4. These relatively high fertilization rates compared with rates applied per ha in the field were found to be necessary to prevent

* Dissolve 4 g paraffin wax in 100 ml benzol (care: toxic) and mix with 1 kg dry quartz sand. Leave to evaporate in a fume hood, and sterilize in bottles in a hot air oven at 160°C for 2 h.

nutrient deficiencies in plants grown in soil cores, which presumably result from exploitation by roots of a larger soil volume in the field than in the cores. Each week the cores were weighed and the weights adjusted with water.

After 10 weeks, plant tops were cut, dried, and analysed for % N content. The soil and roots were allowed to dry partially and then loosened by beating the cores with a wooden stick. The roots were removed and stored in plastic bags in the freezer until nodules could be counted.

Some further strains were screened in experiments not described here, but which are included in the catalogue of strain performance (CIAT 1986b).

For statistical analysis of the data, the nitrogen fertilized controls were excluded, since in some of them nitrogen yields were extremely high, causing a bias in the analysis.

Field evaluation of inoculation response during establishment

Savanna vegetation at Carimagua was burned at the end of the dry season and furrows 1.5 m apart, about 40 cm wide, and 10 cm deep were made using two tynes and a sweep between them. This was sufficient to cut the roots of the savanna vegetation, and leave sizeable sods to prevent excessive soil erosion during rainstorms.

Six treatments (uninoculated, inoculated with individual strains 3810, 3806, 3101, and 2138, and a mixture of the four strains) were established by sowing inoculated or uninoculated seeds of *Arachis pintoi* in the furrows (10 g seeds/furrow), using a completely randomized block design with three replicates. Four 10 m long furrows per plot were fertilized before sowing with 120 Ca, 22 P, 40 S, 33 K, 20 Mg, 5 Zn, 2 Cu, 1 B, and 0.4 Mo (g/10 m furrow). Inoculants were prepared as described above. In this experiment, the number of rhizobia/seed was not counted, although in general, inoculants for such experiments are made with sterile peat and the number of rhizobia are counted using the plate count method. Also, since it was already known that *A. pintoi* required inoculation, no N fertilized control was included. In an unknown situation, a treatment with 20 kg N/ha added 4 weeks after planting and every following 2 weeks is included. In the inoculated treatments, the seeds for each plot (40 g) were weighed and inoculated separately (50 g peat-based inoculant/kg seed, 2 g inoculant/plot). This avoids the inconvenience of separating seeds for each plot after inoculating on a weight or volume basis, as obviously inoculated seeds weigh more, and care must be taken to use the same number of seeds/plot in all treatments. The inoculant for each plot was mixed with 1.3 ml 40% gum arabic and the seed in a disposable cup. The seeds were then coated with about 15 g rock phosphate, added immediately after coating with inoculant, by gently rotating the cup. Where counts are to be made, sterilized gum arabic, rock phosphate, and cups are used and the number of seeds to be removed for the count is added to the seeds to be planted in each plot before inoculation. Once planted, the seeds were compacted by trampling. To reduce contamination between plots new plastic bags were worn over the feet inside the plots, and hands and implements used for weeding were washed with alcohol. Ditches were made in the 3 m wide alleyways between plots to prevent rainwater flooding across them.

After 10, 12, and 14 weeks, three 2 m long samples were cut at different random sites within each plot to ground level to evaluate the rate of establishment of the legume in each treatment. The three samples from each cut within each plot were pooled and weighed as soon as possible. A 100 g subsample was dried at 60°C to constant weight to determine percent humidity. The samples were ground and analysed for % N and the nodulation of randomly selected plants was evaluated.

Results and Discussion

Nitrogen yield of *Arachis pintoi* was increased twofold or more, when inoculated with the most effective rhizobial strains and compared with the uninoculated controls, both in the greenhouse and the field (Figs. 1 and 2). This occurred although abundant, apparently active (internally red) nodules were formed in the uninoculated controls. Numbers of nodules formed did not correlate with nitrogen yields (Fig. 1).

The strains used in the field trial all gave marked responses. Strain 3810 is an isolate from a 4 year old, uninoculated *A. pintoi* plot at Carimagua, and 3806 is from *A. pintoi* growing in soil from a nearby farm (Guayabal). However, other strains which have a similar history were eliminated in the pre-selection trial in soil cores (e.g. strain 3812 from *A. pintoi* at Carimagua). This indicates that effective strains for *A. pintoi* were present in Carimagua soil, but they did not dominate the population, even where *A. pintoi* had been growing for some time. Internationally recommended strains for the inoculation of *A. hypogea* (3823, 3824, and 3825 = TAL 169, USDA 3183, and Nitragin 8A 11T1, respectively) were relatively ineffective in the pre-screening trial.

Two of the selected strains, 2138 and 3101 (which was selected in a trial not shown in this paper), are isolates from *Stylosanthes capitata* in Nopolis, Brazil, and *Centrosema macrocarpum* in Santa Marta, Colombia, respectively. They had previously been selected as effective on their original hosts. This shows that *A. pintoi* could be considered as promiscuous, since it nodulates with rhizobia from other legumes. However, although it nodulates without inoculation, the population of indigenous rhizobia in Carimagua soil is not sufficiently effective for the plant to reach its full yield potential. Thus care must be taken in interpreting the statement that tropical legumes are promiscuous and therefore less likely to need inoculating. As Date (1977) points out, the need to inoculate not only depends on legume specificity but also on soil conditions.

The results reported here are an example of similar results obtained with

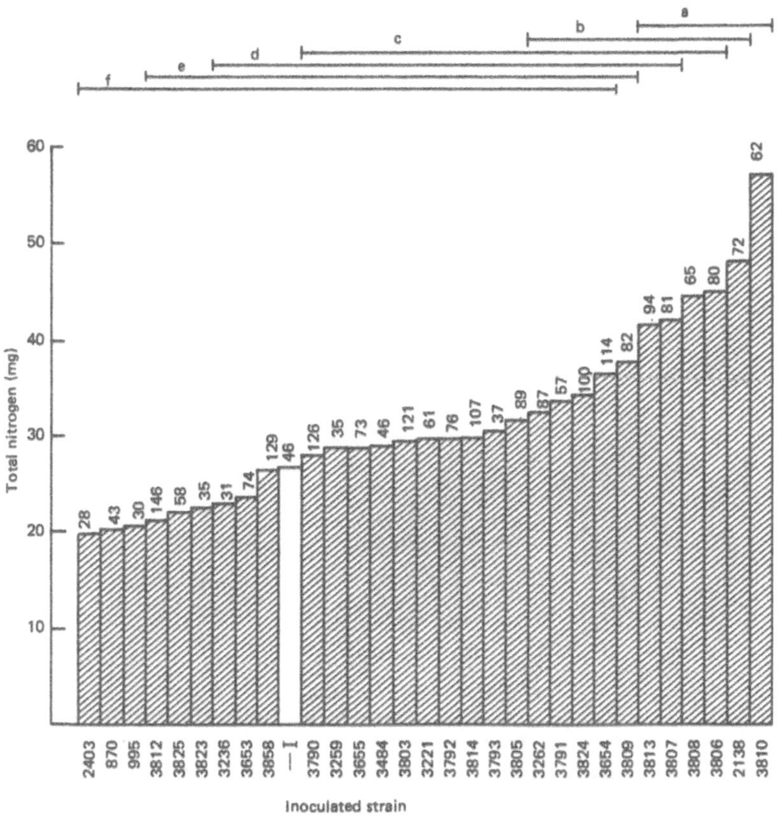

Fig. 1. Total nitrogen in tops/core (two plants) and nodule numbers (above bars) of *Arachis pintoi* 17434 uninoculated (−I) or inoculated with rhizobial strains (CIAT strain nos.) in cores of Carimagua soil. Duncan's MRT (P = 0.05).

other legumes, especially *Centrosema* spp. and *Pueraria phaseoloides*. In some cases, *Desmodium* spp. and *Stylosanthes capitata* have also shown inoculation responses in Carimagua soil. The results of trials with 35 legume ecotypes are summarized in the catalogue of strain performance (CIAT 1986b).

The results show that the simple methods described here can be used to select rhizobial strains which cause inoculation responses in the field, even when a population of indigenous rhizobia able to nodulate the legume in question is present in the soil. Under these circumstances it seems preferable to use soil rather than sterile support materials for strain selection, since this helps to ensure that competitive strains are selected.

One criticism of the method could be that selected strains may be site specific. Tests within the Carimagua farm have shown inoculation responses with the same strains at three different sites (Sylvester-Bradley and Mosquera 1985), and strain 2434, selected for inoculation of *Pueraria phaseoloides*, was effective at five widely separated sites in Colombia (unpublished results and

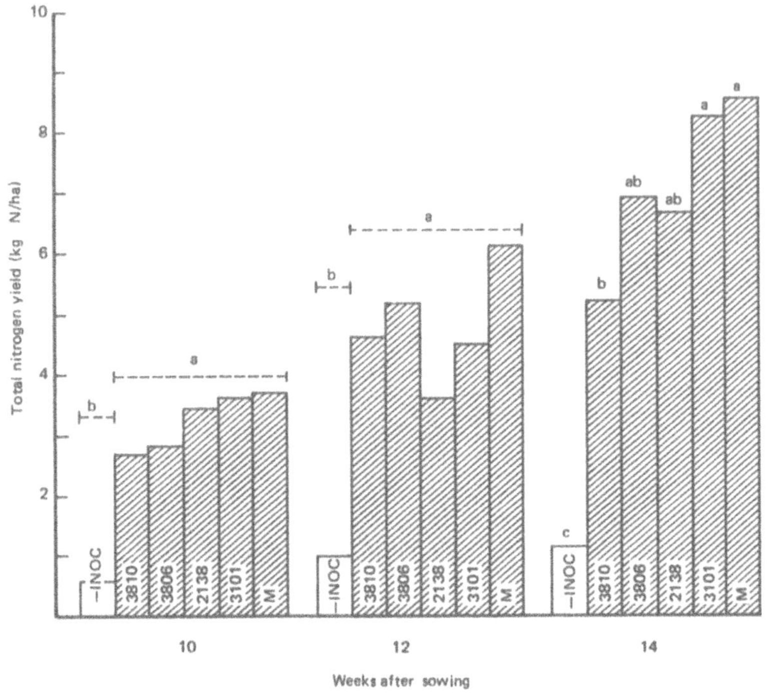

Fig. 2. Total nitrogen yield at three dates during establishment of *Arachis pintoi* at Carimagua; uninoculated (−inoc.), inoculated with four separate rhizobial strains (CIAT strain nos.), or inoculated with a mixture of the four strains (M).

CIAT 1985). However, to ensure that the most widely adapted strains are being selected, it is necessary to test them at as many different sites as possible. Such experiments would also serve to determine differences in the effectivity of native rhizobial populations between sites.

With this aim in mind, microbiologists who in many cases have already been trained in techniques for studying nitrogen fixation, and the agronomists who run the RIEPT trials in six countries, were contacted. These pairs of agronomists and microbiologists are participating in a collaborative training and research project funded by the United Nations Development Programme to evaluate field responses to rhizobial inoculation under local conditions. By using the same control strains in all the trials, an establishment method which minimizes nitrogen mineralization such as that described in this paper, and a nitrogen fertilized control where high nitrogen availability is ensured through repeated applications during the evaluation period, the effectivity of local native and inoculated strains can be assessed by means of yield and nodulation evaluations. Counts of rhizobia/seed at planting may also be necessary. Although the collaborators may choose to study nodule occupancy serologically, this is not required to achieve the aims of the project. Technical assistance in setting up experiments is provided, and participants are expected to present results to the group after one or two growing seasons.

It is hoped that the equal involvement of the two professionals in the trials will help to ensure that they are conducted in a way which is both agronomically and microbiologically sound. It is the experience of the authors that, in many cases, field inoculation trials may fail or be irrelevant either because the agronomist is not aware of the special precautions which are needed when evaluating inoculants, or because the microbiologist does not take into account the agronomic experience of the region, such as locally adapted varieties or fertilization rates. This may lead to an erroneous conclusion that inoculation is not necessary or is too difficult a technology to be useable under local conditions.

In fact, far more complex technologies than those described in this paper are being adopted in the countries under consideration without any problems. Possibly, more collaborative training and research programmes such as the one described here would provide the information needed to justify and support the more widespread adoption of rhizobial inoculants in the tropics.

References

CIAT (Centro Internacional de Agricultura Tropicale) 1985. Page 161 *in* Informa Anual 1984, Programa de Pastos Tropicales, Documento de Trabajo No. 5, 1985.

CIAT (Centro Internacional de Agricultura Tropicale) 1986a. Manual for the Evaluation, Selection and Management of Legume-*Rhizobium* Systems for Increasing Nitrogen Fixation. Vol. A, Theory; Vol. B, Methods. CIAT, Cali, Colombia.

CIAT (Centro Internacional de Agricultura Tropicale) 1986b. Catalogue of *Rhizobium* Strains for Tropical Forage Legumes. Working Document No. 13. CIAT, Cali, Colombia.

Date, R.A. 1977. Inoculation of tropical forage legumes. Pages 293-311 *in* Exploiting the Legume-Rhizobium Symbiosis in Tropical Agriculture (Vincent, J.M., Whitney, A.S. and Bose, L.J., eds.). Micellaneous Publication 145. University of Hawaii, College of Tropical Agriculture, Hawaii, USA.

Sylvester-Bradley, R., Ayarza, M.A., Mendez, J.E. and Moriones, R. 1983. Use of undisturbed soil cores for evaluation of *Rhizobium* strains and methods for inoculation of tropical forage legumes in a Colombian Oxisol. Plant and Soil 74: 237-247.

Sylvester-Bradley, R. and Mosquera, D. 1985. Nitrification and responses to *Rhizobium* inoculation in tropical savanna as affected by land preparation. Pages 167-183 *in* Nitrogen Management in Farming Systems in the Humid and Subhumid Tropics (Kang, B.T. and van der Heide, J., eds.). Institute for Soil Fertility (IB), 9750, R.A. Haren, The Netherlands.

Vincent, J.M. 1970. A Manual for the Study of Root Nodule Bacteria. IBP Handbook 15. Blackwell Scientific, London, UK. 164 pp..s 2;.lm 10; · iscussion

Discussion

Lie

How do you use your freeze-dried inoculants?

Sylvester-Bradley

They are resuspended either in boiled water and injected into sterile peat or in mineral oil which can be applied directly to the seeds. We are doing research on oil-suspended inoculants which has indicated that oil-based, freeze-dried inoculants survive better on seeds than peat-based inoculants under dry conditions.

Eaglesham

In the field studies where you obtained large positive responses to inoculation, do you know if the indigenous rhizobia were low in numbers or poorly effective? Have you scored nodule occupancy of introduced strains?

Sylvester-Bradley

In the counts we have done in savanna soil there were around 10^4 rhizobia/g soil, which implies that they are ineffective on the legumes which respond to inoculation. Possibly this is because the nitrogen demand is less in savanna than improved pastures, so native rhizobia don't have to be so effective. In some legumes the nodule numbers increase with inoculation and in others not, which implies that at least in the latter case the selected strains are competitive. However, we have not studied nodule occupancy. More work on the diversity of native strains is needed.

D.P. Beck and L.A. Materon (editors), Nitrogen Fixation by Legumes in Mediterranean Agriculture
© 1988 ICARDA – Printed in the Netherlands. ISBN 90 247 3624 2

Field Performance of Rhizobial Inoculants for Some Important Legumes in Egypt

HASSAN MOAWAD, S.M.S. BADR EL-DIN and M.A. KHALFALLAH
Soil Microbiology Laboratory, Cairo, Egypt

Summary. The response of lentils, soybeans, faba beans, and peanuts to nitrogen fertilization and inoculation with rhizobia was measured. With mixed strain inoculation of peanut, there were significant increases in dry matter (DM) and nitrogen contents, which were significantly different from those of the nitrogen fertilized treatment. Differences in response were recorded between peat-based and liquid inoculants. In soybean, there was a significant yield response with inoculation which varied with the rhizobial strain. For lentils, there was a significant increase in yield with strains TAL 638 and Nit. 92A3 and with nitrogen fertilization, although differences between the two treatments were not significant. There was a significant response in faba beans to inoculation with strain TAL 1400.

Introduction

Faba bean is, by area, the most common grain legume cultivated in the Nile Valley of Egypt during winter. Lentil was previously planted under basin irrigation in upper Egypt and has been recently introduced to the Nile Valley. Both faba beans and lentils fix atmospheric nitrogen in symbiosis with *Rhizobium leguminosarum* but studies in Egypt to increase faba bean yields through inoculation with *Rhizobium* were inconclusive (Hamdi 1982).

Soybean is the major summer grain legume which was introduced to Egypt recently. This crop fails to nodulate in Nile Valley soils due to the lack of *Rhizobium japonicum* among the indigenous soil rhizobia.

Peanut is another summer legume cultivated in the sandy soil of newly reclaimed areas. This crop fixes nitrogen in symbiosis with cowpea rhizobia (*Rhizobium* spp.), a group of specific rhizobia found in high numbers in sandy soils repeatedly cultivated with peanut. Nodulation with native soil rhizobia in such soils is common.

This study aims to evaluate the performance of rhizobial inoculants for faba bean, lentil, soybean, and peanut in some Egyptian soils.

Materials and Methods

Lentils, soybeans, and faba beans were grown in clay loam Nile Delta soils, while peanuts were grown in a virgin sandy soil at Salhia mechanized project,

Egypt. The treatments were uninoculated, nitrogen fertilized, and inoculated with specific rhizobia. All treatments were fertilized with 40 kg N/ha; the nitrogen treatment had an additional 80 kg N/ha.

Peat-based inoculants containing 10^8 cells/g were used at a rate of 5 g/kg of seeds with gum arabic as the adhesive. Liquid inoculant containing 10^9 cells/ml was injected at the rate of 595.25 ml inoculant/ha with irrigation water 5 days after planting.

Strain-specific fluorescent antibodies were used for nodule typing. Nitrogen determinations were done by the semi-micro Kjeldhal method (Jackson 1958).

Results

Response of peanut to inoculation

Data in Table 1 show that inoculation with a mixture of the strains TAL 1000, TAL 1371, and TAL 169 produced significant increases in dry matter and nitrogen contents of peanut plants compared to the uninoculated control. The peat-based inoculant was superior for plant dry weight accumulation and quantity of nitrogen fixed over the liquid inoculant. Burton (1976) reported that the method of inoculation determines nodulation and nitrogen fixation, since the number of viable cells/seed is obviously critical (Nambiar *et al.* 1981; Roughley 1981). Nambiar *et al.* (1981) reported that peanut requires a heavy inoculation rate for adequate nodulation and nitrogen fixation. In the present work, the liquid inoculant contained 10 times more *Rhizobium* cells than the peat-based inoculant, but the latter was more effective in nodulation and nitrogen fixation (Table 1). However, the liquid inoculant produced significantly higher plant dry weight and nitrogen content than the nitrogen fertilized treatment. Both inoculation techniques seem to successfully place rhizobia in adequate numbers to nodulate peanut roots. The superiority of the peat-based inoculant over the liquid inoculant may be attributed to the direct

Table 1. Nodulation and nitrogen fixation by peanut plants inoculated with peat-based and liquid inoculants in a virgin sandy soil (75 days after sowing).

Treatments		Number of nodules/ plant	Nodule dry wt. (mg)	Plant dry wt. (g)	N content (mg/pl)	N fixed (mg/pl)
Inoculation	Fertilizer (kg N/ha)					
Uninoculated	48	0	0	10.9	204	
Uninoculated	143	0	0	14.4	322	
Peat inoc.	48	61	868	16.9	391	187
Liquid inoc.	48	51	754	15.2	378	174
LSD (5%)				0.3	5	
(1%)				0.4	7	

Table 2. Nitrogen fixation and yield responses of peanut to inoculation with peat-based and liquid inoculants.

Treatments		Seed yield (kg/ha)	Nitrogen fixed (kg/ha)
Inoculation	Fertilizer (kg N/ha)		
Uninoculated	48	1202	
Uninoculated	143	1583	
Peat-based inoculant	48	3626	177.8
Liquid inoculant	48	2919	160.7
LSD (5%)		1359	
(1%)		1907	

exposure of rhizobia in the liquid inoculant to the sun and leaching with irrigation water. Improved rhizobial longevity on seeds in the presence of peat and gum arabic has been previously reported (Burton 1976; Strijdom and Deschadt 1976; FAO 1984).

Data in Table 2 show the significant yield responses to inoculation in newly reclaimed desert soil, with inoculation technique influencing seed yield. For example, there was a highly significant response in seed yield with peat-based inoculant and a smaller but significant response with liquid inoculant application. There was however no significant difference between yields obtained with the peat-based inoculant and the liquid inoculant.

The nitrogen fixed represented 177.8 kg/ha with the peat-based inoculant and 160.7 kg/ha with the liquid inoculant. Similar results were obtained by Zayed *et al.* (1973) in a lysimeter experiment. They estimated a gain of nitrogen in virgin sandy soil under peanut of 178 kg/ha. Dart and Krantz (1977) reported that peanut can fix appreciable amounts of nitrogen (up to 240 kg N/ha).

It is clear that both inoculation techniques, under field conditions, can supply peanut with enough nitrogen to support good growth and yield, with high net nitrogen gains. Although less nitrogen was fixed with the liquid inoculant, this technique was more convenient for production and application using the machinery available in large-scale farming systems.

Field performance of the R. japonicum*/soybean symbiotic system*

Data in Table 3 show the significant yield response in soybean to inoculation with *R. japonicum* strains. Strain USDA 110 was more efficient in nitrogen fixation and in increasing seed yield than the other strains. Significant increases in seed yield were observed with strains USDA 110 and USDA 138 or their combination compared with the nitrogen fertilized treatment. However, there were no significant differences in seed yield between the

Table 3. Effect of inoculation with *R. japonicum* strains on yield and nitrogen fixation of soybean.

Treatments	Seed yield (kg/ha)	Nitrogen fixed (kg/ha)
Uninoculated	1807	
Fertilized (143 kg N/ha)	2743	
USDA 110	3129	200
USDA 138	2986	171
TAL 379	2895	157
USDA 110 + USDA 138 + TAL 379	3048	176
LSD (5%)	193	
LSD (1%)	281	

fertilized treatment and the treatment inoculated with TAL 379. Strain USDA 110 was previously found to be one of the most effective strains for nitrogen fixation and increasing seed yield of soybean (Abel and Erdman 1964; West 1971). In our experiment, the amount of nitrogen fixed by USDA 110, USDA 138, and TAL 379 was 200, 171, and 157 kg/ha, respectively. Similar results were obtained by Nassib *et al.* (1983) who stated that about 60% of the total soybean nitrogen was derived from nitrogen fixation in Egyptian soils.

The recovery of the inoculant strains from the soybean nodules is given in Table 4. Strain USDA 110 was more competitive than either USDA 138 or TAL 379 as it occupied 76% of the nodules when applied with strain USDA 138 and 72% of the nodules when applied with strain TAL 379. However, strain USDA 138 was more competitive than TAL 379, as it was found in 76% of the total nodules when applied as a double strain inoculum. Kosslak *et al.* (1983) found that strain USDA 110 was a better competitor than either USDA 138 or SM-5 on cultivars Lee and Peking. Doubly-infected nodules amounted to 0-36%. The incidence of double-strain occupancy in nodules has been reported for soybean and lentil (Lindemann *et al.* 1974; May and Bohlool 1982; Moawad *et al.* 1984).

Table 4. Competition between *R. japonicum* strains for nodulation of soybean in Nile Valley clay loam soil.

Inoculant strains	% Recovery from nodules		
	USDA 110	USDA 138	TAL 379
USDA 110	88 (12) *	0 (12)	0 (0)
USDA 138	0 (8)	2 (8)	0 (0)
TAL 379	0 (8)	0 (4)	88 (12)
USDA 110 + USDA 138	40 (36)	4 (28)	0 (8)
USDA 138 + TAL 379	0 (4)	44 (32)	20 (36)
USDA 110 + TAL 379	48 (24)	0 (0)	28 (24)
USDA 110 + USDA 138 + TAL 379	32 (28)	24 (20)	4 (32)

* Figures in parentheses are percent of doubly-infected nodules.

Field performance of R. leguminosarum *with two lentil cultivars*

Inoculation of lentil cultivars Giza 9 and Giza 370 with *R. leguminosarum* strains TAL 638 and Nit. 92A3 in the Nile Delta alluvial soil induced significant increases in seed yield for both cultivars under regular or frequent irrigation regimes (Table 5). However, no significant differences in seed yield were recorded between different inoculant strains, indicating the similarity between strains and their mixture in nitrogen fixing capacity. Previous research on lentil has shown a growth response to rhizobial inoculation (Taha *et al.* 1967; Islam and Afandi 1980; Mukhtar and Noural 1985). In our study, a significant response to nitrogen fertilization was also recorded compared to the control. However, no significant differences were observed between the nitrogen fertilized and the inoculated treatments.

Significant differences in seed yield were observed with each of the inoculant strains and their mixture, and it appears that lentil cultivar Giza 370 is more efficient in nitrogen fixation with any of the inoculant strains than cultivar Giza 9. The differences between cultivars in their ability to fix nitrogen in symbiosis with *Rhizobium* has been demonstrated in a number of legumes (Gibson 1968; Hardy *et al.* 1973; Graham 1981). In this work there was no strain/cultivar interaction, which may be due to the limited numbers of the cultivars and strains used.

Our results show that the seed yields of both cultivars were significantly affected by watering regime. Frequent irrigation resulted in higher seed yield compared with regular irrigation. Sprent (1976) and Abdel-Ghaffer *et al.* (1981) reported that water stress depressed nitrogen fixation and seed yield of grain legumes. The results also show that the irrigation/strain or irrigation/cultivar interactions were not significant, indicating that strains and cultivars were similar in their sensitivity to water stress.

The competition pattern of strain TAL 638 and Nit. 92A3 on two lentil cultivars in soil with native *R. leguminosarum* is given in Table 6. The

Table 5. Seed yields (kg/ha) of two field grown lentil cultivars (Giza 9 and Giza 370) with rhizobial inoculation and two irrigation regimes.

Treatments	*Regular irrigation*		*Frequent irrigation*	
	Giza 9	Giza 370	Giza 9	Giza 370
Uninoculated	2220	2580	2760	3050
Fertilized	3150	3150	3300	3640
TAL 638	3100	3100	3290	3550
Nit. 92A3	3080	3360	3740	3990
TAL 638 + Nit. 92A3	2920	3280	3460	3610
LSD		5%	1%	
between treatments		307	447	
between cultivars		157	223	
between irrigation regimes		264	357	

Table 6. Percent recovery of introduced and native strains of *R. leguminosarum* from nodules of two lentil cultivars (Giza 9 and Giza 370) in clay soil.

Strain in *inocula*	Strains recovered from nocules					
	Giza 9			Giza 370		
	Tal638	*Nit.92A3*	*Unidentified*	*TAL638*	*Nit.92A3*	*Unidentified*
		Regular irrigation				
Uninoculated	12 (0) *	4 (0)	84	8 (0)	12 (0)	80
TAL 638	80 (0)	0 (0)	20	72 (0)	0 (0)	28
Nit. 92A3	0(0)	92 (0)	8	0 (0)	80 (0)	20
TAL 638 + Nit. 92A3	32 (16)	28 (16)	24	32 (28)	24 (28)	16
		Frequent irrigation				
Uninoculated	0 (0)	8 (0)	92	16 (0)	4 (0)	80
TAL 638	76 (0)	0 (0)	24	80 (0)	0 (0)	20
Nit. 92A3	0 (0)	80 (0)	20	0 (0)	84 (0)	16
Tal 638 + Nit. 92A3	40 (20)	20 (20)	20	40 (36)	16 (36)	8

* Figures in parentheses are percentages of doubly infected nodules.

inoculant strains outcompeted the native rhizobia and occupied 72-92% of nodules formed on the two cultivars inoculated with single strain inocula. With double-strain inocula, strain TAL 638 formed 48% of the total nodules on cultivar Giza 9 and 60% on cultivar Giza 370 under the regular irrigation regime. Strain Nit. 92A3 occupied 44% of Giza 9 nodules and 52% of Giza 370 for the same irrigation regime. The same competition pattern was observed with frequent irrigation. These results show that both strains were equally competitive in this experiment. The native population of *Rhizobium* sp. occupied 8-84% of the total nodules, depending on the inoculation treatments. The nodules from the control plot did not cross-react with fluorescent antibodies specific for TAL 638 or Nit. 92A3. May and Bohlool (1982) found that the pattern of competition of inoculant strains in lentil depended on the host variety. In the present work, however, there was no marked difference in competition between strains on the two lentil cultivars. However, both of the strains showed high competition with the native strains.

The incidence of double strain occupancy in nodules was studied by dual examination of the same microscope fields using fluorescent and phase contrast microscopy (Lindermann *et al.* 1974). Doubly-infected nodules were 0-36% of the total nodules. May and Bohlool (1982) reported the incidence of double strain occupancy in nodules of field grown lentil to be 0-38%, depending on the strain.

These data show the importance of considering the competitive ability of the strain in selecting rhizobia. In this study, strains TAL 638 and Nit. 92A3 were effective nitrogen fixers and they out-competed the native strains (Table 6). This resulted in significant increases in seed yield for both lentil cultivars (Table 5). Moreover, by adding extra water cycles the improvement of soil-

Table 7. Response of faba bean to inoculation with three strains of *R. leguminosarum*.

Treatments	Nodule mass (g/plant)	Plant dry weight (g)	Nitrogen content (mg/plant)
Uninoculated	0.33	8.8	195
TAL 634	0.40	8.9	200
NRC 65	0.46	12.4	272
TAL 1400	1.73	16.5	413
TAL 634+NRC 65	0.48	8.6	194
NRC 65+TAL 1400	1.58	16.0	392
TAL 634+TAL 1400	1.41	13.7	363
NRC 65+TAL 634+TAL 1400	1.65	15.3	384
LSD (5%)	0.14	0.7	17
(1%)	0.19	1.0	23

water-plant relationships further strengthened plant growth and nitrogen fixation.

Response of faba bean to inoculation

The response of faba beans to inoculation with three strains of *R. leguminosarum* in fertile Nile Delta soil is given in Table 7. One of the strains, TAL 1400, is the strain recommended by NifTAL for INLIT trials (International Network of Legume Inoculation Trials) for faba bean; another strain, TAL 634, is also recommended for lentil inoculation; and the third strain, NRC 65, is a local strain isolated from lentil grown in Egyptian soils.

There was a highly significant response to inoculation with strain TAL 1400 and significant response with local isolate NRC 65. There was no significant response in nodulation or plant dry weight with TAL 634. This shows the wide differences in nitrogen fixing capacity of strains of some species of *Rhizobium*. In this particular experiment strain TAL 634, originally isolated from *Lathyrus hirsutus* (Halliday and Somasegaran 1984), was ineffective in nodulating faba beans. The results in Table 7 also show that the presence of strain TAL 1400 in the inoculant mixtures resulted in significant nodulation and dry matter accumulation increases, suggesting a higher competitive ability in this highly effective strain over the two other strains.

References

Abdel-Ghaffar, A.S., El-Attar, H.A., El-Halafawi, M.H., and Abdel-Salam, A.A. 1981. Effect of inoculation, nitrogen fertilizer, salinity and water stress on symbiotic N_2 fixation by *Vicia faba* and *Phaseolus vulgaris*. Pages 153-159 *in* Biological Nitrogen Fixation Technology for Tropical Agriculture (Graham, P.H. and Harris, S.C., eds.). CIAT, Cali, Colombia.

Abel, G.H. and Erdman, L.W. 1964. Response of Lee soybeans to different strains of *Rhizobium japonicum*. Agronomy Journal 56: 423-424.

Burton, J.C. 1976. Methods of inoculating seeds and their effect on survival of rhizobia. Pages 175-190 *in* Symbiotic Nitrogen Fixation in Plants (Nutman P.S., ed.). Cambridge University Press, Cambridge, England.

Dart, P.J. and Krantz, B.A. 1977. Legumes in the semi-arid tropics. Pages 119-154 *in* Exploiting the Legume-*Rhizobium* Symbiosis in Tropical Agriculture (Vincent, J.M., Whitney, A.S. and Bose, J., eds.). Miscellaneous Publication 145. University of Hawaii, College of Tropical Agriculture.

FAO (Food and Agriculture Organisation). 1984. Legume inoculants and their use. NifTAL Project, USA and FAO, Rome, Italy.

Gibson, A.H. 1968. Nodulation failure of *Trifolium subterraneum* L.C.V. Woogenellup (syn. Marrar). Australian Journal of Agricultural Research 19: 907-918.

Graham, P.H. 1981. Some problems of nodulation and symbiotic nitrogen fixation in *Phaseolus vulgaris* L. A review. Field Crops Research 4: 93-112.

Hamdi, Y.A. 1982. Symbiotic nitrogen fixation in faba beans. Pages 127-138 *in* Faba Bean Improvement: Proceedings of the Faba Bean Conference (Hawtin, G. and Webb, C., eds.), ICARDA/IFAD Nile Valley Project, 7-11 March 1981, Cairo, Egypt. Martinus Nijhoff Publishers, The Hague, The Netherlands.

Hardy, R.W.F., Burns, R.C. and Hostern, R.D. 1973. Applications of the acetylene-ethylene assay for measurement of nitrogen fixation. Soil Biology and Biochemistry 5: 47-82.

Halliday, J. and Somasegaran, P. 1984. The *Rhizobium* germplasm resource at NifTAL. Catalogue of Strains. University of Hawaii NifTAL Project and Mircen, USA.

Islam, R. and Afandi, F. 1980. Response of lentil cultivars to *Rhizobium* inoculation and nitrogen fertilization. LENS 7: 50-51.

Jackson, M.L. 1958. Soil Chemical Analysis. Prentice Hall Inc., Englewood Cliffs, New Jersey, USA.

Kosslak, R., Bohlool, B.B., Dowdle, S. and Sadowsky, M.J. 1983. Competition of *Rhizobium japonicum* strains in early stages of soybean nodulation. Applied and Environmental Microbiology 46: 870-873.

Lindemann, W.C., Schmidt, E.L., and Ham, G.E. 1974. Evidence for double infection within soybean nodules. Soil Science 118: 274-279.

May, S.N. and Bohlool, B.B. 1982. Competition among *Rhizobium leguminosarum* strains for nodulation of lentil (*Lens esculenta*). Applied and Environmental Microbiology 45: 960-965.

Moawad, H., Ellis, W.R., and Schmidt, E.L. 1984. Rhizosphere response as a factor in competition among three serogroups of indigenous *Rhizobium japonicum* for nodulation of field-grown soybeans. Applied and Environmental Microbiology 47: 607-612.

Mukhtar, N.O. and Noural, A.H. 1985. Lentil inoculation and fertility trial. ICARDA/IFAD Nile Valley Project on Faba Bean Improvement. Sixth Annual Coordination Meeting, Cairo, 9-13 September 1985. ICARDA, Aleppo, Syria.

Nambiar, P.T.C., Dart, P.J., Srinivasa Rao, B. and Ravishankar, H.N. 1981. Response of groundnut (*Arachis hypogaea*) to inoculation. Pages 241-248 *in* Biological Nitrogen Fixation Technology for Tropical Agriculture (Graham, P.H. and Harris, S.C., eds.). CIAT, Cali, Colombia.

Nassib, A.M., Cassman, K.G., Hassan, M.Z., El-Sherbeeny, M.H. and Abdalla, S.T. 1983. Irrigated soybean production in Egypt: past, present and future trends. Research report. EMCIP Publication 69: 1-14. Ministry of Agriculture, Cairo, Egypt.

Roughley, R.J. 1981. The storage, quality control, and use of legume seed inoculants. Pages 115-126 *in* Biological Nitrogen Fixation Technology for Tropical Agriculture (Graham, P.H. and Harris, S.C., eds.). CIAT, Cali, Colombia.

Sprent, J.I. 1976. Nitrogen fixation by legumes subjected to water and light stresses. Pages 405-420 *in* Symbiotic Nitrogen Fixation in Plants. IBP 7. (Nutman, P.S., ed.). Cambridge University Press, Cambridge, England.

Strijdom, B.W. and Deschadt, C.C. 1976. Carriers of rhizobia and the effects of prior treatment on the survival of rhizobia. Pages 151-168 *in* Symbiotic Nitrogen Fixation in Plants. IBP 7. (Nutman, P.S., ed.). Cambridge University Press, Cambridge, England.

Taha, S.M., Mahmoud, S.A.Z. and Salem, S.H. 1967. Effect of inoculation with rhizobia on some leguminous plants in the United Arab Republic. II. Nitrogen fertilization. Journal of Microbiology of the United Arab Republic 2: 31-41.

West, G. 1971. Nitrogen increases in a non-nodulating soybean genotype grown with nodulating genotypes. Agronomy Journal 63: 356-359.

Zayed, M.N., Taha, S.M., Saber, M. and Badr El-Din, S.M.S. 1973. Nitrogen transformation in soils. I. In sandy soil under barley and peanut in a two years rotation. Zbl. Bakt. II. 116-125..lm 10;

Discussion

Roughley

Would you please describe your carrier material used in the peat cultures and what were the numbers of rhizobia per gram of carrier? How was the liquid inoculum prepared and applied? What was the number of rhizobia applied by both inoculation methods in terms of number of rhizobia per m^2?

Moawad

The carrier we used was prepared in our laboratory. It consists of a mixture of Irish peat + peanut husks + $CaCO_3$ and charcoal. The carrier was able to support good growth of rhizobia; the number of rhizobia was 10^8 cells/g. The liquid inoculum was prepared in the laboratory on mannitol-sucrose medium. The amount of rhizobia added by this method was 10 times as much as the peat-based carrier. Each hectare received 595.25 ml of 10^9-10^{10} cells/ml rhizobia. The injection of the inocula was done with water during the night to prevent the direct harmful effect of the sun.

Danso

How did you determine nitrogen fixed in soybean? Assume you had 6% nitrogen in your seed, then the 3129 kg/ha seed yield, for example, would give you a total nitrogen of approximately 140.5 kg/ha. Can you please explain this?

Moawad

We measured the nitrogen fixed in soybean by the difference method. The calculation of the nitrogen fixed was based on the actual amounts of nitrogen in the crop but not on the hypothetical N%.

Eaglesham

The positive response to TAL 1400 in comparison to the failure of the other two strains is interesting. Have you checked the effectiveness of the three strains in the greenhouse?

Moawad

No we have not, since this is a current experiment which has not been harvested yet. We plan to follow this up next season.

Mytton

What genetic markers were used for strain identification?

Moawad

We did not use any genetic markers. The nodules were serotyped using immunofluorescence to identify the inoculant strains.

Saxena

Your results with faba beans are very interesting. Could you tell us what is the source of the strains TAL 634 and TAL 1400? I also wonder why you did not use the local *Rhizobium* because the faba beans in the traditional production areas in Egypt are well nodulated and capable of producing large yields.

Moawad

The source of the strains was the NifTAL project in Hawaii. Strain 1400 was isolated from *Vicia faba* and strain 634 from *Lathyrus*. We have isolated a large collection of *R. leguminosarum* from Egypt. These strains will be screened for nitrogen fixing capacity.

D.P. Beck and L.A. Materon (editors), Nitrogen Fixation by Legumes in Mediterranean Agriculture
 ISBN 90 247 3624 2

Effect of *Rhizobium* Inoculation, Lime and Addition of Molybdenum on the Yield of *Vicia faba*

FATIHA BRHADA

Universite Mohammed V, Rabat, Morocco

Summary. A study was carried out to determine the cause of low faba bean yields in the central region of Morocco. *Rhizobium leguminosarum* was shown to be present in the soil of Sidi-Bettache area and 18 isolates were found to be infective and effective. In an attempt to increase yield, faba bean variety F269 was inoculated with strains of *R. leguminosarum* and treated with molybdenum and lime. Addition of molybdenum increased yield compared to the control, while inoculation alone did not increase yield. However, inoculation with the addition of molybdenum and lime increased yield. The indigenous *R. leguminosarum* was highly competitive as few nodules were produced by the introduced strain.

Introduction

In Morocco legumes are grown on less than 10% of the cultivated land. Faba bean, the predominant legume, is grown on 40–50% of the land under legumes and comprises 50–60% of total legume production. Average annual yield for faba beans is 900 kg/ha compared to 500–600 kg/ha for peas, chickpeas, and lentils (Cottet 1981). However, yields are low and there has been no increase during the past few years. The Ministry of Agriculture of Morocco now has a policy for intensification, to which the improvement of nitrogen fixation by faba bean could contribute.

Objectives

Our objective was to find out the reasons for the low yield of faba beans in Sidi-Bettache, where climatic conditions are satisfactory. Several hypotheses have been advanced for the low yields, such as absence of rhizobia, presence of non-infective rhizobia or infective but inefficient rhizobia, lack of soil micronutrients or toxicity of some soil elements, and extremes of pH.

Region of Study

The work described here was carried out in the Zaers region of central Morocco from 1982 to 1984. This is a semi-arid region with 350–400 mm annual rainfall. Cereals are the staple food and legumes are grown on a small scale although yields are variable: they are good in Merchouch area where the predominant soils are clayey (Ghanem 1978) and average in the Ezzhiliga area where clay/sand soils are predominant. However, in Sidi-Bettache area where the soils are muddy, less than 200 kg/ha was obtained and farmers were forced to abandon cultivation (Ghanem 1978).

Indigenous *Rhizobium* Strains

Nodulation occurred on plants grown in pots containing unsterilized soil indicating that *Rhizobium leguminosarum* exists in the soil of Sidi-Bettache. There were 9.5×10^4 cells of *R. leguminosarum* per gram of dry soil which was considered satisfactory for nodulation. Amarger (1980) suggested that 10^6 cells of *Rhizobium* gives good nodulation.

Eighteen strains were isolated from this soil and 5 ml of a 3-day old culture of each strain were inoculated on four faba bean plants with four replicates in greenhouse pot culture. A study of their infectivity and nitrogen fixation capacity showed that they were infective and efficient (Table 1).

Table 1. Infectivity and efficiency of *Rhizobium* strains isolated from Sidi-Bettache soil (mean of four replicates).

Strain	Number of nodules	Dry weight of nodules (g)	Dry weight of tops (g)	Relative efficiency *
RLF22	195.25	0.163	4.779	59.92
RLF23	131.25	0.175	4.102	51.43
RLF24	245.75	0.159	3.822	47.92
RLF25	122.25	0.159	4.581	57.43
RLF26	260.50	0.163	3.650	45.76
RLF27	279.75	0.155	4.390	55.04
RLF28	330.75	0.227	4.462	55.94
RLF29	200.50	0.168	3.900	48.90
RLF30	326.50	0.242	4.479	56.15
RLF31	201.25	0.184	5.009	62.80
RLF32	31.50	0.07	4.232	53.06
RLF33	180.25	0.187	4.321	54.17
RLF34	39.50	0.069	4.030	50.53
RLF35	250.75	0.218	4.714	59.10
RLF36	328.00	0.182	4.556	57.12
RLF37	223.25	0.166	4.016	50.33
RLF38	177.75	0.178	4.644	58.22
RLF39	188.50	0.186	4.361	54.68
Control	0	0	3.469	43.59
Nitrogencontrol	0	0	7.976	100

$$* \text{ Relative efficiency} = \frac{\text{Dry weight of tops of inoculated plants}}{\text{Dry weight of plant tops of nitrogen control}} \times 100$$

Field Assay Treatment

A field assay was carried out at Sidi-Bettache, using faba bean variety F269, to study the effect on yield of soil pH and inoculation with *R. leguminosarum*.

For each treatment the soil was fertilized with 60 units P_2O_5/ha, 40 units K_2O/ha, and 20 units N/ha, applied as ammonium sulphate. As the soil at Sidi-Bettache is slightly acid (pH 5.4), it was thought that there may be a lack of molybdenum, which is not readily available in acid soils (Mulder *et al.* 1969). Therefore molybdenum was applied in some of the treatments.

To determine if soil acidity was the cause of low yield the following treatments were applied:

T_1 control, without additional nitrogen,
T_2 nitrogen control (total 100 units),
T_3 lime (400 kg $CaCO_3$/ha) (Finck 1979),
T_4 molybdenum (2.2 kg $Na_2MoO_4.2H_2O$/ha) (Tisdale and Nelson 1975), and
T_5 molybdenum + lime.

There were four other treatments, all of which included inoculation at a rate of 10^7 cells/seed with an infective and efficient strain of *R. leguminosarum* (RLF_2) from the Merchouch area:

T_6 inoculation,
T_7 inoculation + lime,
T_8 inoculation + molybdenum, and
T_9 inoculation + molybdenum + lime.

In the inoculated and molybdenum treatments, a coating method was used, with gum arabic as adhesive (Marrett 1969). For each treatment, there were four replicates distributed within four blocks. Plots were 2 m long and 4 rows wide with 50 cm between rows.

Results of the Field Assay

At the 50% blossoming stage, the percentage of nodules induced by strain RLF_2 was determined using the immunofluorescence technique. For all of the inoculated treatments this percentage was low (Table 2), probably because of the poor competitive capacity of RLF_2 against indigenous strains or its poor capacity to survive in this soil. At the same time, total number and dry weight of nodules and the dry weight of the aerial plant parts were determined and compared with the values obtained for the control (Table 2). Only two of the treatments (lime only and inoculation only) did not increase seed yield. The other treatments gave a significant increase in yield (Table 2), the greatest

Table 2. Symbiotic response and relative nodulation success of strain RLF2 in faba beans *.

Treatments	Number of nodules	Dry weight of nodules (g)	Percentage nodules formed by RLF2	Dry weight of tops (g) **	Yield (kg/ha)
T_1	677.5	1.141	0	77.09	1347
T_2	509.75	0.818	0	73.15	1896
T_3	521.25	0.930	0	51.44	1337
T_4	545.75	0.742	0	77.90	1832
T_5	335.00	0.490	0	59.82	1498
T_6	397.50	0.343	14.06	64.07	1308
T_7	602.00	0.786	12.5	62.99	1742
T_8	376.25	0.753	6.25	71.33	1699
T_9	288.75	0.577	15.62	75.10	1522

* Means of four replicates
** Weight of 10 plants.

increase being in treatment 2 (100 units of N), followed by treatment 4 (addition of molybdenum).

Conclusion

This work confirmed that current faba bean yields in the region studied (200 kg/ha) could be improved through adequate fertilization and periodic weed control. The yield of our control plot (1300 kg/ha) is evidence of that fact. Addition of molybdenum can also increase yield, probably by alleviating constraints to nitrogen fixation with indigenous *Rhizobium* strains.

Indigenous *R. leguminosarum* strains seem to be very competitive with the inoculated strain as shown by the low percentage of nodules induced by the strain used in our experiments (RLF_2). This high competitiveness may be due to the adaptation of these strains to the acid soil conditions.

References

Amarger, N. 1980. Aspect microbiologique de la culture des legumineuses. La Selectionneur Francais 28: 61-66.

Cottet, C. 1981. Caracteres succints de la production des legumineuses a graines au Maroc. Ecole National de l'Agriculture, Meknes, Morocco.

Finck, A. 1979. Pages 182-185 *in* Dunger and Dungung Verlay Chemie Weinhein, New York.

Ghanem, H. 1978. Contribution a l'etude des sols du Moroc. Genese, classification et repartition des sols des regions des Zaers et de ses bordures. Ph.D. thesis. GAND University, Morocco.

Marrett, P.L. 1969. Legume inoculation. Journal of Agriculture, South Africa 72: 232-237.

Mulder , E.G., Lie, T.A., and Woldendorp, J.W. 1969. Fixation symbiotique d'azote *in* Biologie des Sols. Compte rendu de recherche de l'UNESCO.

Saito, S.M.T. 1982. Avaliacao em campo da capacidade de fixacar simbiotica de estripes de *Rhizobium phaseoli* Pesq. Agropec. Bras., Brasilia, 17, 7, 999–1006.
Tisdale, S.L. and Nelson, W.L. 1975. Molybdenum *in* Soil Fertility and Fertilizers, Third Edition. Macmillan, New York, USA.

Discussion

Weaver

You listed 27 strains of rhizobia in your first table. Do you know if they were different strains or simply different isolates?

Brhada

They were 18 strains of rhizobia from Sidi-Bettache soils. They were different strains as shown by differences in the effects of pH and temperature and by the immunofluorescence technique.

D.P. Beck and L.A. Materon (editors), Nitrogen Fixation by Legumes in Mediterranean Agriculture
 ISBN 90 247 3624 2

Seed Inoculation and Inoculant Technology

J.C. CLEYET-MAREL

INRA Laboratoire de Recherches sur les Symbiotes des Racines, Montpellier, France

Summary. Legume inoculation with *Rhizobium* spp. ensures nodulation and efficient nitrogen fixation. However, the need for inoculation must first be evaluated. This paper discusses this evaluation and methods of inoculant production, application of rhizobia to seed, production of pelleted seed, and the response to inoculation.

Introduction

The soil-enriching quality of leguminous crops was known during the time of the Roman Empire when rotational cropping systems were introduced. Perhaps they used a crude method of inoculation by spreading trash or soil from the field where the same crop had previously been grown.

With the isolation of the causal organism, *Rhizobium*, by Beijerinck in the late 19th century, microbiologists and agronomists began to devise more efficient ways of achieving inoculation of seeded legumes.

The requirements for a successful symbiosis are definable and outwardly simple. They are:

* a reasonably healthy host legume, capable of synthesizing enough carbohydrate to satisfy its own and the bacterial energy requirements;
* a strain of *Rhizobium* compatible with the host, effective in assimilating atmospheric nitrogen; and
* an environment conducive to the vigorous growth of both organisms, resulting in symbiosis.

Legume inoculation is one way of ensuring that a strain of *Rhizobium* appropriate for the cultivar being seeded is present in the soil at the proper time and in sufficient numbers to assure a quick and effective infection and efficient subsequent nitrogen fixation.

Evaluation of the Need for Inoculation

The presence of rhizobia in the field can be detected easily by merely looking at the roots of legumes growing at the site. The presence of nodules is the first evidence of bacterial presence and the position and size of the nodules are an indication of the presence, number, and effectiveness of the bacteria.

Many large nodules on the main root, close to the crown, indicate abundant bacteria, while small nodules scattered widely on the secondary or finer roots indicate late nodulation, often delayed because of low bacterial populations. These norms should not be interpreted too rigidly and there are many significant exceptions.

The physical appearance of the host plant is also a good indicator of the presence of effective bacteria. If the legume is robust and green and the roots are nodulated, one may assume there is an adequate population of efficient rhizobia. However, if the plant is pale green and not vigorous but the roots are well nodulated, the possibilities are that (1) the rhizobia that infected the root are ineffective or (2) that some other environmental factor is limiting growth. If the legume is not vigorous and the roots are not nodulated, it may be that infective rhizobia are not present in adequate numbers or at a time appropriate to cause nodulation. In either case, further investigation is necessary, but inoculation should increase plant growth unless a factor such as pH is limiting the crop, the bacteria, or both. Ability to distinguish between nodulation due to introduced bacteria and that due to naturally-occurring rhizobia is essential.

There are several tests such as phage-typing, spontaneous antibiotic resistance, and serology, which are extremely specific and are excellent for establishing the identity of a particular strain. Immunofluorescence has been successfully used to show which strain occupies a nodule (Trinick 1969) and is a basis for quantitative determination of specific rhizobia in the soil (Schmidt 1974). It is now possible to readily identify the occupants of a large number of nodules by means of an enzyme-linked immunosorbent assay (ELISA) (Olsen and Rice 1984). With these two serological techniques there is no need to purify rhizobia from nodules and make subcultures as for phage-typing and spontaneous resistance to an antibiotic.

How to Produce Inoculum

Inoculation is usually accomplished by treating the seed or the soil with a rhizobial suspension or a mixture of bacteria and a carrier, called an inoculum. Several factors lead to the choice of rhizobial culture and carrier.

The rhizobia may be used as a suspension of a pure culture grown on agar or in a liquid medium. Alternatively, the rhizobia may be intimately mixed with a carrier. For instance, agar cultures are less troublesome to prepare in small quantities, whereas for a large volume, liquid cultures are more suitable. These two forms are useful when the inoculum is to be used immediately and near to the place of production. For distribution over long distances and for long-term storage, peat-based inocula are especially convenient.

Agar cultures are convenient where demand is low, particularly if this is combined with a need for diverse and more specialized cultures. The yield of a well-grown agar culture is about 10^{10} rhizobia per 10 ml agar slope (Vincent

1970). This is sufficient to provide inoculum for preliminary field strain selection. When there is good liaison between the producing laboratory and researchers in the field, and where a seed inoculation program is sufficiently intensive, liquid cultures are useful (Van Scheven 1958; Lie pers. comm.).

In industrial countries, industry provides the farmer with the *Rhizobium* strains specific to the leguminous crop he wishes to grow, whereas in developing countries where there is a great need for legume inoculants, numerous problems are associated with inoculant production. A frequent problem is the availability of a suitably-sized and low-cost fermenter. For instance, in Vietnam, this problem was solved by using a local adaptation of the rotary shaker. In this way it is possible to produce 100 litres (100×1 litres) of inoculum per week. This system has the advantage of permitting individual inspection of each batch of inoculant with less chance of contamination. However it is time consuming.

Scientists from the NifTAL Project propose an inoculant production system based on a stainless steel pressure vessel which can be sterilised with a small gas or propane burner. An inexpensive adaptation of this system is proposed by Lie and Houwers (pers. comm.).

During growth in broth medium, contamination must be avoided. A few fast-growing contaminants present early in the process can suppress growth of rhizobia. This can occur with fast-growing rhizobia but the slow growers are more susceptible. Several legumes, such as *Glycine max* and *Cicer arietinum* can be nodulated by both fast- and slow-growing strains, so when we have the choice between fast- and slow-growing strains, the fast one will reduce the time needed to obtain sufficient growth and the risks associated with an occasional contaminant.

Nowadays, most common inocula consist of a moist suspension of bacteria in some sterile carrier. Peat is commonly used but materials such as bagasse, coconut waste, and rice hulls have been used. A remarkable survey of growth and survival of *Rhizobium* in peat culture has been published by Roughley and Vincent (1967). Most peat is acidic and must be neutralized (pH 6.5-7.0) by the addition of $CaCO_3$. Generally, slight differences in quality between inoculants produced with autoclaved and irradiated peat should have no practical drawbacks (Roughley and Vincent 1967; Somasegaran 1985).

In the usual method of peat-based inoculant production, rhizobia are multiplied in fermenters and the bacterial culture added to the peat. In the pure culture or aseptic technique, a definite volume (30-50% of the total final weight) of a fully grown culture (10^9 cells per ml) is added aseptically to a relatively dry peat sample (8-10% moisture) which has been presterilized in polyethylene bags by gamma irradiation. As the peat is presterilized, the added rhizobia multiply further, but minimal to no multiplication may take place if the peat is not sterilized.

Recently, Somasegaran (1985) proposed the use of steam-sterilized peat and diluted liquid cultures of *Rhizobium* spp. He tested the influence of various diluents on the subsequent multiplication and long-term survival of *R. phaseoli* and *R. japonicum*, and found that sterile, deionized water is not significantly different from the other nutrient-enriched diluents in supporting multiplication and survival of both species. Peat from diverse sources is compatible with diluted liquid cultures of *Rhizobium* spp. (Somasegaran 1985). The use of diluted cultures and presterilized peat (autoclaved in polypropylene bags) may be a cheaper option for small to medium scale inoculant production compared with the use of undiluted cultures and gamma irradiated peat.

Dommergues *et al.* (1979) entrapped rather than adsorbed *Rhizobium* cells by incorporating the bacteria in a polymeric gel. The first experiments using polyacrylamide as the entrapping agent showed that the resulting inoculant (PER) compared favorably with peat-based inoculant, provided PER was maintained at a suitable moisture content. Polyacrylamide can be replaced by other polymers, and xantham significantly outranked polyacrylamide. Nevertheless, this inoculant has not been used extensively.

Inoculation and Application of *Rhizobium*

Agronomic utilization of inoculants requires a production technology capable of processing rhizobial strains into inoculants with a high viable cell density. Current methods involve some form of cultured suspension of bacteria applied to the seed just before planting. The number of viable rhizobia that should be applied to the seed varies with the size of the seed. In France, recommendations are 5000 rhizobia per seed for lucerne and 10^6 per seed for soybean. For lucerne, inoculation is carried out at sowing by mixing seeds (20 kg/ha) with 200 g of peat inoculant containing at least 2.5×10^8 rhizobia per gram. For soybean, 400 g of peat inoculant containing 1×10^9 rhizobia per gram are mixed with 80 kg of seeds. Seeds should be inoculated just before planting and no more than 4-6 hours should elapse before inoculated seed is used. The seed should be visibly moist but should not clump (for small seed) or drip (for large seed). A sticking agent, such as gum arabic or molasses, causes the inoculum carrier to stick better to the seeds and each seed should have some black specks on it, indicating the presence of the carrier. Direct sunlight on inoculated seeds must be avoided because of the sterilizing effect of the ultraviolet rays which are lethal to the rhizobia.

Treatment of seed with biologically active materials such as fungicides, insecticides, and herbicides could destroy *Rhizobium* and must be avoided when their use is not essential. Various authors have suggested that the inoculant for chemically-protected seeds should be added directly to the soil before planting or at the time of planting. However, broadcasting the mixture over the soil surface and then covering it lightly by cultivation necessitates two special field operations. This is not feasible in a hot, dry climate, since it is likely to result in rapid drying of the scattered mixture and a high dispersion and mortality of bacteria before they are incorporated into the moist soil.

Application of a rhizobial suspension to the planting furrows is more promising. The diluted suspension of rhizobia is introduced into a tank and dropped through tubes which end in seed furrow openers approximately 4 cm above the bottom of the planting furrow. An adaptation for dry areas is the placement of inoculum 12 cm below the planting furrow to prevent eventual drying of the inoculum between two consecutive irrigations. At the same time, seeds are placed at the usual depth of 4-5 cm (Schiffmann and Alper 1968). Unfortunately, for this inoculant application we need special equipment which is not available everywhere. Granular inocula (peat or polymer-entrapped *Rhizobium*) for direct application to the soil at seeding time have recently come onto the market, and seeding machines equipped for fertilizer application at sowing can be used for application of granular inocula. For experimental field sowing of inoculated legume seed, a hand-operated aseptic seeding device was proposed by Ireland (1969).

Pelleted Seed

Many attempts have been made to develop a successful method of pre-inoculating seed. With this method, the inoculum is protected from the toxic effects of acid soil and the appropriate rhizobia are added to the seed at the factory prior to distribution (Loneragan *et al.* 1955; Brockwell 1962; Mackinnon *et al.* 1977). Enclosing the seed in a coating of lime or phosphate, along with an inoculum, protects the bacteria and the seed from adverse conditions in the soil surrounding the seed.

Because low pH has often been implicated in legume and inoculation failures under temperate conditions, limestone has been used to form the pellet. However, lime itself has an alkaline reaction that is lethal for rhizobia. The fact that the alkalinity of lime varies from source to source may explain some of the conflicting results that have been published. Therefore, it may be reasonable to consider rock phosphate unless a definite gain from limestone can be demonstrated. Pelleting is a good alternative to broadcast lime or rock phosphate only when inoculation and pelleting techniques are satisfactory (Lowher 1975).

Response to Inoculation

Population densities in soil and rhizospheres

If inoculation has been done properly and the seed sown promptly, the bacteria will begin to multiply shortly after they are placed in the soil. Fluorescent antibody (FA) techniques, which allow the identification and enumeration of specific strains of *Rhizobium* directly in the environment, provide a unique tool for the study of population changes of rhizobia in soil and in the rhizosphere (Reyes and Schmidt 1979).

By applying the FA technique to survival kinetics of *Bradyrhizobium japonicum* strains introduced into soils, we observed that all soil-strain combinations during incubation reached the same survival balance level, generally between 10^3 and 10^4 *B. japonicum* per gram of soil. The equilibrium threshold may correspond to the number of ecological sites available to rhizobia. This number depends on the strain's saprophytic ability (Parker *et al.* 1977), and on physical, chemical, and biological factors, such as prey and predator relations (Alexander 1977).

The strategy whereby the appropriate *Rhizobium* symbiont is selectively favoured by its host root and responds with overwhelming growth in the host rhizosphere may be a prerequisite to successful nodulation. This is illustrated by the occurrence of cells and microcolonies in immunofluorescence microscopic observations of the root systems of plants grown in controlled conditions.

The same factors which influence rhizobia in the soil are likely to be found in the rhizosphere. Direct examination of the population dynamics of a *Rhizobium* strain in developing rhizospheres under natural conditions showed that the stimulation was only slightly, if at all, greater for host than for nonhost plants. Observed stimulation was not at all suggestive of a massive population expansion in the immediate vicinity of the roots (Reyes and Schmidt 1979). However, additional studies are necessary and would provide good model systems for the study of prenodulation events in soil to understand and resolve the inconsistencies of inoculation.

Nodulation Capabilities in the Field

Determining the nodulation capabilities of a strain involves the selection of a site or sites for field work. To evaluate nitrogen fixation, a site should be selected where there are no nodulating strains for the particular legume under study. If the competitiveness and persistence of a test strain are to be tested, a native population of rhizobia is needed.

Nodulation is assessed between 6 and 20 weeks, depending on the legume and the seasonal conditions, by sampling a number of plants in each treatment. Nodule number is a valuable criterion in the first 6-8 weeks of plant development, but it is of very little use thereafter. At that point, the best criteria are nodule dry weight and dry weight of the whole plant.

Conclusions

Some of the basic principles concerning inoculation of legumes are well documented (Brockwell 1962; Vincent 1970; Burton 1976).

If rhizobial strains are not present in the soil, inoculation will be beneficial. If rhizobial strains are present and infection occurs, but plant yields show parasitic nodulation, then inoculation with effective strains should be beneficial.

Effectivity still represents the fundamental criterion for selecting strains. However, it is now considered only part of an overall selection program. Current thought establishes three criteria as basic to a strain selection program:

(1) efficiency in nitrogen fixation,
(2) competitive ability in the presence of a native rhizobial population, and
(3) ability to survive in various environmental conditions.

The latter two points are extremely important. If a strain fails to compete successfully with the native rhizobial population for infection sites and does not survive environmental conditions where it is introduced, inoculation is unsuccessful.

However, it must be kept in mind that legume inoculation is not a panacea. It represents only one aspect of a total management program. Where inoculum is inexpensive relative to the expected gain, inoculation is often practiced as a form of "insurance". In tropical or subtropical conditions, the result may not be success or failure, but modest gains in infectivity of an introduced strain or efficiency in fixing a greater quantity of nitrogen. The final decision on the value of inoculation at any given site must be an economic one.

References

Alexander, M. 1977. Ecology of nitrogen fixing organisms. Pages 99-114 *in* Biological Nitrogen Fixation in Farming Systems of the Tropics (Ayanaba, A. and Dart, P., eds.). Wiley Interscience, New York, USA.

Brockwell, J. 1962. Studies on seed pelleting as an aid to legume seed inoculation. I. Coating materials, adhesives, and methods of inoculation. Australian Journal of Agricultural Research 13: 638-649.

Burton, J.C. 1976. Methods of inoculating seeds and their effect on survival of rhizobia. Pages 175-189 *in* Symbiotic Nitrogen Fixation in Plants. IBP Vol. 7 (Nutman, P.S., ed.). Cambridge University Press, Cambridge, England.

Dommergues, Y.R., Diem, H.G. and Divies, C. 1979. Polyacrylamide entrapped *Rhizobium* as an inoculant for legumes. Applied and Environmental Microbiology 37: 779-781.

Ireland, J.A. 1969. A hand operated seeding device for experimental field sowing of inoculated legume seeds. Australian Journal of Experimental Agriculture and Animal Husbandry 9: 209-212.

Loneragan, J.F., Meyer, D., Fawcett, R.G. and Anderson, A.J. 1955. Lime pelleted clover seeds for nodulation in acid soils. Journal of the Australian Institute of Agricultural Science 21: 264-265.

Lowher, W.L. 1975. Pelleting materials for oversown clover. New Zealand Journal of Experimental Agriculture 3: 121-125.

Mackinnon, P.A., Robertson, J.G., Scott, D.J. and Hale, C.N. 1977. Legume inoculant usage in New Zealand. New Zealand Journal of Experimental Agriculture 5: 35-39.

Olsen, P.E. and Rice, W.A. 1984. Minimal antigenic characterization of eight *Rhizobium meliloti* strains by indirect enzyme linked immunosorbent assay (ELISA). Canadian Journal of Microbiology 30: 1093-1099.

Parker, C.A., Trinik, M.S. and Chatel, D.L. 1977. Rhizobia as soil and rhizosphere inhabitants. Pages 311-352 *in* A Treatise on Dinitrogen Fixation (Hardy, R.W.F. and Gibson, A.H., eds.). Wiley Interscience, New York, USA.

Roughley, R.J. and Vincent, J.M. 1967. Growth and survival of *Rhizobium* spp. in peat culture. Journal of Applied Microbiology 30: 362-376.

Reyes, V.G. and Schmidt, E.L. 1979. Population densities of *Rhizobium japonicum* strain 123 estimated directly in soil and rhizosphere. Applied and Environmental Microbiology 37: 854-858.

Schiffmann, J. and Alper, Y. 1968. Inoculation of peanuts by application of *Rhizobium* suspension into the planting furrows. Experimental Agriculture 4: 219-226.

Schmidt, E.L. 1974. Quantitative autocological study of microorganisms in soil by immunofluorescence. Soil Science 118: 141-149.

Somasegaran, P. 1985. Inoculant production with diluted liquid cultures of *Rhizobium* spp. and autoclaved peat: evaluation of diluents, *Rhizobium* spp., peats, sterility requirements, storage, and plant effectiveness. Applied and Environmental Microbiology 50: 398-405

Trinick, M.J. 1969. Identification of legume nodule bacteria by the fluorescent antibody reaction. Journal of Applied Bacteriology 32: 181-186.

Van Schreven, D.A. 1958. Methods used in the Netherlands for the production of legume inoculants. Pages 328-333 *in* Nutrition of the Legumes (Hallsworth, ed.). London, England.

Vincent, J.M. 1970. A Manual for the Practical Study of the Root-Nodule Bacteria. IBP Handbook No. 15. Blackwell Scientific Publishers, Oxford, England.

D.P. Beck and L.A. Materon (editors), Nitrogen Fixation by Legumes in Mediterranean Agriculture
 ISBN 90 247 3624 2

Legume Inoculants; Their Technology and Application

R.J. ROUGHLEY

72 Bradleys Road, Terrigal, NSW 2260 Australia

Summary. In this paper, the various stages of inoculant production at which improvements could be made or future research carried out are discussed, with particular reference to broth cultures of rhizobia, inoculant carriers, and application.

Introduction

In a series of experiments spanning some 20 years J.M. Vincent, with various co-workers, put the ecology of rhizobia in soil and legume inoculants and their survival on seed on a quantitative basis. This allowed both a logical explanation for inoculation failures and the setting of goals for inoculum quality and seed inoculation rates. A particularly striking demonstration (Ireland and Vincent 1968) set the problem of competition in soil in terms of numbers of soil rhizobia and the size of the population on seed needed to establish an elite strain. Although later research (Robinson 1969; Franco and Vincent 1976) showed that factors other than simple population dynamics may determine nodule occupancy by a strain, numbers of rhizobia applied to the seed or introduced into the soil compared with the number already in the soil remains the most practical way of predicting the success of inoculation.

Clearly, the inoculum potential of a strain introduced into soil by seed inoculation has two components, the number applied to each seed and the seeding rate. Thus, although high numbers of rhizobia can be applied to large-seeded legumes, fewer seeds are sown per hectare. Numbers of rhizobia in soybean rhizospheres responded to a 100-fold increase over normal inoculation (Herridge *et al.* 1984) illustrating the need for yet higher inoculum potentials.

To produce inoculants with high numbers of rhizobia requires the evaluation of the raw materials and their processing at each separate stage of production. These aspects have been reviewed extensively (Roughley 1970; Date and Roughley 1977; Thompson 1980; Roughley and Pulsford 1982). Therefore, I intend to make only a few observations on what I see as the stages offering room for improvement or future research. Methods of using legume inoculants have also been reviewed recently by Brockwell (1980; 1982) so my treatment of the topic will be selective rather than comprehensive.

Inoculant Production

Broth culture of rhizobia

Perhaps for too long there has been general satisfaction with producing broth cultures containing 5-10 $\times 10^8$ viable cells/ml. This satisfaction may derive in part from studies indicating that an inoculum of 10^3 cells/ml will, after injection and incubation in sterilized peat, produce inoculants containing 10^9 rhizobia/g (Roughley 1968). In Australia, this has led to the dilution of broths with sterile water prior to inoculation of the carrier. Although the bulk of inoculants are currently not produced in sterile carriers, the greater availability of autoclavable films and an increase in irradiation plants in, for example Brazil and Thailand, could change this situation rapidly. The method developed by Somasegaran and Halliday (1982) to dilute broth cultures could then be used, making smaller and cheaper simple fermentors more feasible. Still further dilution would be possible with higher count broth cultures.

Balatti (1982) described media and growing conditions capable of producing *R. japonicum* cultures containing 1.5×10^{10} viable cells/ml. It is surprising that this work has not been verified and, if found generally applicable, not adopted by both research workers and industry.

A 100 l fermentor is capable of producing 80 l of broth containing 10^9 cells/ml using commonly accepted procedures. This is sufficient to prepare 800 $\times$ 250 g packets of inoculum based on the carrier used in Australia. Balatti's procedure would produce 8000 $\times$ 250 g packets when the broth is diluted 10-fold with UV sterilized water. Even greater dilution would be feasible.

Since its inception, the inoculant industry in Australia has modified small containers of 20-50 l capacity to serve as fermentors. These suffer the disadvantage of requiring large autoclaves for sterilization and a long period for cooling. At the other extreme are the excellent facilities of a cooperative in Brazil, Fecotrigo at Cruz Alta, where a battery of 1600 l stainless steel fermentors and associated smaller units to produce starter broth grow sufficient broth to inject 1 million packets in 3 months. K.S.L. Engineering * in Thailand now produces a 100 l stainless steel fermentor, costing approximately 50,000 Baht (in 1986). The fermentor is itself a pressure vessel and may be sterilized directly by steam under pressure passed through the cooling coil or by heating from below with a gas burner. The design includes a cooling coil to reduce prolonged heating of the medium after sterilization and may be inoculated in less than 3 hr from when the gas burner was first lit. It is superior in many respects, including cost, to the impractical FAO – sponsored fermentor.

Cost of the medium is of concern in some countries and has led to the general use of sucrose, even if it is not metabolised by the strain cencerned.

* K.S.L. Engineering Co. Ltd., 275/1 Soil Thavi-Wathana, Sathupradit Rd, Yannava, Bangkok 12, Thailand.

Yeast extract is universally expensive. Cheap alternatives such as malt sprout extract (Boiardi and Ertola 1985) point the way to investigating cheap, local alternatives.

Carriers for Inoculants

Powdered inoculants

Finely-milled peat still remains the standard by which other carriers are judged, but its deficiencies stimulate a continuing search for an alternative carrier. While availability is one reason for such studies, there is a general problem that the variability of peat, even within a deposit, poses an ever-possible hazard. Clearly, the botanical composition of the original flora, the conditions both past and present prevailing in the swamp, the possibility of pollution from drainage, and the accumulation of salt in unusually dry periods, all may affect the suitability of peat. Most researchers have concentrated on high organic matter material such as coals, filter mud, bagasse, coir dust, charcoal, diatomaceous earth, composted maize cobs, and various organic supplements to soil, either pre-composted or not. Clay minerals, talc, and vermiculite have also been used.

It is often instructive to re-examine the original literature on alternative carriers rather than consult reviews. The standards by which they may have been judged at the time might not be acceptable today. For example, bagasse was found by Leiderman (1971) to provide only 7.2×10^6 cells at inoculation and 4.6×10^6 cells/g after 30 days. Does such a population represent the potential of bagasse? We do not know, for example, what the population of contaminants was in this material at 30 days, nor do we know whether the moisture relations were optimal. There still appears to be a lack of realisation that when comparing carriers it is essential to first determine the water potential curve for the material and to adjust the potential of each to that optimal for rhizobia. This could vary according to strain, but as a guide could be taken as between -14 bars and -2.6 bars for unsterile carriers and -2.6 bars and -0.5 bars in sterile carriers (Roughley 1968).

Research into the suitability of different carriers has generally ended with the production of an inoculant with a specified population of rhizobia. Research which ends at this point fails to recognize that the achievement of large populations of rhizobia in an inoculant is not an end in itself but rather a means of introducing rhizobia into soil to nodulate the host. Surprisingly, this appears not to be widely acknowledged. Consequently, while the ability of peat to protect organisms on the seed is well established, this requirement of alternative carriers has largely been ignored.

Granular inoculants

Inoculant granules were prepared by Nitragin Co., USA to solve a nodulation problem in groundnuts growing in hot, sandy soils in Florida. The granules are naturally-occurring in the peat and are separated by sieving milled peat and collecting the 16-40 mesh fraction. The granules are neutralised by adding extra $CaC0^3$ to the broth before mixing with the granules, taking care to keep them intact. The normal sowing rate is 5 kg granules/ha, which costs some 20 times more than conventional inoculants (Burton personal communication).

Non-powder type inoculants

Two new formulations for inoculants have been reported in recent years. One is a new type of inoculant based on entrapping rhizobia in polyacrylamide gel (Dommergues *et al.* 1979), and the other uses oil to improve the survival of freeze-dried rhizobia on seed (Kremer and Peterson 1982a). Polyacrylamide gel was reported to allow improved survival of *R. japonicum*, as with peat-based inoculants, but the data presented show an unexplained, unusually high death rate in sterile peat. The method has not been widely utilised.

McLeod and Roughley (1961) found that lyophilised cultures could provide good nodulation in the field but this is likely to be restricted to favourable conditions, as survival of lyophilised rhizobia on seed is poor (Vincent 1965). Peterson and co-workers appear to have overcome this problem in an investigation into the protective action of peanut oil and soybean oil for rhizobia (Kremer and Peterson 1982a; 1982b). Their claim is that freeze-dried rhizobia suspended in dried oil are protected, particularly against high soil temperatures. Used at the rate necessary to provide 10^8 rhizobia/seed as in their experiments, and given a concentration of rhizobia in oil of 10^{11}/ml (Kremer and Peterson 1983), a volume of 200 ml would be required to inoculate 1 ha of soybeans.

The Application of Inoculants

Seed inoculation

Seed inoculation remains the most widely used method of inoculating legumes and is the almost exclusive method for pasture and forage legumes. Methods vary from the simplest but least effective dusting of the seed with powdered inoculant, through the use of adhesives to stick the inoculant onto seed, to the quite complex pelleting techniques used to pre-inoculate pasture seed.

For some 30 years, there has been a quest for methods to improve survival on seed with the ultimate aim of some to increase the effective inoculum potential. For others, the goal was to market pre-inoculated seed. To these ends, the protective action of maltose was demonstrated by Vincent *et al.* (1962), and the suitability of adhesives and coating materials by Brockwell (1962). Using the information available at that time and trial and error experimentation, Roughley *et al.* (1966) described a suitable, basic, pelleting recipe. However, the resulting survival of rhizobia by this method and later modifications gave little hope of prolonged adequate survival on seed prior to sowing (Brockwell and Roughley 1967).

Seed coat toxicity (Thompson 1960) was considered to be a major problem and led some to try to separate rhizobia from the seed coat using a protective layer, and others (Hale and Mathers 1976) to investigate the possibility of using polyvinylpyrrolidone (PVP) to render such toxins unavailable. Small improvements in survival were achieved.

Materon and Weaver (1984), using PVP and peat inoculant to coat seed, found that PVP improved the survival of *R. trifolii* on arrowleaf clover seed. The use of PVP as a protectant for rhizobia against seed coat toxicity was patented by Coated Seed Ltd, Christchurch, New Zealand in 1981.

The suitability of pelleting techniques is not solely a question of the survival of rhizobia, it also depends on the physical quality of the pellet. Recipes such as those proposed by Roughley *et al.* (1966) usually result in a somewhat powdery surface which will abrade when transported in bags or when sown with machinery. Pellets with good physical properties may be prepared in a fluid bed but, without protectants, survival of rhizobia is poor (Herridge *et al.* 1974). The Australian Inoculants Research and Control Service (AIRCS) has modified previous recipes to include a final overspray of polyvinyl acetate (PVA) which binds the outside of the pellet. More sophisticated pelleting methods have been developed in New Zealand and Canada which have significantly improved the survival of rhizobia. Such pelleted seed pre-inoculated 1 month prior to testing by the AIRCS, indicated that numbers of *R. meliloti* on lucerne seed exceeded 2500/seed, and of *R. trifolii* on white clover seed exceeded 1000/seed. Numbers well in excess of these minimum figures for *R. meliloti* are not uncommon.

Pellets now prepared by the AIRCS include a pre-coating of the seed to provide a better surface for the coating. The use of PVP as an adhesive rather than as a protectant only, and the use of oversprays to bind the pellet achieve similar survival. The excess moisture in the pellet may be removed by using a fluid bed dryer.

Direct Inoculation of Soil

Granules

Rhizobia in granules may be introduced directly into the drill row with the seed or placed below the seed when soil temperatures are high. Granules therefore allow seed treated with toxic chemicals to be physically separated from the rhizobia.

Natural peat granules are produced in the United States or they may be prepared by inoculating, with peat cultures, any non-toxic material which can serve as a nucleus to form a pellet, using $CaC0^3$ as a final coating. Marble chips and coarse sand are suitable. Such granules may be prepared by farmers immediately prior to use, but a suitable sowing rate must be determined for each type of granule.

Inoculation of soil using a non-legume crop

Experiments to investigate the practice of pre-inoculating fields with rhizobia by inoculating seed of the crop sown before the legume are currently being conducted in Australia by Brockwell, Gault, Roughley and Gemell. This interest stems from the finding that *R. japonicum* inoculated into new soybean land when sowing the first crop, formed more nodules than the same strain inoculated onto the second-year crop in the same soil (Brockwell *et al.* 1985).

Liquid inoculation

This method is particularly suited to large-seeded grain legumes which, because of their bulk, make seed inoculation a formidable task. Such seeds may often be subject to damage by moist inoculation thus reducing germination.

Liquid inoculation was used by Schiffmann and Alper (1968) and developed and evaluated by Gault *et al.* (1982). They reported comparable nodulation to that achieved by seed inoculation without increasing the inoculation rate. Nodulation of soybeans may be somewhat slower when a standard inoculum rate is used but, by flowering, nodulation was unaffected (Brockwell personal communication).

The technique of liquid inoculation involves preparing a suspension of peat inoculant in water which is pumped at a pressure of 140 kPa through fine plastic tubing mounted behind the seed boot. The outlet should be 2.5-5.0 cm above the bottom of the furrow; a broader spray may be obtained using a deflector. A rate of 112 l/ha is suitable and the pump should have excess capacity to enable 75% of its output to be bypassed back to the tank to keep the peat in suspension. Small areas can be inoculated using a knapsack.

Concluding Comments

A significant increase in the inoculum potential of an elite strain of rhizobia must depend on increasing their numbers in the inoculant and prolonging their survival on the seed prior to sowing and in the soil. Both growth and subsequent survival should be considered concurrently by research workers by ensuring that formulations which promote high count inoculants also promote good survival on seed or granules.

There is considerable scope for modifying known suitable carrier materials with substances which protect rhizobia such as clay minerals, carbohydrates, and adhesives. Such additions will have greatest scope when used with sterilized carriers. The search for a defined carrier should continue and the claim that one with a wide availability has already been identified (Lloyd personal communication, 1983) should give extra stimulus.

One factory in Brazil produces one million packets of inoculum in a 3 month period by broth injection of individual packets. Problems with large-scale production seem, therefore, a poor excuse for the continued production of second-class inoculants in unsterilized carriers. Countries without suitable carriers or where autoclavable films are unavailable should consider importing uninoculated packets.

We can look with confidence to inoculants which will regularly contain 10^{10} rhizobia/g. To increase still further the inoculum potential we must look towards improving inoculation methods.

References

Balatti, A.P. 1982. Culturing *Rhizobium* in large-scale fermentors. Pages 127-132 *in* Biological Nitrogen Fixation Technology for Tropical Agriculture (Graham, P.H. and Harris, S.C., eds.). Centro Internacional de Agricultura Tropical, Cali, Colombia.

Boiardi, J.L. and Ertola, R.J. 1985. *Rhizobium* biomass production in batch and continuous culture with malt-sprouts medium. MIRCEN Journal of Applied Microbiology and Biotechnology 1: 163-173.

Brockwell, J. 1962. Studies on seed pelleting as an aid to legume seed inoculation. 1. Coating materials, adhesives, and methods of inoculation. Australian Journal of Agricultural Research 13: 638-649.

Brockwell, J. 1980. Experiments with crop and pasture legumes-principles and practice. Pages 417-488 *in* Methods for Evaluating Biological Nitrogen Fixation (Bergersen, F.J., ed.). John Wiley and Sons, New York, USA.

Brockwell, J. 1982. Inoculation methods for field experimenters and farmers. Pages 211-227 *in* Nitrogen Fixation in Legumes (Vincent, J.M., ed.). Academic Press, New York, USA.

Brockwell, J. and Roughley, R.J. 1967. An examination of the number of nodule bacteria associated with legume seed following commercial multiple inoculation. Journal of the Australian Institute of Agricultural Science 33: 204-207.

Brockwell, J., Roughley, R.J. and Herridge, D.F. 1985. Ecological studies of *Rhizobium japonicum*: relationships between inoculum establishment and nodule occupancy in competitive and non-competitive soils. Page 397 *in* Nitrogen Fixation Research Progress (Evans, H.J., Bottomley, P.J. and Newton, W.E., eds.). Martinus Nijhoff, The Hague, The Netherlands.

Date, R.A. and Roughley, R.J. 1977. Preparation of legume seed inoculants. Pages 253-275 *in* A Treatise on Dinitrogen Fixation, Section IV (Hardy, R.W.F. and Gibson, A.H., eds.). John Wiley and Sons, New York, USA.

Dommergues, Y.R., Diem, H.G. and Divies, C. 1979. Polyacrylamide-entrapped *Rhizobium* as an inoculant for legumes. Applied and Environmental Microbiology 37: 779-781.

Franco, A.A. and Vincent, J.M. 1976. Competition amongst rhizobial strains for the colonization and nodulation of two tropical legumes. Plant and Soil 45: 27-48.

Gault, R.R., Chase, D.L. and Brockwell, J. 1982. Effects of spray inoculation equipment on the viability of *Rhizobium* spp. in liquid inoculants for legumes. Australian Journal of Experimental Agriculture and Animal Husbandry 22: 299-309.

Hale, C.N. and Mathers, D.J. 1976. Toxicity of white clover seed diffusate and its effect on the survival of *Rhizobium trifolii*. New Zealand Journal of Agricultural Research 20: 69-73.

Herridge, D.F., Robinson, F.D., Roughley, R.J., Mullett, J.H. and Cook, L.J. 1974. Inoculating and lime-pelleting legume seed using a fluid bed. Pages 124-133 *in* Plant Introduction Breeding and Seed Production. Proceedings of the Twelfth International Grass Congress, Moscow, Soviet Union.

Herridge, D.F., Roughley, R.J. and Brockwell, J. 1984. Effect of soil nitrate on the establishment and functioning of the soybean symbiosis in the field. Australian Journal of Agricultural Research 35: 149-161.

Ireland, J.A. and Vincent, J.M. 1968. A quantitative study of competition for nodule formation. Transactions of the Ninth International Congress on Soil Science 2: 85-93.

Kremer, R.J. and Peterson, H.L. 1982a. Effect of inoculant carrier on survival of *Rhizobium* in inoculated soil. Soil Science 134: 117-125.

Kremer, R.J. and Peterson, H.L. 1982b. Effects of carrier and temperature on survival of *Rhizobium* spp. in legume inocula: development of an improved type of inoculant. Applied and Environmental Microbiology 45: 1790-1794.

Kremer, R.J. and Peterson, H.L. 1983. Field evaluation of selected *Rhizobium* in an improved legume inoculant. Agronomy Journal 75: 139-143.

Leiderman, J. 1971. El bagazo como excipiente de inoculantes para leguminosas. Revista Industrial of Agricultura de Tucuman 48: 51-58.

McLeod, R.W. and Roughley, R.J. 1961. Freeze-dried cultures as commercial legume inoculants. Journal of Experimental Agriculture and Animal Husbandry 1: 29-33.

Materon, L.A. and Weaver, R.W. 1984. Survival of *Rhizobium trifolii* on toxic and non-toxic arrowleaf clover seeds. Soil Biology and Biochemistry 16: 533-535.

Robinson, A.C. 1969. Competition between effective and ineffective strains of *Rhizobium trifolii* in the nodulation of *Trifolium subterraneum*. Australian Journal of Agricultural Research 20: 827-841.

Roughley, R.J., Date, R.A. and Walker, M.H. 1966. Inoculating and lime pelleting legume seed. Agricultural Gazette of New South Wales 77: 142-146.

Roughley, R.J. 1968. Some factors influencing the growth and survival of root nodule bacteria in peat culture. Journal of Applied Bacteriology 31: 259-265.

Roughley, R.J. 1970. The preparation and use of legume seed inoculants. Plant and Soil 32: 675-701.

Roughley, R.J. and Pulsford, D.G. 1982. Production and quality control of legume inoculants. Pages 193-209 *in* Nitrogen Fixation in Legumes (Vincent, J.M., ed.). Academic Press, New York, USA.

Schiffmann, J. and Alper, Y. 1968. Inoculation of peanuts by application of *Rhizobium* suspensions into planting furrows. Experimental Agriculture 4: 219-226.

Somasegaran, P. and Halliday, J. 1982. The dilution of liquid cultures of *Rhizobium* for increasing the production capacity of inoculant plants. Applied and Environmental Microbiology 44: 330-333.

Thompson, J.A. 1960. Inhibition of nodule bacteria by an antibiotic from legume seed coats. Nature 187: 619-620.

Thompson, J.A. 1980. Production and quality control of legume inoculants. Pages 489-553 *in* Methods for Evaluating Biological Nitrogen Fixation (Bergersen, F.J., ed.). John Wiley and Sons, New York, USA.

Vincent, J.M. 1965. Environmental factors in the fixation of nitrogen by the legume. Pages 384-435 *in* Soil Nitrogen (Bartholomew, W.V. and Clark, F.E., eds.). American Society of Agronomy, Madison, Wisconsin, USA.

Vincent, J.M., Thompson, J.A. and Donovan, K.O. 1962. Death of root-nodule bacteria on drying. Australian Journal of Agricultural Research 13: 258-270.

Discussion

Lie

I wonder about the price of your simplified fermentor. We developed a system for about US$ 1000 (40 l fermentor).

Roughley

Concerning the price of the fermentor made in Thailand of approximately US$ 2000 this should be considered in the light of the price of the FAO fermentor which I understand is US$ 6000 and requires a large autoclave for sterilization.

Lie

How long does it take for the fermentor to be sterilized?

Roughley

The time for sterilization is 1 hr after reaching 121° C. However, the time to reach this temperature is considerable and depends on the methods of heating which can be a gas burner, line steam, or a combination of both. This can be of the order of 1 hr and a further period of 45 min is required to cool the medium to inoculation temperature. This time is dependent on the temperature of the cooling water and may be reduced by injecting air through the sparger once the pressure has returned to normal.

Lie

Can you use freeze-dried bacteria as a starter for your fermentor? That may simplify the system.

Roughley

I believe that freeze-dried cultures may be used to inoculate a starter flask rather than the fermentor direct. This allows for a check on contamination of the starter before use.

Witty

You said that the "super pellets" were dried at 32° C. Is that correct?

Roughley

Seed pellets may be dried at temperatures up to 32° C provided they do not remain long at this temperature. Presumably latent heat of evaporation keeps the microenvironment inside the pellet well below the air temperature.

Danso

The results you showed on the effect of ten-fold dilutions of *Rhizobium* on the number of nodules on lupin has implications for the accuracy of the most probable number (MPN) technique for counting *Rhizobium*. Would you please comment on this?

Roughley

The results presented showed that nodule dry weight and nodule number varied directly with the number of rhizobia applied to the seed, between 10 and 10^6/seed. In fact these numbers were determined by plate counts but could well have been done by MPN counts. The confidence that can be placed in MPN counts, having chosen the dilution intervals, e.g. ten-fold, depends on the number of plants inoculated at each dilution. Biological factors also affect the count e.g. conditions of plant growth and whether other organisms on the seed or soil being counted interfere with nodulation. Compensation for such "skips" which may occur may be made by using discretion in allocating to the non-nodulated tube (the skip) a + or − value.

Solh

In some areas in Saudi Arabia where the temperature may be 35 °C, inoculating alfalfa is done by pre-irrigation, then planting inoculated seed and continuing irrigation at a low flow rate through center pivots for 4 – 5 days. This is possible to follow in sandy soils but probably not in other types of soils.

Eaglesham

Has any comparison been made between the standard peat method and the liquid inoculum method in very dry soils?

Roughley

I have not made such a comparison directly but related studies show that death of the inoculant using both methods can be very rapid and when temperatures are also in the region of 37 °C, over 99% of the rhizobia may die in 24 hours.

Thompson

Would you please comment on the role of acetylene reduction for selection of strains for effectiveness or the comparison of the relative effectiveness of a group of strains.

Roughley

Any doubts we may have had regarding the limitations of the acetylene reduction technique for strain evaluation have been dispelled by Dr. Witty's paper at this conference. There seems to be no place for a technique which suffers from such major errors, and only provides us with acceptable data when it is correlated with other more readily measured variables.

D.P. Beck and L.A. Materon (editors), Nitrogen Fixation by Legumes in Mediterranean Agriculture
© 1988 ICARDA – Printed in the Netherlands. ISBN 90 247 3624 2

Inoculant Production in Turkey

MUJGAN ENGIN
University of Cukurova, Adana, Turkey

Summary. A number of legume crops are grown in Turkey but good rotation systems with legumes are not common. To reduce the dependence on nitrogen fertilizers, the use of inoculants has been developed. This paper reviews the state of the inoculant industry and methods of inoculant production in Turkey.

Introduction

The climate and topographic characters of Turkey are not uniform so various crops of diverse geographical origin can be grown, ranging from subtropical to cool. Of the agricultural land, 70% is used for field crops, with leguminous crops covering 9% of this area (Die 1984). Land use for agricultural purposes in Turkey has not yet reached its limit. A new irrigation project will undoubtedly increase the area under production and the use of leguminous crops. At present, peas and faba beans are common in the west and south coastal regions, while beans, chickpeas, and lentils are common in southeast and central Anatolia. Soybean is popular in the west and south, and will extend southeast after the introduction of irrigation. In many areas outside the traditional legume growing areas, grain legumes are becoming more profitable and popular compared to wheat. In the past 5 years, production and export figures for grain legumes have risen sharply, especially for lentil in 1984 (Table 1) (FAO 1984).

In 1984, Turkey produced more than one third of the world's lentils. Despite the increase in acreage and production, good rotation systems with legumes are not common. Meanwhile, the use of nitrogen fertilizers for field crops has increased. In general, Turkey's highest foreign exchange expenditure is for petroleum followed by fertilizers.

The value of biological nitrogen fixation by legumes in Turkish agricultural systems has so far not been appreciated. The need for inoculation of legumes

Table 1. Area and production of legumes in Turkey (FAO 1984).

	Lentil	Chickpea	Bean	Faba bean	Pea	Cowpea	Soybean
Area (1000 ha)	620	345	112	42	1.8	3.14	28
Production (1000 m ton)	570	335	164	76	3.7	3.10	60

in all areas is not clear, and has consistently been overlooked. Nevertheless, the few existing studies for some legumes (Gurbuzer 1978; Ulgen 1978) have shown benefit from inoculation in many cases. Decision-making bodies for agricultural policies naturally give priority to the immediate need to increase yield and quality in the short term. Industrial fertilizers meet this need in practice.

Most farmers in the south prefer growing crops such as cotton, which have an established market and good price. However, farmers are considering their incomes from profitable crops in a stable market but are not considering soil fertility in the long term. In some regional agricultural projects, farmers are being encouraged to grow legumes after wheat instead of leaving fields fallow for a year. In some parts, where the climate is favorable for a second annual crop, legumes are recommended. For example, growing soybean in the south after winter wheat is becoming more common.

Use of inoculants for legumes is not a common practice except for soybean, on which inoculants were first used in Turkey. At the end of the 1970's there was a severe oil shortage and animal feed was very expensive. In some institutes, including our university, soybean trials were conducted in collaboration with INTSOY (Atakisi and Arioglu 1983a; 1983b). The first experimental results coincided with the authorities deciding to import some soybean varieties. When the crop was first introduced, seed and inoculant were imported and subsidized. At present the majority of the seed and approximately 12 tons of inoculant are being produced in Turkey while the rest is imported.

Inoculant is produced by the Soil and Fertilizer Research Institute in Ankara. Annual capacity is approximately 12 tons, with most of the inoculant produced for soybean. Inoculants for lentil, chickpea, bean, pea, faba bean, and some forage crops are also being produced on a small scale. The facilities for production are generally adequate, the exception being fermentation vessels.

The peat used as a carrier comes from Karasaz, near Kayseri in central Anatolia. It is rather powdery and like charcoal dust after grinding. Raw peat has a high salt content and pH (Gurbuzer 1982), so to reduce the salt and pH, the peat is washed several times (Table 2). Following washing, the peat is dried to 4–5% moisture and ground to 0.2 mm particle size. The peat is then sterilized for 3 hrs at 121° C.

A broth culture of rhizobia is prepared from a mother culture in YMA growth medium and incubated at 28° C in big glass balls. When rhizobial

Table 2. Some characteristics of Krasaz-Kayseri peat (Gurbuzer 1982).

	Organic matter	Salt (%)	$CaCO_2$ (%)	P_2O_5 (kg/da)	K_2O (kg/da)	Water saturation (%)	pH
Before washing	14.4	1.8	6.9	10.53	172.8	198	7.95
After washing *	10.2	0.3	2.7	7.31	117.0	147	7.60

* 2–3 days washing with 5–6 water changes/day.

numbers are high, the broth is mixed with sterile peat in a modified cement mixer. This mixing is the most difficult part of the procedure. The broth-peat mixture is packed in 0.5–1 kg quantities in polyethylene bags and sealed immediately. Bags of freshly inoculated peat are incubated on shelves in controlled temperature rooms until the final high number of rhizobia is reached. The maximum number of rhizobia at dispatch is around 10^8–10^9 per gram. For storage and transportation, the polyethylene bags are packed ten to a box.

The institute is renewing and extending their production facilities, and new fermentors may be in use by 1987.

References

Atakisi, I.K. and Arioglu, H., 1983a. International Soybean Variety Experiment. Seventh Report of Results 1979. INTSOY. No.24. Illinois, USA.

Atakisi, I.K. and Arioglu, H. 1983b. Cukurova kosullarinda farkli soya cesitlerinin 2. urun olarak yetistirme olanaklari uzerinde bir arastirma. C.U. Zir. Fak. Yilligi, Ayri Baski No.2. Adana.

DIE (Devlet Istatistik Enstitusu). 1984. Ankara, Turkey.

FAO (Food and Agriculture Organisation). 1984. Production Yearbook. FAO, Rome, Italy.

Gurbuzer, E. 1978. En fazla azot tespit etme ozelligi gosteren soya fasulyasi nodozite bakterilerinin secilmesi. Top. Gub. Aras. Enst. Yayinlari. No.78. Ankara.

Gurbuzer, E. 1982. Nodozite bakteri kulturlerinin degisik sicaklik derecelerinde kullanilabilme sureleri., Top. Gub. Aras. Enst. Yayinlari, Ankara.

Ulgen, H. 1978. Orta anadolu kosullarinda fig bitkisinin urun miktarinda ve azot kapsaminda en fazla artis saglayan nodozite bakteri suslarinin sera ve tarla sartlarinda secilmesi. Top. Gub. Aras. Enst. Yayinlari, No.79. Ankara.

Part IV

Nitrogen Fixation Limitations and Potential

D.P. Beck and L.A. Materon (editors), Nitrogen Fixation by Legumes in Mediterranean Agriculture
 – Printed in the Netherlands. ISBN 90 247 3624 2

The Role of Legume, *Rhizobium*, and Environment in Nitrogen Fixation: Constraints on Symbiotic Potential and their Removal

J.M. VINCENT

Emeritus Professor, University of New South Wales, Australia

Summary. Any research on the legume-*Rhizobium* symbiosis should involve rhizobiologists and agronomists. In this paper, the need for close cooperation is stressed. The effect of each component of the symbiosis – legume, rhizobia, environment – and their interaction on the success of the symbiosis is discussed. The constraints to N_2 fixation and a strategy for their removal are outlined.

Introduction

The effectively nodulated legume is the major biological system for reductive conversion of N_2 to a utilisable combined form (NH_4^+). This account will be directed to well established aspects that bear directly on the agronomic usefulness of the symbiosis. The list of reference books and review articles at the end will be useful for methodologies and concepts and will be referred to in the various sections, where relevant.

The "Productivity Triangle" (Dart and Krantz 1977; Herridge 1982; Nutman 1965; Vincent 1984)

Operative factors

There are three major components of the legume symbiotic system which determine its contribution to productivity; legume, *Rhizobium*, and environment. The achievement of a satisfactory symbiosis does not depend on either symbiont alone, but on the inherent characteristics of each, on their interaction with each other, and on the influence of outside factors (environment). The last reacts with the two symbionts separately and with the symbiosis.

Need for cooperation between plant and Rhizobium workers

The *Rhizobium* worker who fails to take sufficient account of the host, the agronomist who ignores the rhizobial factor, and either of the two who neglects the over-riding influence of environmental factors is equally culpable. Unfortunately such tunnel-vision is all too common.

One has only to look at reports from agronomists and other plant workers, even from well reputed international institutes where interdisciplinary cooperation is to be expected, to find an extraordinary willingness to ignore the rhizobial factor or take it for granted. Why is it so often assumed that all is well with the nodulation of a legume, which is expected to be independent of soil nitrogen?

At the same time, it is common for a rhizobiologist, even when involved in what purports to be a practical project, to seize upon a rhizobial investigation suitable for laboratory or greenhouse, with scant attention to the agronomic realities which will determine its applicability. This approach may be justified in a laboratory funded to make fundamental contributions to our knowledge of the symbiosis, but it is hard to justify in a setting which purports to achieve higher productivity.

In this paper, both symbionts and their interaction with and reaction to the environment will be discussed.

The Legume (Allen *et al.* 1981; Kay 1979; NAS 1979; Vincent 1984)

Intrinsic properties

The intrinsic genetic complement of the host species or variety is a major determinant of the amount of N_2 which will be fixed and its agronomic usefulness. As such, it is the concern of the agronomist, introduction officer, and plant breeder.

Besides overall yield, these workers are concerned with the way in which gross productivity is partitioned between vegetative growth (for pasturage and fodder) and grain. They must also consider characteristics which will determine agronomic acceptability, such as quality of the product, profitability, and potential contribution to soil fertility.

Reaction to local environment (Andrew 1978; Franco 1978; Munns 1977)

As important as a plant's intrinsic properties is its suitability to the environment in which it will be grown. Environmental factors include soil (structure, H^+ concentration, nutritive status, and toxic elements), climate (particularly temperature and rainfall), and day length.

Aspects of the legume-environment interaction include the legume's ability to survive during and beyond a growing season, its recovery from stress (due to water shortage, grazing, or hay-making), its ability to produce sufficient seed for re-establishment, and its resistance to disease and pests (e.g. ascochyta blight of faba bean, lentil, and chickpea, and nematode infection of chickpea). Mycorrhizal association, if occurring, will be an advantage to the host.

The rhizobial factor must be considered as part of the plant's environment.

For this reason, the plant worker should take into account the selected host's compatibility with rhizobia already in the soil and/or those available for introduction.

The *Rhizobium*

Symbiotic capacity in relation to the chosen host

Rhizobia able to nodulate a particular legume may be absent, sparse, or abundant.

In the first case successful nodulation and the establishment of a suitable, effective, rhizobial population through seed or soil inoculation can be relatively straightforward, provided critical environmental factors such as soil pH, moisture, and nutrient status are already favourable or subject to improvement.

If naturally occurring, specifically nodulating rhizobia are present but sparse, initial nodulation success can be readily achieved with an inoculant, subject to suitable soil conditions. However, long-term persistence of a superior introduced strain in competition with an increasing natural, possibly better adapted population cannot be guaranteed. If the introduced strain does not persist and the increased natural population is fully effective, no harm will have been done and inoculation will have taken care of the host's initial nodulation requirement. However, a build-up of ineffective, or only partially effective strains can lead to inferior plant performance. The extent of problems due to failure to establish an introduced inoculum strain will depend on the effectiveness of those rhizobial strains already in the soil.

The genetic stability of any introduced rhizobial strain is important. Strains differ in the incidence of ineffective forms and many instances of lost or reduced symbiotic capacity have been reported with cultured rhizobia under controlled conditions which permit detection. Changes in symbiotic capacity could occur amongst natural populations, although field conditions would probably select against such variants, at least so long as available nitrogen is limiting plant growth. Demonstration that DNA sequences which control some aspect of symbiotic capacity can be located in an exchangeable plasmid makes it not surprising that variability in this property is seen from time to time. Temperate rhizobial phage causing lysogenisation can also play a role in the loss of effectivity.

Environmental determinants of rhizobial growth and survival (Bushby 1983; Chowdhury 1977; Munns 1977; Vincent 1965; Vincent 1974; Vincent 1977)

Environmental factors such as soil pH, available water, temperature, soil antagonists, toxic chemicals, and the presence or absence of plants are important in determining rhizobial growth and survival. Species and strains within species differ in sensitivity to acidity, desiccation, and elevated temperature. For example, the rhizobia able to nodulate faba beans and lentils are likely to be more tolerant of low pH (about 5) than those of medics. However, the latter are likely to be more tolerant of high temperature and salinity.

Interaction Between Legumes and *Rhizobium* (Parker and Chatel 1982; Quispel 1974)

The symbiosis (Dart 1974; Trinick 1982a; Vincent 1974)

An established symbiosis is a close association between host and *Rhizobium* in a morphologically defined, cytologically specialised nodule which is the site of N_2-fixation and transfer to the plant. In broad terms, the steps to nodule formation are invasion, nodule formation, and nodule function.

Roots of legumes may stimulate rhizobia to a greater degree than those of other plants. However, the seed coats of some host species carry a water-soluble factor toxic to rhizobia.

Specificity between host and *Rhizobium* is the major factor which determines, according to the partners involved, a result ranging from failure to form nodules or complete nodule ineffectivity, through degrees of N_2-fixing effectiveness to full effectivity.

Species of the genus *Medicago* demonstrate host-*Rhizobium* specificity which ranges from a wide tolerance of *Rhizobium meliloti* strains in *Medicago sativa* (lucerne, alfalfa) to a very specialised host response in *M. laciniata*. Other species fall between these extremes.

Amongst any collection of rhizobia able to nodulate a particular host, there is a wide range of effectiveness in N_2-fixation for that legume, which is not necessarily paralleled with other species or varieties.

Environmental factors affecting the symbiosis (Andrew 1977; 1978; Dart 1974; Dart *et al.* 1976; Franco 1977; 1978; Gibson 1976; Hardy and Havella 1976; Lie 1974; Lie *et al.* 1976; Nosse 1977; Munns 1977; Pate *et al.* 1981; Phillips *et al.* 1981; Sprent 1976; Trinick 1982b; Vincent 1965

Some environmental factors operate directly on the symbiosis. These include chemical elements such as molybdenum (part of the Mo - Fe protein of nitrogenase), cobalt (needed for vitamin B_{12} synthesis by rhizobia), and boron (required for a normal vascular system). There is a narrower optimal temperature range for the symbiotically-dependent plant than for the host with access to adequate combined nitrogen. Sufficient soil water, but not to the exclusion of essential O_2, the level of combined nitrogen available to the plant, and the proper functioning of its photosynthetic system (affected by light and CO_2 supply) are also important in determining symbiotic performance.

Constraints to N_2-Fixation (Vincent 1982)

We can now list in broad terms the limitations which may constrain legume productivity.

With the host: intrinsic difficulties in yield capacity and/or unsuitability to the environment.

With the rhizobia: absence or scarcity of specific rhizobia able to provide effective N_2-fixing nodulation.

With the symbiosis: environmental factors which limit nodule formation or function.

A Strategy for the Removal of Constraints

This can be considered in four stages.

Stage 1: Choice of legume and exploratory field trial

The first step is the selection and supply of seeds of legume species or varieties to serve the needs of a particular area, with regard to suitability to the regional environment (actual or subject to improvement). This will be an agronomic responsibility.

Trial sowings of the selected legumes could then be undertaken with and without fertilizer nitrogen and under conditions in which other nutrients (especially P, K, Mg, S, Mo, and other trace elements) and soil pH are non-limiting. At this stage the legume might be sown without inoculation to assess the local rhizobia, but if an inoculation treatment is included, care should be exercised to avoid confusion between the native population and the introduced strain. The agronomist could play a major role at this stage with the collaboration of the rhizobiologist.

Inspection of nodule form and degree of nodulation in the non-N treatment is likely to indicate the effectiveness of naturally occurring rhizobia. This will, in the first instance, be the responsibility of the rhizobiologist, with the agronomist regularly inspecting the legume's roots and nodules.

Stage 2: Assessment of the local rhizobial population (Date 1982a)

The rhizobiologist will be able to assess the symbiotic effectiveness of the local rhizobia during stage 1 by testing isolates from representative nodules for specific effectiveness in the greenhouse. Information likely to be useful for strain identification (such as serological characteristics) can be obtained at the same time.

An alternative, or additional, more quantitative estimate of the local rhizobial population can also be obtained by soil sampling and counts using a plant infection method.

Stage 3: Inoculation trials

The benefit achievable by using one or more selected strains (introduced or developed from effective local isolates) can be determined experimentally with seed or soil inocula. Again, care must be taken to ensure sufficient nutrient elements, a suitable soil pH, and if appropriate, sufficient moisture.

These experiments can be interpreted using yield measurements and macroscopic examination of nodulation when local nodulating strains are absent, sparse, and/or clearly ineffective. However, when such a sharp distinction is not possible, full interpretation of results will require identification of the strains actually responsible for nodulation - whether introduced or native. Such identification could involve serological identification (against a background of known and checked native serotypes), or the experimental use of inoculant strains labelled with antibiotic resistance.

Continued collaboration between rhizobiologist and agronomist is desirable at this stage although greater responsibility will rest with the former.

Stage 4: On-farm inoculation; inoculant quality and use (Batthyany 1977; Date 1977; 1982b; Sahni 1977; Vincent 1974)

When inoculation is essential or advantageous, a reliable supply of inoculant containing a large number of appropriate rhizobia is vital for on-farm application. This involves selection of suitable strains (Brockwell *et al.* 1982) and a sound system of quality control from manufacture through to supply to the farmer (Roughley 1976; Roughley and Pulsford 1982; Strijdom and Deschodt 1976; Thompson 1980). Details of such control cannot be dealt with here but its importance cannot be over-emphasised (Roughley 1976; Roughley and Pulsford 1982; Strijdom and Deschodt 1976; Thompson 1980).

Proper use of the inoculant is also important as is any necessary amelioration of soil conditions. Many unsuccessful seed inoculation reports have been attributable to failure at this level (Brockwell 1982; Burton 1976; 1979).

This is a stage when all participants (farmer, agronomist, and rhizobiologist) can play an important part in determining any deficiencies, their cause, and their remedy.

Methodology

A very useful manual has been produced by the NifTAL Project (Somasegaran and Hoben 1985) which up-dates an earlier text (Vincent 1970). Some specific procedures are described in more detail in other accounts, for example;

(1) Measurement of N_2-fixation: direct (Bergersen 1980), and indirect (Herridge 1982; Turner and Gibson 1980).

(2) Methods of plant experimentation (Brockwell 1980; 1982b; 1982c; Gibson 1980).
(3) Direct enumeration and plant infection counts (Brockwell 1982a; Vincent 1982b).
(4) Strain identification and direct recovery of labelled rhizobia (Bushby 1982; Schwinghammer and Dudman 1980).

References

Source Books and Monographs

Allen, O.N. and Allen, Ethyl K. 1981. The Leguminosae - A Source Book of Characteristics, Uses and Nodulation. University of Wisconsin Press, Wisconsin, USA.

Bartholemew, W.V. and Clark, F.E. (eds.). 1965. Soil Nitrogen. American Society of Agronomy, Madison, Wisconsin, USA.

Bergersen, F.J. (ed.). 1980. Methods for Evaluating Biological Nitrogen Fixation. John Wiley and Sons, Chichester, England.

Broughton, W.J. (ed.). 1982. Nitrogen Fixation. Volume 2. Clarendon Press, Oxford, England.

Dobereiner, J., Burris, R.H. and Hollander, A. (eds.). 1978. Limitations and Potentials for Biological Nitrogen Fixation in the Tropics. Plenum Press, New York, USA.

Gibson, A.H. and Newton, W.E. (eds.). 1981. Current Perspectives in Nitrogen Fixation. Australian Academy of Sciences, Canberra, Australia.

Hardy, R.W.F. and Silver, W.S. (eds.). 1977. A Treatise on Dinitrogen Fixation, Section III, Biology. John Wiley and Sons, New York, USA.

Kay, D. E. 1979. Food Legumes. TPI Crop and Product Digest No. 3. Tropical Products Institute, England.

NAS (National Academy of Sciences Advisory Panel) 1979. Tropical Legumes: Resources for the Future. National Academy of Sciences, Washington, USA.

Nutman, P.S. (ed.). 1976. Symbiotic Nitrogen Fixation in Plants. Cambridge University Press, Cambridge, England.

Quispel, A. (ed.). 1974. The Biology of Nitrogen Fixation. North-Holland, Amsterdam, The Netherlands.

Somasegaran, P. and Hoben, H.J. 1985. Methods in Legume-*Rhizobium* Technology. University of Hawaii, NifTAL Project and Mircen, Paia, Hawaii, USA.

Subba Rao, N.S. (ed.). 1978. Recent Advances in Biological Nitrogen Fixation. Oxford and IBH Publishing, New Delhi, India.

Vincent, J.M. 1970. A Manual for the Practical Study of the Root-Nodule Bacteria. IBP Handbook No. 15. Blackwell Scientific Publications, Oxford, England.

Vincent, J.M. (ed.). 1982d. Nitrogen Fixation in Legumes. Academic Press, Sydney, Australia.

Vincent, J.M., Whitney, A.S. and Bose, J. (eds.). 1977. Exploiting the Legume-*Rhizobium* Symbiosis in Tropical Agriculture. Miscellaneous Publication 145. University of Hawaii College of Tropical Agriculture, USA.

Vose, P.B. and Blixt, S.C. (eds.). 1984. Crop Breeding: A Contemporary Basis. Pergamon Press, Oxford, England.

Review Chapters

Andrew, C.S. 1977. Nutritional restraints on legume symbiosis. Pages 253-274 *in* Exploiting the Legume-*Rhizobium* Symbiosis in Tropical Agriculture (Vincent, J.M., Whitney, A.S. and Bose, J., eds.). Miscellaneous Publication 145. University of Hawaii College of Tropical Agriculture, USA.

Andrew, C.S. 1978. Legumes and acid soils. Pages 135-160 *in* Limitations and Potentials for Biological Nitrogen Fixation in the Tropics (Dobereiner *et al.*, eds.). Plenum Press, New York, USA.

Batthyany, C. 1977. Legume inoculants for Latin America. Pages 429-439 *in* Exploiting the Legume-*Rhizobium* Symbiosis in Tropical Agriculture (Vincent, J.M., Whitney, A.S. and Bose, J., eds.). Miscellaneous Publication 145. University of Hawaii College of Tropical Agriculture, USA.

Bergersen, F.J. 1977. Physiological chemistry of dinitrogen fixation by legumes. Pages 519-555 *in* A Treatise on Dinitrogen Fixation, Section III, Biology (Hardy, R.W.F. and Silver, W.S., eds.). John Wiley and Sons, New York, USA.

Bergersen, F.J. 1980. Measurement of nitrogen fixation by direct means. Pages *5-110* *in* Methods for Evaluating Biological Nitrogen Fixation (Bergersen, F.J., ed.). John Wiley and Sons, Chichester, England.

Brockwell, J. 1980. Experiments with crop and pasture legumes-principles and practice. Pages 417-488 *in* Methods for Evaluating Biological Nitrogen Fixation (Bergersen, F.J., ed.). John Wiley and Sons, Chichester, England.

Brockwell, J. 1982a. Plant-infection counts of rhizobia in soils. Pages 41-58 *in* Nitrogen Fixation in Legumes (Vincent, J.M., ed.). Academic Press, Sydney, Australia.

Brockwell, J. 1982b. Host plants for symbiotic experiments. Pages *9-83* *in* Nitrogen Fixation in Legumes (Vincent, J.M., ed.). Academic Press, Sydney, Australia.

Brockwell, J. 1982c. Inoculation methods for field experimenters and farmers. Pages 211-227 *in* Nitrogen Fixation in Legumes (Vincent, J.M., ed.). Academic Press, Sydney, Australia.

Brockwell, J., Diatloff, A., Roughley, R.J. and Date, R.A. 1982. Selection of rhizobia for inoculants. Pages 173-191 *in* Nitrogen Fixation in Legumes (Vincent, J.M., ed.). Academic Press, Sydney, Australia.

Burton, J.C. 1976. Methods of inoculating seeds and their effect on survival of rhizobia. Pages 175-189 *in* Symbiotic Nitrogen Fixation in Plants (Nutman, P.S., ed.). Cambridge University Press, Cambridge, England.

Burton, J.C. 1979. New developments in inoculating legumes. Pages 380-405 *in* Recent Advances in Biological Nitrogen Fixation (Subba Rao, N.S., ed.). Oxford and IBH Publishing, New Delhi, India.

Bushby, H.V.A. 1983. Ecology. Pages 37-75 *in* Nitrogen Fixation. Volume 2. (Broughton, W.J., ed.). Clarendon Press, Oxford, England.

Bushby, H.V.A. 1982. Direct quantitative recovery of *Rhizobium* from soil and rhizosphere. Pages 59-67 *in* Nitrogen Fixation in Legumes (Vincent, J.M., ed.). Academic Press, Sydney, Australia.

Chowdhury, M.S. 1977. Effects of soil antagonists on symbiosis. Pages 385-411 *in* Exploiting the Legume-*Rhizobium* Symbiosis in Tropical Agriculture (Vincent J.M., Whitney, A.S., and Bose, J., eds.). Miscellaneous Publication 145. University of Hawaii College of Tropical Agriculture, USA.

Dart, P.J. 1974. Development of root-nodule symbiosis: the infection process. Pages 381-429 *in* The Biology of Nitrogen Fixation (Quispel, A., ed.). North-Holland, Amsterdam, The Netherlands.

Dart, P.J., Day, J., Islam, R. and Dobereiner, J. 1976. Symbiosis in tropical grain legumes: some effects of temperature and the composition of the rooting medium. Pages 361-384 *in* Symbiotic Nitrogen Fixation in Plants (Nutman, P.S., ed.). Cambridge University Press, Cambridge, England.

Dart, P.J. and Krantz, B.A. 1977. Legumes in the semi-arid tropics. Pages 119-154 *in* Exploiting the Legume-*Rhizobium* Symbiosis in Tropical Agriculture (Vincent, J.M., Whitney, A.S., and Bose, J., eds.). Miscellaneous Publication 145. University of Hawaii College of Tropical Agriculture, USA.

Date, R.A. 1976. Principles of *Rhizobium* strain selection. Pages 137-150 *in* Symbiotic Nitrogen Fixation in Plants (Nutman, P.S., ed.). Cambridge University Press, Cambridge, England.

Date, R.A. 1977. Inoculation of tropical pasture legumes. Pages 293-311 *in* Exploiting the Legume-*Rhizobium* Symbiosis in Tropical Agriculture (Vincent, J.M., Whitney, A.S., and Bose,

J., eds.). Miscellaneous Publication 145. University of Hawaii College of Tropical Agriculture, USA.

Date, R.A. 1982a. Assessment of rhizobial status of the soil. Pages 85-94 *in* Nitrogen Fixation in Legumes (Vincent, J.M., ed.). Academic Press, Sydney, Australia.

Date, R.A. 1982b. Collection, isolation, characterisation and conservation of *Rhizobium*. Pages 95-109 *in* Nitrogen Fixation in Legumes (Vincent, J.M., ed.). Academic Press, Sydney, Australia.

Franco, A.A. 1977. Nutritional restraints for tropical grain legume symbiosis. Pages 237-252 *in* Exploiting the Legume-*Rhizobium* Symbiosis in Tropical Agriculture (Vincent, J.M., Whitney, A.S., and Bose, J., eds.). Miscellaneous Publication 145. University of Hawaii College of Tropical Agriculture, USA.

Franco, A.A. 1978. Micronutrient requirements of legume-*Rhizobium* symbiosis in the tropics. Pages 161-171 *in* Limitations and Potentials for Biological Nitrogen Fixation in the Tropics (Dobereiner, J., Burris, R.H., and Hollander, A., eds.). Plenum Press, New York, USA.

Gibson, A.H. 1976. Recovery and compensation by nodulated legumes to environmental stress. Pages 385-403 *in* Symbiotic Nitrogen Fixation in Plants (Nutman, P.S., ed.). Cambridge University Press, Cambridge, England.

Gibson, A.M. 1980. Methods for legumes in glasshouses and controlled environment cabinets. Pages 140-184 *in* Methods for Evaluating Biological Nitrogen Fixation (Bergersen, F.J., ed.). John Wiley and Sons, Chichester, England.

Hardy, R.N.F. and Havella, U.B. 1976. Photosynthate as a major factor limiting nitrogen fixation by field-grown legumes with emphasis on soybeans. Pages 421-439 *in* Symbiotic Nitrogen Fixation in Plants (Nutman, P.S., ed.). Cambridge University Press, Cambridge, England.

Herridge, D.F. 1982. Assessment of nitrogen fixation. Pages 123-136 *in* Nitrogen Fixation in Legumes (Vincent, J.M., ed.). Academic Press, Sydney, Australia.

Herridge, D.F. 1982. Crop rotation involving legumes. Pages 253-261 *in* Nitrogen Fixation in Legumes (Vincent, J.M., ed.). Academic Press, Sydney, Australia.

Lie, T.A. 1974. Environmental effects on nodulation and symbiotic nitrogen fixation. Pages 555-582 *in* Biology of Nitrogen Fixation (Quispel, A., ed.). North Holland, Amsterdam, The Netherlands.

Lie, T.A., Hille, L., Lambers, R. and Houwers, A. 1976. Symbiotic specialisation in pea plants: some environmental effects on nodulation and nitrogen fixation. Pages 319-333 *in* Symbiotic Nitrogen Fixation in Plants (Nutman, P.S., ed.). Cambridge University Press, Cambridge, England.

Mosse, B. 1977. The role of mycorrhiza in legume nutrition on marginal soils. Pages 275-292 *in* Exploiting the Legume-*Rhizobium* Symbiosis in Tropical Agriculture (Vincent, J.M., Whitney, A.S. and Bose, J., eds.). Miscellaneous Publication 145. University of Hawaii College of Tropical Agriculture, USA.

Munns, D.N. 1977. Soil acidity and related factors. Pages 211-236 *in* Exploiting the Legume-*Rhizobium* Symbiosis in Tropical Agriculture (Vincent, J.M., Whitney, A.S. and Bose, J., eds.). Miscellaneous Publication 145. University of Hawaii College of Agriculture, USA.

Nutman, P.S. 1965. Symbiotic nitrogen fixation. Pages 360-383 *in* Soil Nitrogen (Bartholemew, W.V. and Clark, F.E., eds.). American Society of Agronomy, Madison, Wisconsin, USA.

Parker, C.A. and Chatel, D.L. 1982. Factors determining success or failure in legume establishment. Pages 145-153 *in* Nitrogen Fixation in Legumes (Vincent, J.M., ed.). Academic Press, Sydney, Australia.

Pate, J.S. 1976. Physiology of the reaction of nodulated legumes to environment. Pages 335-360 *in* Symbiotic Nitrogen Fixation in Plants (Nutman, P.S., ed.). Cambridge University Press, Cambridge, England.

Pate, J.S. 1977. Functional biology of dinitrogen fixation by legumes. Pages 473-517 *in* A Treatise on Dinitrogen Fixation (Hardy, R.W.F. and Silver, W.S., eds.). John Wiley and Sons, New York, USA.

Pate, J.S., Atkins, C.R. and Rainbird, R.W. 1981. Theoretical and experimental costing of nitrogen fixation and related processes in nodules of legumes. Pages 105-116 *in* Current

Perspectives in Nitrogen Fixation (Gibson, A.H. and Newton, W.E., eds.). Australian Academy of Science, Canberra, Australia.

Phillips, D.A., De Jong, T.M. and Williams, L.S. 1981. Carbon and nitrogen limitations on symbiotically grown soybean seedlings. Pages 117-120 *in* Current Perspectives in Nitrogen Fixation (Gibson, A.H. and Newton, W.E., eds.). Australian Academy of Science, Canberra, Australia.

Quispel, A. 1974. Prerequisites for biological nitrogen fixation: in root-nodule symbioses. Pages 719-745 *in* The Biology of Nitrogen Fixation (Quispel, A., ed.). North-Holland, Amsterdam, The Netherlands.

Roughley, R.J. 1976. The production of high quality inoculants and their contribution to legume yield. Pages 125-136 *in* Symbiotic Nitrogen Fixation in Plants (Nutman, P.S., ed.). Cambridge University Press, Cambridge, England.

Roughley, R.J. and Pulsford, D.J. 1982. Production and control of legume inoculants. Pages 193-209 *in* Nitrogen Fixation in Legumes (Vincent, J.M., ed.). Academic Press, Sydney, Australia.

Sahni, V.P. 1977. Inoculants for India. Pages 413-427 *in* Exploiting the Legume-*Rhizobium* Symbiosis in Tropical Agriculture (Vincent, J.M., Whitney, A.S. and Bose, J., eds.). Miscellaneous Publication 145. University of Hawaii College of Agriculture, USA.

Schwinghamer, E.A. and Dudman, W.F. 1980. Methods for identifying strains of diazotrophs. Pages 337-365 *in* Methods for Evaluating Biological Nitrogen Fixation (Bergersen, F.J., ed.). John Wiley and Sons, Chichester, England.

Smith, F.W. 1982. Mineral nutrition of legumes. Pages 155-172 *in* Nitrogen Fixation in Legumes (Vincent, J.M., ed.). Academic Press, Sydney, Australia.

Sprent, J.I. 1976. Nitrogen fixation by legumes subjected to water and light stress. Pages 405-420 *in* Symbiotic Nitrogen Fixation in Plants (Nutman, P.S., ed.). Cambridge University Press, Cambridge, England.

Strijdom, B.W. and Deschodt, C.C. 1976. Carriers of rhizobia and the effects of prior treatment on the survival of rhizobia. Pages 151-168 *in* Symbiotic Nitrogen Fixation in Plants (Nutman, P.S., ed.). Cambridge University Press, Cambridge, England.

Thompson, J.A. 1980. Production and quality control of legume inoculants. Pages 489-533 *in* Methods for Evaluating Biological Nitrogen Fixation (Bergersen, F.J., ed.). John Wiley and Sons, Chichester, England.

Trinick, M.J. 1982a. Host-*Rhizobium* associations. Pages 111-122 *in* Nitrogen Fixation in Legumes (Vinçent, J.M., ed.). Academic Press, Sydney, Australia.

Trinick, M.J. 1982b. Competition between rhizobial strains for nodulation. Pages 229-238 *in* Nitrogen Fixation in Legumes (Vincent, J.M., ed.). Academic Press, Sydney, Australia.

Turner, G.L. and Gibson, A.H. 1980. Measurement of nitrogen fixation by indirect means. Pages 111-138 *in* Methods for Evaluating Biological Nitrogen Fixation (Bergersen, F.J., ed.). John Wiley and Sons, Chichester, England.

Vincent, J.M. 1965. Environmental factors in the fixation of nitrogen by the legume. Pages 384-435 *in* Soil Nitrogen (Bartholemew, W.V. and Clark, F.E., eds.). American Society of Agronomy, Madison, Wisconsin, USA.

Vincent, J.M. 1974. Root-nodule symbiosis with *Rhizobium*. Pages 265-341 *in* the Biology of Nitrogen Fixation (Quispel, A., ed.). North-Holland, Amsterdam, The Netherlands.

Vincent, J.M. 1977. *Rhizobium*: general microbiology. Pages 277-366 *in* A Treatise on Dinitrogen Fixation (Hardy, R.F.W. and Silver, W.S., eds.). John Wiley and Sons, New York, USA.

Vincent, J.M. 1977. Quality control of inoculants. Pages 447-456 *in* Exploiting the Legume-*Rhizobium* Symbiosis in Tropical Agriculture (Vincent, J.M., Whitney, A.S. and Bose, J., eds.). Miscellaneous Publication 145. University of Hawaii College of Tropical Agriculture, USA.

Vincent, J.M. 1982a. Nature and basic properties of rhizobia. Pages 5-11 *in* Nitrogen Fixation in Legumes (Vincent, J.M., ed.). Academic Press, Sydney, Australia.

Vincent, J.M. 1982b. Enumeration and determination of growth. Pages 35-40 *in* Nitrogen Fixation in Legumes (Vincent, J.M., ed.). Academic Press, Sydney, Australia.

Vincent, J.M. 1982c. Role, needs and potential of the nodulated legume. Pages 263-284 *in* Nitrogen Fixation in Legumes (Vincent, J.M., ed.). Academic Press, Sydney, Australia,
Vincent, J.M. 1984. Potential for enhancing biological nitrogen fixation. Pages 185-215 *in* Crop Breeding: a Contemporary Basis (Jose, P.B. and Blixt, S.C., eds.). Pergamon Press, Oxford, England..lm 10;

Discussion

Weaver

How do you feel about applying fertilizer nitrogen during legume screening and avoiding the complexity of inoculation?

Roughley

This may be the only practical way to ensure good plant growth particularly in the absence of appropriate strains of rhizobia. It must be borne in mind that the response of the legume to environmental stress may differ when using inorganic nitrogen than when depending on fixed nitrogen. It is important that selection in breeding programmes is based on symbiotic plants to ensure that the character of nitrogen fixation is expressed.

Broughton

You mentioned the possibility that genetic exchange occurs amongst rhizobia in the soil. Do you know if Prof. Vincent has any data to support this hypothesis?

Roughley

I don't know if Prof. Vincent has any data from his own experiments but he would be aware of the Australian experience with *R. trifolii* strain CC2480a. Two to three years after inoculation into soil it was reisolated from nodules and shown to vary from ineffective through the spectrum to full effectiveness. These reisolates were serologically indistinguishable from the parent strain using gel diffusion.

Brockwell

The pathological effects of the environment on the symbiosis have not been mentioned. It is clear that a plant without healthy roots cannot be expected to nodulate properly and fix nitrogen. This means, of course, that root health is the number one priority. So when it is necessary to use seed dressings of, say, a fungicide that is incompatible with *Rhizobium*, there are obvious implications for inoculation method.

D.P. Beck and L.A. Materon (editors), Nitrogen Fixation by Legumes in Mediterranean Agriculture
© 1988 ICARDA – Printed in the Netherlands. ISBN 90 247 3624 2

Influence of some Agroclimatic Factors and Agronomic Practices on Nitrogen Nutrition of Chickpea (*Cicer arietinum* L.)

J. WERY, M. DESCHAMPS and N. LEGER-CRESSON

ENSAM – INRA (LECSA) Equipe Legumineuses a Graines, Place Viala, 34060 Montpellier Cedex, France

Summary. The nitrogen nutrition of field-grown chickpeas was studied using *in situ* measurement of acetylene reduction activity and nitrate reductase activity. The influence of drought stress, sowing date, and mineral nitrogen supply on growth, nitrogen fixation (ARA), and nitrogen assimilation (NRA) was studied.

Drought stress, frequently experienced by spring-sown chickpeas, depressed nitrogen fixation more than nitrogen assimilation. The decrease of ARA appeared to be related to poorer growth of the vegetative parts of the non-irrigated plants.

Winter sowing improved grain and total dry matter production and stimulated ARA to a greater extent than total NRA. These effects could be explained by better growth conditions and less competition between pods and nodules in winter-sown chickpeas. Although nitrogen assimilation is essential to start plant growth and for subsequent establishment of good fixation activity, in fertile soils nitrate supply simultaneously increased assimilation and decreased fixation.

As previously shown in other legumes, nitrogen nutrition of chickpea appears to be the result of a necessary but unstable equilibrium between atmospheric nitrogen fixation and mineral nitrogen assimilation.

The implications for improving the production and nitrogen balance of chickpea crops are discussed in this paper.

Introduction

Increasing biological nitrogen fixation in cropping systems is one way to increase productivity and improve the farming systems of rainfed areas in the Mediterranean region. In addition to the potential offered by genetic improvement of the *Rhizobium*-legume symbiosis, rapid improvements in the amounts of nitrogen fixed could be obtained by:

(a) Increasing the importance of legumes in rotations. In the Mediterranean basin, grain legumes are cultivated on only a small part of the arable land area: 3.5% compared to 52.5% for cereals. This percentage could easily be increased.

(b) Increasing the amounts of nitrogen fixed by legumes.

To achieve these two objectives, the production and nitrogen balance of legume crops must be increased and regularized, making these crops more economically attractive to farmers.

First, a better understanding is required of nitrogen fixation activity in the field and of the influence of various factors (associated with the environment or with farmers' practices) on the efficiency of the symbiosis. For this, we must consider the complexity of legume nitrogen nutrition, which is a combination of atmospheric nitrogen fixation and mineral nitrogen assimilation (Wery *et al.* 1986).

In this study, the relationships between these two nitrogen reduction processes in the plant were analysed in field-grown chickpea by studying the influence of the mineral nitrogen level in the soil, water supply, and time of sowing. For each treatment we analysed a set of three curves representing the evolution of shoot dry matter production, acetylene reduction activity, and nitrate reductase activity during the growth cycle.

Materials and Methods

The results presented in this paper were obtained from two trials.

Trial 1: Effect of mineral nitrogen applied to winter- or spring-sown chickpeas (carried out during 1983/84) (Leger-Cresson 1984).

Trial 2: Effect of mineral nitrogen applied to spring-sown chickpea, with or without a water deficit (carried out during 1985) (Deschamps 1985).

Experiments were carried out at the experimental station of l'Ecole National Superieure Agronomique de Montpellier (ENSAM) on the following soils.

Trial 1: sandy-clay silt, 1m deep, with moderate fertility.
Trial 2: clay-sandy silt, very deep, with good fertility.

The chickpea cultivar INRA 199 was used. It is a desi type, with black seed. Its productivity is acceptable for spring sowing, but its level of frost and ascochyta blight resistance allows winter sowing under the conditions at ENSAM.

Using the results of previous experiments (Leger-Cresson 1984; Wery 1985), the chickpeas were cultivated intensively. In trial 1 there were 30 plants/m^2 with an inter-row spacing of 40 cm, while in trial 2, with the more fertile soil, there were 60 plants/m^2 with an inter-row spacing of 30 cm. Plants were fertilized with 100 kg P_2O_5/ha and 100 kg K_20/ha, and Herbalt S (Nitrofene + Neburon) was applied pre-emergence for weed control, which was completed by a hand weeding. Inoculation was not necessary as the soils contained an efficient population of *Rhizobium cicerii* (Leger-Cresson 1984).

The following treatments were used.

Sowing date

In trial 1 a spring sowing (6 March 1984) was compared with a winter sowing (24 November 1983). Treatments were labelled S and W respectively. In trial 2, the crop was sown on 28 February 1985.

Level of mineral nitrogen in the soil

In both trials, a control without nitrogen application (labelled O N) was compared with a treatment receiving 120 kg N/ha as ammonium nitrate (labelled N) (Table 1).

Water deficit

In trial 2 an irrigated chickpea crop (I) was compared with a crop planted on previously stored moisture and additional rainfall received during development (NI). For treatment I, irrigation was scheduled according to tensiometric information, using the method previously used for soybean (Wery *et al.* 1983; Roux 1985). Soil water content was measured every 10 days using a neutron probe. During the season, the irrigated treatment received 60 mm (three times), for a total water consumption of 327mm (Deschamps 1985).

To summarize, the following combinations were studied:

Trial 1	S–O N	Trial 2	I–O N
	S–N		I–N
	W O N		N I–O N
	W–N		N I–N

Table 1. Timing of mineral nitrogen supply.

Trial	Days after emergence	Crop stage	Nitrogen (kg/ha)	
1	Spring sowing	7	5-6 Leaf stage	60
		39	Early flowering	60
	Winter sowing	10	1-2 Leaf stage	30
		58	5-6 Leaf stage	30
		118	Early grain filling	60
2		3	Emergence	50
		13	7-8 Leaf stage	50
		34	13 Leaf stage	20

The trials were arranged in a split plot design with four replications. Sowing dates were whole plot treatments in trial 1 and water level whole plots in trial 2. For both trials, nitrogen application level was assigned to subplots of 16 m^2.

Measurements

Growth and yield

Growth was followed by measuring dry matter or nitrogen accumulation in aerial parts of three plants per replication. When pods appeared the reproductive parts (seeds and pods) were separated from the vegetative parts of the plant. At maturity, yield and yield components were measured on the central part (1 m^2) of each plot.

Nitrogen fixation

Atmospheric nitrogen fixation was evaluated by weekly measurements of acetylene reduction activity (ARA), using an *in situ* method (Wery 1983; Wery *et al.* 1986). This method allows measurements to be made on the same plants throughout the season. Measurements were made on four cylinders (30 cm diameter) per treatment. As in previous experiments on alfalfa (Wery 1983) and soybean (Roux 1985), ARA evolution was similar for the four cylinders of a given treatment even if spatial variability led to differences in absolute values (data not shown). We have reported the average values of the four cylinders, associated with the confidence interval at $P = 0.05$, on graphs. This ARA is expressed in μmole C_2H_4/hr/plant. By integrating each ARA curve on the whole cycle we obtained a comparative value of the total fixation of each treatment.

Nitrogen assimilation

Nitrogen (NO_3) assimilation was followed using measurements of nitrate reductase activity (NRA), using the *in situ* method of Robin *et al.* (1983), adapted to field-grown chickpea (Leger-Cresson 1984; Deschamps 1985).

In trial 1, assimilation activity of the crop (total NRA) was obtained by mutiplying NRA per gram of dry matter (measured on the four upper leaves of the primary stem) by the total plant dry matter. For each treatment 6–8 samples were used.

In trial 2, NRA and total NRA were directly measured on aerial parts of four plants per treatment.

NRA was expressed in u mole NO_2/hr/g DM and total NRA in u mole NO_2/hr/plant. Integration of the total NRA curves over the growth cycle gave a comparative value of total nitrogen assimilation between treatments.

Results and Discussion

Effect of drought on growth and yield

Irrigation delayed flowering by 1 week and this shift persisted until maturity (Fig. 1a). Water stimulated growth and caused lodging due to the uncontrolled development of vegetative plant parts (Fig. 1a). Although irrigation clearly increased total production and number of pods per plant (Table 2), it caused a significant decrease in grain yield (Table 3), probably due to increased competition between the vegetative and reproductive parts of the plant.

Effect of drought on nitrogen fixation

Nitrogen fixation (represented by ARA) increased quickly during the active growth phase (Fig. 1b). ARA decreased as soon as grains began to fill and became the dominant sink for photosynthates. The ARA peak corresponded to the beginning of remobilization from leaves (Figs 1a and 1b). According to Evans (1982), after a brief period of simultaneous increase in nodule mass and activity (with rapid development of fixation; a consequence of nodule mass increase), nodule specific activity decreases during the remainder of the cycle. Nitrogen deficiency resulting from this decrease in fixation at a stage when the grains' needs are higher, accelerates remobilization, carrying the plant towards what Sinclair and De Witt (1976) called self destruction.

The first irrigation stimulated nitrogen fixation, leading to a 43.3% increase in the maximum value of ARA (Fig. 1b). This resulted from improved nodule mass associated with the supplemental growth, and from a prolonged fixation period due to the delay of seed filling in the irrigated crop (Fig. 1b). Water supply may also have increased the acetylene reducing activity of the nodule tissues by improving their water status. Irrigation increased the amount of nitrogen fixed.

Effect of drought on nitrogen assimilation

Nitrate assimilation–as represented by total NRA–was maximum at the beginning of the growth cycle, ensuring nitrogen nutrition when fixation was still low or nil (Fig. 1c). Assimilation then decreased during the active growth phase as nitrogen fixation increased.

Irrigation slightly increased total NRA by stimulating the shoot specific activity (represented by NRA/g DM) (Fig. 1c), probably because of improved nitrate flow associated with increased transpiration (Triboi-Blondel 1978).

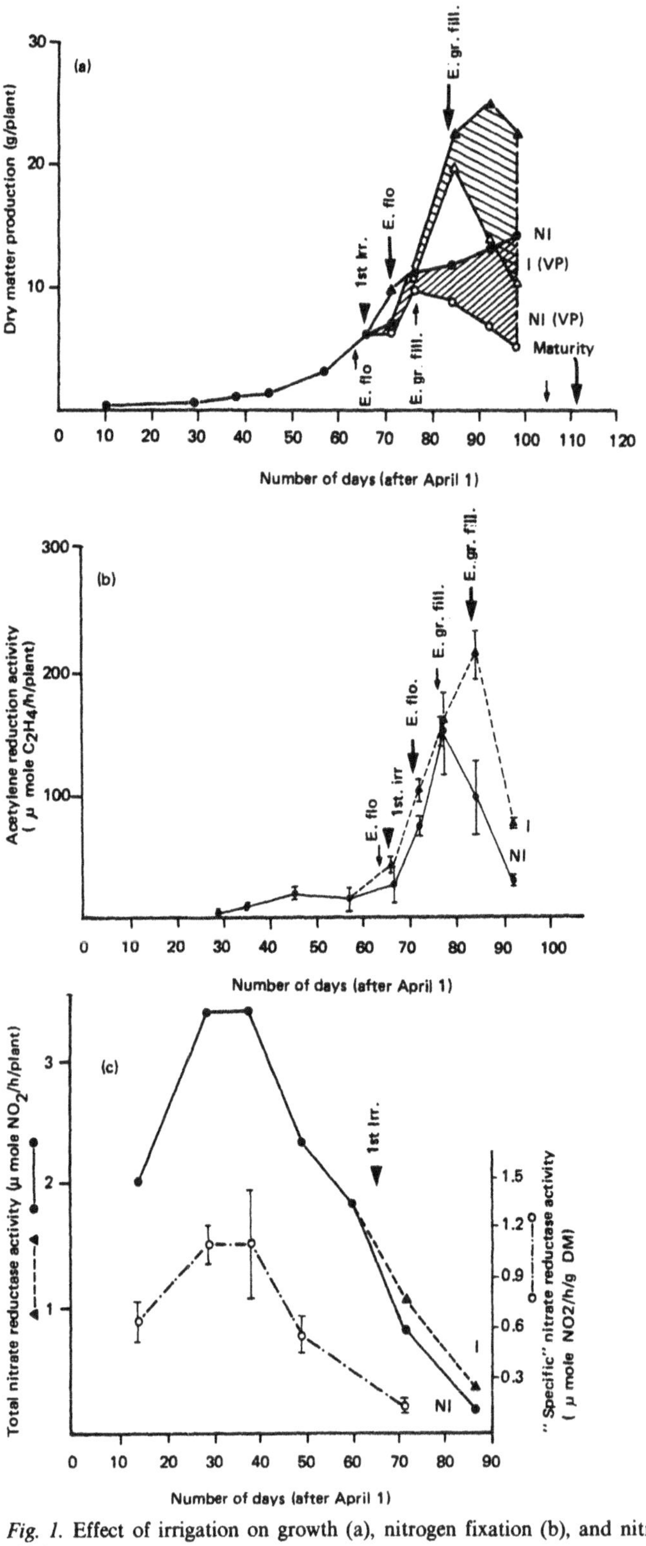

Fig. 1. Effect of irrigation on growth (a), nitrogen fixation (b), and nitrogen assimilation (c) in chickpea.
I = irrigated; NI = not irrigated; E. flo. = early flowering; E.gr.fill. = early filling. Open symbols are for total plant, closed symbols for vegetative parts. Thick and thin arrows for irrigated and nonirrigated crop.

Table 2. Effect of irrigation on number of pods per plant.

Date	No. of pods/plant	
	Non-irrigated	Irrigated
June 5	12 (1.3) *	1 (1.3)
June 11	21 (1.1)	8 (1.3)
June 17	24 (1.7)	31 (1.4)
June 25	31 (1.9)	44 (0.7)
July 2	42 (2.1)	60 (1.7)

* Figures in parentheses are confidence intervals at P = 0.05.

Conclusion

In agreement with the work of Obaton *et al.* (1982) on soybean, these results show that for chickpea, atmospheric nitrogen fixation is more sensitive to drought stress than nitrate assimilation, as seen from the area of the ARA curve which was 70.2% lower for the non-irrigated treatment. For total NRA however, this area was reduced by only 2.1%. The drought stress applied to the crop in our experiment was probably later and smaller than actually experienced by most chickpea crops. However, the high level of water storage in the soil and the rainy spring did not allow extreme drought stress in the non-irrigated treatment, as shown by the good yields obtained in this trial (Table 3), and from the small differences in soil water status between the two treatments (Deschamps 1985). Nevertheless it seemed that such a small water deficit is essential to obtain a high grain yield, as an unlimited water supply totally disrupted the plant to the benefit of the vegetative parts.

Finally, as shown for soybean (Bouniols *et al.* 1985), drought stress in chickpea depressed the percentage of nitrogen fixed more than grain production. It is possible that for most spring-sown chickpea crops when production is strongly limited by water deficit the amounts of nitrogen fixed could be very low.

In addition, these results could explain why, in poor soils, mineral nitrogen supply, which has no effect on irrigated chickpea, improves the yield of a non-irrigated crop (ICRISAT 1981). In this case, improved nitrogen assimilation compensated for part of the nitrogen deficiency resulting from decreased nitrogen fixing activity. This regulating role of nitrogen assimilation under drought stress was also observed in alfalfa (Wery *et al.* 1986).

Table 3. Effect of irrigation on yield and some yield components of chickpea.

Treatment	Biological yield DM (t/ha)	Grain yield DM (t/ha)	Harvest index	100-seed weight (g)	Protein (%)	Plants/m^2
Non-irrigated	8.46	2.64 (0.19) *	0.31	22.64 (1.43)	23.0	60
Irrigated	13.61	1.79 (0.45)	0.13	21.84 (0.74)	22.3	60

* Figures in parentheses are confidence intervals at P = 0.05.

Effect of sowing date on growth and yield

Winter sowing significantly increased chickpea yield which agreed with the results of several workers (Hawtin and Singh 1984; Keatinge and Cooper 1983; Saxena 1984). This is mainly due to an increase in the number of seeds per plant, better plant development, and lengthening of the growth cycle (173 days between emergence and maturity compared to 91 days with spring sowing). Winter sowing significantly increased dry matter production by 58.8% (Fig. 2a). In addition, the seed filling phase started 1 month earlier making water supply (Keatinge and Cooper 1983) and temperature conditions more

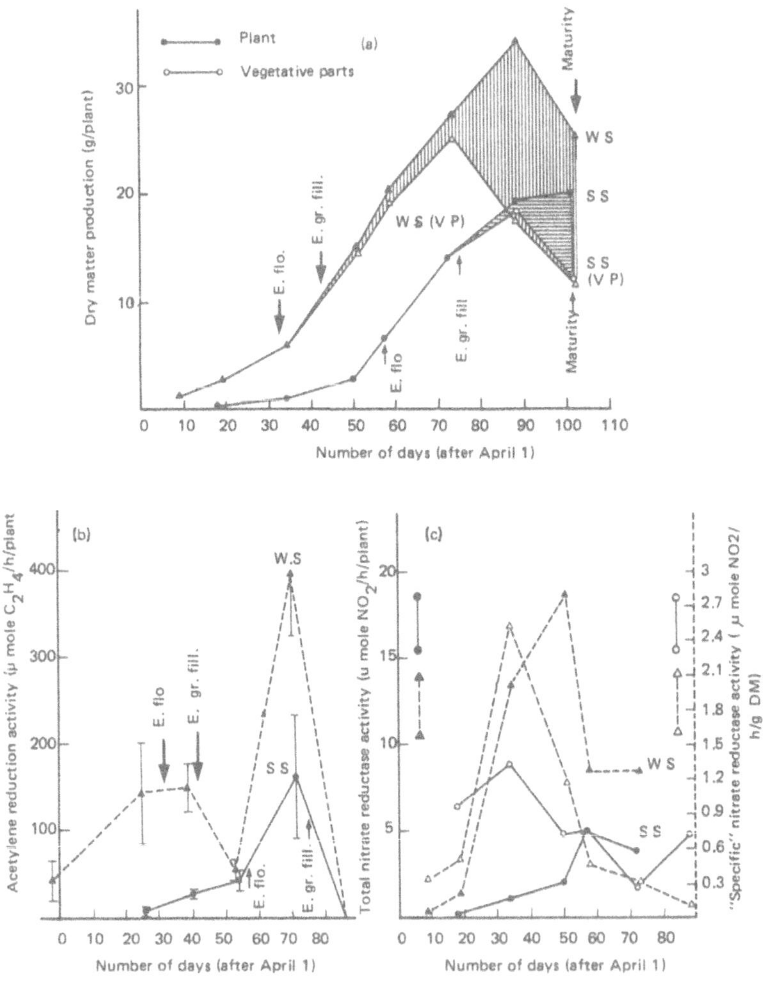

Fig. 2. Effect of sowing date on growth (a), nitrogen fixation (b), and nitrogen assimilation (c) of winter (WS, ↓) and spring sown (SS, ↓) chickpea.

Table 4. Effect of winter sowing on yield and some yield components of chickpea.

Treatment	Grain yield DM (t/ha)	100-seed weight (g)	Seeds/plant	Plants/m^2
Spring sowing	1.86 (0.20) *	25.7	13.3 (3.8)	21.1 (3.0)
Winter sowing	3.57 (0.13)	22.1	48.3 (15.2)	26.3 (3.0)

* Figures in parentheses are confidence limits at P = 0.05.

favorable for grain yield during this crucial period. Nevertheless, according to Saxena (1984) this earlier sowing has no effect on harvest index.

Effect of sowing date on nitrogen fixation

For spring-sown chickpea, ARA followed a similar pattern to that observed previously with a sharp decline as soon as seeds began to fill (Fig. 2b). With winter sowing, the fixation period was longer (100 days compared to 67 with spring sowing) and ARA values were clearly higher than for spring sowing. In the same type of comparisons, Islam (1984) observed an increase in the number of nodules which correlated with improved total dry matter production. Therefore, the earlier sowing produced a strong stimulation of nitrogen fixation, which could be accounted for by the 90% increase in area of the ARA curve. The pattern of the curve is also different. In winter-sown chickpea, the maximum ARA occurred at a later stage (mid-seed filling). This peak in fixing activity occurred at the same time (June 10) for the two sowing dates, which was the time when nitrogen accumulation was near its maximum value and when the curves representing total dry matter and vegetative dry matter began to separate due to remobilization towards pods (Fig. 2a).

It seemed therefore that winter sowing, by altering the beginning of the seed filling phase, reduced the competition between pods and nodules. For the two sowing dates, fixation fell as soon as the rate of nitrogen accumulation decreased, probably due to the combined action of drought stress and high temperature. Moreover these two factors directly inhibited nitrogen fixation, according to our results (for drought stress) and to the experiment of Rawsthorne *et al.* (1985) (for high temperatures).

Effect of sowing date on nitrogen assimilation

Winter sowing did not change the pattern of the NRA curve (Fig. 2c), which was maximum at the beginning of the cycle (early May), when fixation was still low. This activity decreased thereafter, during the active growth phase. Total NRA i.e., the product of NRA and dry matter production, presented different values for the two sowing dates. With winter sowing, the total NRA peak preceded the ARA peak: fixation gradually replaced assimilation as previously observed for soybean (Obaton *et al.* 1982). The same was observed for spring-sown chickpea with a good water supply. On the other hand, for spring sowing when conditions are less favourable for fixation, nitrogen assimilation continuously increased during the cycle (Fig. 2c).

Finally, by increasing growth winter sowing increased the amount of assimilated nitrogen, but in a lower proportion than fixed nitrogen.

Conclusion

Winter sowing lengthened the chickpea growth cycle so that the reproductive phase occurred in more favourable hydric and thermic conditions and grain and straw production were stimulated. Nitrogen fixation was improved due to an increase in the photosynthetic area and reduced competition between pods and nodules. Winter sowing simultaneously improved nitrogen production and the proportion of nitrogen coming from fixation, which was more enhanced than nitrogen assimilation. According to Keatinge *et al.* (1985), spring-sown chickpea fixed less than 50% of its nitrogen, while this proportion was clearly increased by winter sowing.

Effect of mineral nitrogen supply on growth and yield

In the spring sowing of trial 1, supplying mineral nitrogen had no effect on nitrogen accumulation in shoots (Fig. 3a: treatments S-ON and S-N). For all other treatments, supplying nitrogen slightly increased growth (Figs 3a and 4a). Increases occurred mainly in the winter sowing of trial 1 (Fig. 3a) and to a greater degree for the irrigated crop in trial 2 (Fig. 4a), indicating that under conditions of high growth potential nitrogen fixation, even when complemented by soil nitrogen, cannot ensure unlimited nitrogen nutrition. Therefore, nitrate supply is needed for optimal growth.

Table 5. Effect of mineral nitrogen supply on chickpea grain yield.

Trial	Treatment	Seed yield DM (t/ha)	100-seed weight (g)	Protein (%)
1	S–ON	1.86 (0.19) *	25.7	25
	S–N	1.46 (0.25)	27.1	25
	W–ON	3.57 (0.13)	22.1	25
	W–N	3.38 (0.31)	23.7	25
	NI–ON	2.64 (0.19)	22.63 (1.43)	26.13
2	NI–N	2.79 (0.19)	23.64 (0.52)	26.25
	I–ON	1.79 (0.45)	21.97 (0.74)	25.31
	I–N	1.92 (0.62)	22.74 (1.09)	26.31

* Figures in parentheses are confidence intervals at P = 0.05.

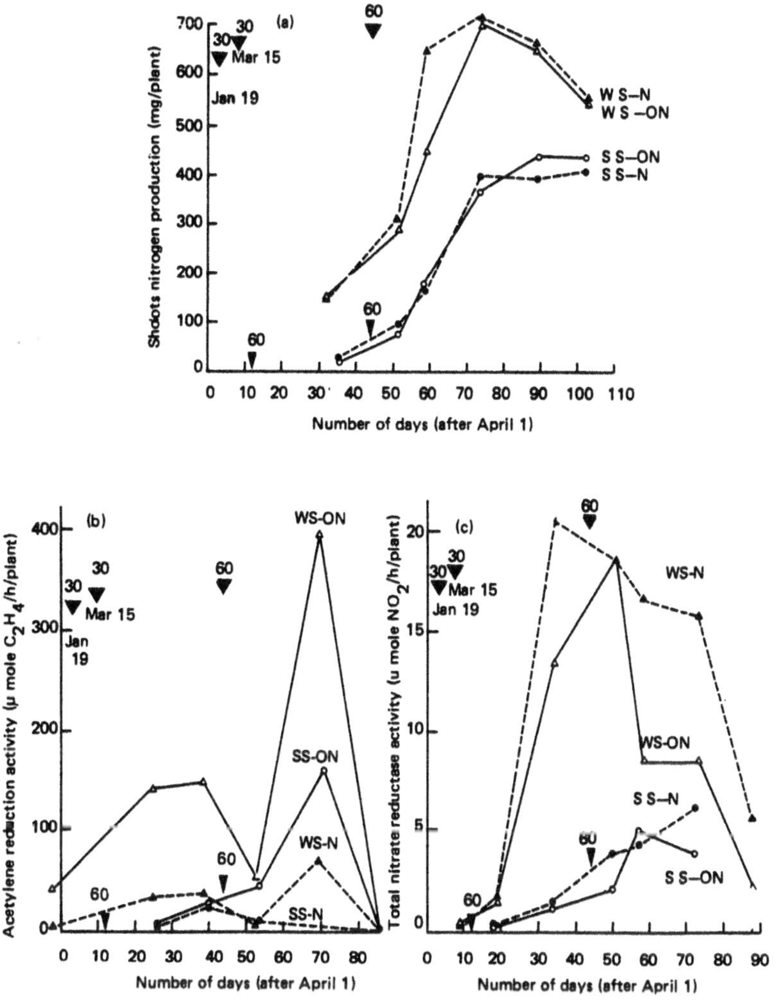

Fig. 3. Effect of mineral nitrogen supply on growth (a), nitrogen fixation (b), and nitrogen assimilation (c) of winter (WS) and spring sown (SS) chickpea.
▼ 60 and ▼ 60 indicate dates and rates (kg/ha) of nitrogen applied to WS and SS chickpea.

On the other hand, in the four treatments in each of the two trials, there were no significant increases in grain yield (Table 5). Nevertheless, grain mass and nitrogen content (in trial 2) were slightly increased with mineral nitrogen supply, indicating improved seed filling conditions.

For the non-irrigated crop (trial 2; Fig. 4a) growth stimulation with nitrate application occurred mainly during the period of drought stress, when nitrogen fixation was limited. Experiments in India showed that on poor soil the application of 20 kg N/ha improved the grain yield of a non-irrigated chickpea crop, but it had no effect on an irrigated crop (ICRISAT 1981).

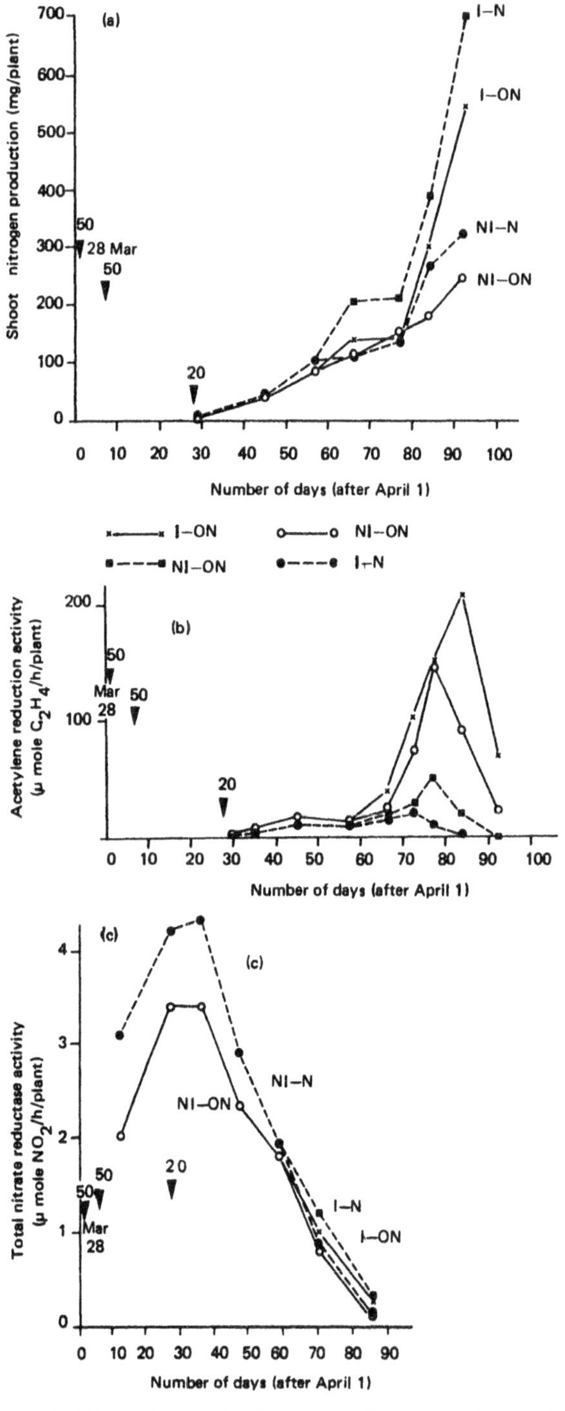

Fig. 4. Effect of mineral nitrogen supply on growth (a), nitrogen fixation (b), and nitrogen assimilation (c) of irrigated (I) and nonirrigated (NI) chickpea. N and ON = with and without nitrogen, respectively, 20 indicates dates and amounts (kg/ha) of nitrogen applied.

Effect of mineral nitrogen supply on nitrogen nutrition

For all treatments, supplying nitrate completely changed the nitrogen nutrition of chickpea. Nitrate assimilation (NRA; Figs 3c and 4c) was improved, probably because of the increase in amounts of nitrate absorbed by the plant. The pattern of the NRA curves was unchanged, so we can assume that evolution of this specific activity during the cycle was controlled more by the plant than by the nitrate availability in the soil.

On the other hand, nitrogen fixation (ARA; Figs 3b and 4b) decreased, with no change in the curve pattern. This inhibition of ARA by nitrate is considered to be a consequence of the direct effect on nodule formation and functioning, and of competition between the two pathways of nitrogen reduction (Wery *et al.* 1986).

Conclusion

These results, confirming those previously obtained on alfalfa (Wery *et al.* 1986), show that nitrogen fixation is very sensitive to increased nitrate availability in soil. Stimulation of assimilation is reflected by a decrease in fixation, as long as the nitrate supply does not improve plant growth status.

In soils having a low nitrate level or for plants grown in nutritive solution, chickpea growth could be limited by nitrogen deficiency, especially at the beginning of the growth cycle. In this case, a slight mineral nitrogen supply could improve both nitrogen production and nitrogen fixation, as observed by Jessop *et al.* (1984) and Rawsthorne *et al.* (1985).

Final conclusions

As previously shown for soybean (Obaton *et al.* 1982) and for alfalfa (Wery *et al.* 1986), nitrogen nutrition of chickpea is the result of an unstable equilibrium between N^2 fixation and NO^3 assimilation.

Normally, nitrate assimilation will ensure the start of growth and is then progressively replaced by symbiotic fixation. The latter is especially dependent on plant growth capacity; it falls as soon as the rate of dry matter production decreases with the start of remobilization from the vegetative to the reproductive parts. In most cases, except for winter-sown chickpea, this stage also corresponds with the start of seed filling.

In chickpeas, many factors associated with soil, climate, or cultural practices can totally change this equilibrium. Winter sowing and/or irrigation increase the amount of nitrogen fixed, mainly by stimulating growth. Under these conditions nitrogen fixation is more improved than nitrogen assimilation. Fixation seems to be more sensitive to stresses such as high temperatures (Rawsthorne *et al.* 1985) or drought than assimilation, which is mainly limited by nitrate availability in the soil.

The amount of nitrogen fixed can also be limited by factors which reduce chickpea growth, like phosphorus deficiency (Shukla and Yadav 1982), potassium deficiency (Idris and Sandhu 1983), and salt stress (Balasubramanian and Sinha 1976).

These results lead to two conclusions essential for grain legume improvement:

– Most of the cultural practices ensuring increased chickpea growth can probably improve the amount of nitrogen fixed by the crop. For chickpea in the Mediterranean region, the most promising procedure seems to be winter sowing or early spring sowing, particularly if it is associated with optimal stand, phosphorus fertilization, weed control, and protection against diseases and pests. Indeed this could simultaneously improve the grain yield, straw production, and nitrogen balance of the crop.

– Some factors, such as drought stress, are more limiting for fixation than for grain production, because of compensation by nitrogen assimilation and the remobilization process. These probably cause degradation of the nitrogen balance of the crop, more difficult to resolve as it is not detected by classic agronomic measurements.

In field conditions the *in situ* measurement of nitrate reductase activity and acetylene reduction activity, eventually associated with ^{15}N, could help to improve our knowledge of the efficiency of symbiosis in the field, and to analyse the influence of production factors on the nitrogen nutrition of legumes.

Acknowledgement

We gratefully acknowledge Dr M.S. Mekni (ICARDA) for the English editing of this paper.

References

Balasubramanian, V. and Sinha S.K. 1976. Nodulation and nitrogen fixation in chickpea (*Cicer arietinum*) under salt stress. Journal of Agricultural Science, Cambridge 87;465-466.

Bouniols, A., Puech, J., Chalamet, A. and Mondies, M. 1985. Influence des conditions d'alimentation hydrique ou (et) azotee a differents stades du developpement sur la production de grains et la nutrition azotee du Soja. Eurosoya 3: 55-61.

Deschamps, M. 1985. Influence de l'eau et de l'azote sur la croissance, la formation du rendement et la nutrition azotee de deux legumineuses a graines: *Cicer arietinum* et *Pisum sativum*. DEA Sciences Agronomiques, Option Phytotechnie, Universite des Sciences et Techniques du Langedoc, Ecole Nationale Superieure Agronomique de Montpellier, Montpellier, France. 65 pp.

Evans, J. 1982. Symbiosis, nitrogen and dry matter production in chickpea (*Cicer arietinum*). Experimental Agriculture 18: 339–351.

Hawtin, G.C. and Singh, K.B. 1984. Prospects and potential of winter sowing of chickpeas in the Mediterranean region. Pages 7-16 *in* Ascochyta Blight and Winter Sowing of Chickpeas (Saxena, M.C. and Singh K.B. eds.). World Crops, Vol. 9. Martinus Nijhoff, The Hague, The Netherlands.

ICRISAT (International Crops Research Institute for the Semi-Arid Tropics). 1981. Annual Report. ICRISAT, Hyderabad, India.

Idris, M. and Sandhu, G.R. 1983. The effect of potash fertilizer on the yield and N_2 fixing efficiency of chickpea. International Potash Institute, Section 13, No. 11.

Islam, R. 1984. Nodulation aspects of winter-planted chickpeas. Pages 159-166 *in* Ascochyta Blight and Winter Sowing of Chickpeas (Saxena, M.C. and Singh K.B. eds.). World Crops, Vol. 9. Martinus Nijhoff, The Hague, The Netherlands.

Jessop, R.S., Hetherington, S.J. and Hoult, E.H. 1984. The effect of soil nitrate on the growth, nodulation and nitrogen fixation of chickpeas (*Cicer arietinum*). Plant and Soil 82: 205–214.

Keatinge, J.D.H. and Cooper, P.J.M. 1983. Kabuli chickpea as a winter sown crop in Northern Syria: moisture relations and crop productivity. Journal of Agricultural Science (Camb.) 100: 667–680.

Keatinge, J.D.H., Saxena, M.C., Cooper, P.J.M. and Stephens, J. 1985. Biological nitrogen fixation by food legumes in dry areas. The scope for increase by improved management. Pages 219-226 *in* Proceedings of the International Workshop on Faba Beans, Kabuli Chickpeas and Lentils in the 1980's (Saxena, M.C. and Varma, S. eds.). Martinus Nijhoff, The Hague, The Netherlands.

Leger-Cresson, N. 1984. La formation du rendement chez le pois chiche (*Cicer arietinum*) de printemps et d'hiver. Etude de la nutrition azotee et de quelques aspects agronomiques. DEA Sciences Agronomiques, Option Phytotechnie, Universite des Sciences et Techniques du Langedoc, Ecole National Superieure Agronomique de Montpellier, Montpellier, France. 60 pp.

Obaton, M., Miquel, M., Robin, P., Conejero, G., Domenach, A.M. and Bardin, R. 1982. Influence du deficit hydrique sur l'activite nitrate reductase et nitrogenase chez le soja. Compte Rendus d'Academie de Science, Paris, Serie III, 294: 1007–1012.

Rawsthorne, S., Hadley, P., Roberts, E.H. and Summerfield, R.J. 1985. Effects of supplemental nitrate and thermal regime on the nitrogen nutrition of chickpea (*Cicer arietinum* L.). I. Growth and development. Plant and Soil 83: 265–277.

Rawsthorne, S., Hadley, P. Roberts, E.H. and Summerfield, R.J. 1985. Effects of supplemental nitrate and thermal regime on the nitrogen nutrition of chickpea (*Cicer arietinum* L.). II. Symbiotic development and nitrogen assimilation. Plant and Soil 83: 279–293.

Robin, P., Conejero, G., Tranchant, J.P., Passama, L. and Salsac, L. 1983. Mesure de reduction du nitrate dans les feuilles intactes, Physiologie Vegetale 21: 123–128.

Roux, M. 1985. Influence du mode de conduite de l'irrigation sur la croissance, le rendement et la fixation d'azote chez le soja en culture derobee. Memoire de fin d'etudes BTS CERCA Angers. Ecole National Superieure Agronomique de Montpellier, Montpellier, France.

Saxena, M.C. 1984. Agronomic studies on winter chickpeas. Page 123 *in* Ascochyta Blight and Winter Sowing of Chickpeas (Saxena, M.C. and Singh K.B., eds.). World Crops, Vol. 9 Martinus Nijhoff, The Hague, The Netherlands.

Shukla, U.C. and Yadav, D.P. 1982. Effects of phosphorus and zinc on nodulation and nitrogen fixation in chickpea (*Cicer arietinum* L.). Plant and Soil 65: 239–248.

Sinclair, T.R. and De Witt, C.T. 1976. Analysis of carbon and nitrogen limitations to soybean yield. Agronomy Journal 68: 319–324.

Triboi-Blondel, A.M. 1978. Effects de differents regimes d'alimentation hydriques sur l'activite *in vivo* de la nitrate reductase dans les feuilles de dactyle. Compte Rendus d'Academie de Science, Paris, Serie D, 286: 1795–1798.

Wery, J. 1983. Contribution a l'etude de la nutrition azotee d'une legumise fourragere (*Medicago sativa*) et de legumineuses a graines. These de Docteur Ingenieur en Agronomie. Ecole National Superieure Agronomique de Montpellier, Montpellier, France. 239 pp.

Wery, J. 1985. Le pois chiche: un proteagineux pour les regions mediterraneennes? Pages 24-25 *in* La France Agricole, 5 Juillet 1985.

Wery, J., Moutonnet, P. and Verret, F. 1983. Economie de l'eau d'une culture de soja en irrigation localisee. Bulletin du Groupe Francais d'Humidimetric Neutronique 14: 95–115.

Wery J., Turc, O. and Salsac, L. 1986. Relationships between N fixation, N assimilation and growth in a legume (*Medicago sativa* L.). Plant and Soil. (In press).

D.P. Beck and L.A. Materon (editors), Nitrogen Fixation by Legumes in Mediterranean Agriculture
© 1988 ICARDA – Printed in the Netherlands. ISBN 90 247 3624 2

Effect of Edaphic Factors on Biological Nitrogen Fixation in *Vicia faba* under Egyptian Field Conditions

A.S. ABDEL-GHAFFAR

Soil and Water Science Department, Alexandria University, Alexandria, Egypt

Introduction

Faba bean (*Vicia faba* L.) is the most important grain legume in Egypt. It is produced on more than 50% of the total area under food legumes, including soybeans. It serves as an important source of protein for humans, especially for those with low incomes. In spite of the rapid increase in population, the rise in prices of animal protein, and reclamation of new agricultural land, the area under faba bean decreased from 153420 ha in 1964 to only 113760 ha in 1984. This is mostly due to competition from other winter crops, mainly wheat and clover. However, the average yield for the country increased from 1.8 t/ha in 1964 to 2.4 t/ha in 1984. This poor yield may be the result of (a) abscission of flowers and immature pods, which may reach 75-95% (Salem *et al.* 1983), (b) inadequate measures to prevent or control weeds, diseases, and pests, (c) lack of adequate irrigation at critical stages during the growth period, (d) little or no application of fertilizers, especially phosphates, (e) problems of soil salinity associated with poor drainage, (f) lack of seed inoculation with effective strains of specific root nodule bacteria (*Rhizobium* sp.), (g) early or late planting; in Egypt, recommended planting dates are mid-October in Upper Egypt, late October in Middle Egypt, and early November in the Delta (Ibrahim 1982), and/or (h) losses due to manual harvesting and threshing.

Symbiotic nitrogen fixation by legumes is important as a practical means of improving the yield and quality of these crops. It depends on the interaction between the root-nodule bacteria (*Rhizobium* spp.) and the host plant. This interaction can be affected by environmental conditions which may affect the occurrence, growth, and survival of the rhizobia, modify nodule formation, and influence the function of the formed nodules (Vincent 1965). Also, environmental conditions may directly affect the growth of the plants independently of the biological nitrogen fixation process. To decide which type of reaction is involved, the response of the plants to a particular factor should be ascertained (a) in the absence and (b) in the presence of combined nitrogen. If the factor under consideration affects N_2 fixation, a response will be observed in (a) and not in (b).

Faba beans are efficient nitrogen-fixing legumes. Under Egyptian field conditions, the estimated average amount of atmospheric nitrogen fixed by faba bean is 135 kg/ha, 97 kg/ha for chickpea, 83 kg/ha for lentil, 78 kg/ha for peanut, and 40 kg/ha for soybean (Rizk 1966).

This paper discusses the main factors that may impose stress on or stimulate nodulation and subsequent N_2 fixation by faba beans under Egyptian conditions.

Inoculation

A compatible symbiosis, under field conditions, is established only with the strains of rhizobia that are able to survive in the soil, multiply in the rhizosphere of the legume host, compete successfully with native root nodule bacteria for sites on the roots, resist antagonistic action of soil organisms, infect the host plant, induce nodulation of the plant roots, and fix atmospheric nitrogen. If any of these steps are incomplete, an effective nitrogen-fixing symbiotic association between the rhizobia and the host plant is not established. To raise the nitrogen-fixing ability and yield of a leguminous crop, effective rhizobial strains able to compete successfully with the indigenous strains should be introduced into the soil in adequate quantities. Successful legume inoculation will make the host plant quickly independent of soil nitrogen and will most likely result in increased and better quality yield. In addition, under certain conditions, a net gain in soil nitrogen may result.

Sometimes, legumes do not respond to inoculation. This may be due to one or more of the following reasons (Dawson 1970).

1. The soil may already contain an abundant supply of efficient and effective nitrogen-fixing rhizobia of the proper strain or strains for the cultivated legume.
2. Environmental conditions in the soil, such as moisture, temperature, aeration, pH, etc., may be unfavorable.
3. The inoculant may be inappropriate, non-viable, contaminated with antagonistic organisms, or unsuited to the particular species or cultivar of cultivated legume.
4. Direct contact of inoculated seeds with fertilizers, seed disinfectants, or toxic chemicals.
5. Lack of certain nutrients such as P, Mo, Zn, Co, or B.
6. The soil may be rich in available nitrogen.
7. Presence of specific bacteriophage in the soil.

Egyptian farmers have been producing faba beans without inoculation and with little or no nitrogen fertilizer for hundreds of years. In general, with this practice the crop does not show signs of nitrogen deficiency. Examination of faba beans growing in the traditional areas indicated that the roots of every plant are well nodulated. The average number of nodules per plant in uninoculated faba bean fields varied between 80 and 94 and the dry weight of the nodules was 0.36–1.31 g/plant (Table 1).

This nodulation status of faba bean plants indicates that the soils of the traditional faba bean growing areas have adequate native populations of *Rhizobium leguminosarum*. The observed variation in nodulation in the differ-

Table 1. The status of nodulation of faba bean plants in uninoculated farmers' fields (Saxena and Stewart 1983).

Governorate	Number of nodules/plant		Nodule dry weight (g/plant)	
	Range	Average	Range	Average
Minia (Middle Egypt)	13–170	80	0.02–1.42	0.36
Kafr El-Sheikh (Delta)	30–137	94	0.22–3.14	1.31

ent regions can be attributed to fluctuations in the populations of specific root nodule bacteria. The population density of *R. leguminosarum* varied in Egyptian soils according to location, crop rotation, drying, waterlogging, and time of year (Hamdi 1982). In addition, it is well known that rhizobia are capable of an independent saprophytic existence in soil which does not contain legumes and can survive under these conditions for several years. However, they become drastically reduced in numbers or will disappear in the continued absence of the appropriate host (Hamdi 1982).

In spite of the large indigenous populations of *R. leguminosarum* in Egyptian soils, significant responses to inoculation have been recorded in terms of nodule numbers, weight, and seed and straw yields (Seleem 1947; Hamdi *et al.* 1966; Taha *et al.* 1967a; 1967b; Hamdi 1976; 1982; and Abdel-Ghaffar *et al.* 1982). Seleem (1947) studied the effect of inoculation on leguminous crops in farmers' fields. His data indicated that the inoculated faba bean plants had a faster growth rate, dark green color, more seeds per pod, and higher yields (50%) in comparison with the uninoculated plants. Field experiments conducted during 1979/80 at 10 locations in Middle Egypt (Minia Governorate), revealed significant responses to inoculation with respect to nodule number and weight in plants aged 77–91 days (Table 2). In this investigation, a significant seed yield response was only observed in 3 out of the 10 locations and there was no significant effect on straw yield at all locations (Saxena and Stewart 1983).

However, Hamdi *et al.* (1966) conducted experiments at the experimental farm of the Faculty of Agriculture of Ain Shams University (Cairo) and found that inoculation separately or combined with P and N fertilizers significantly increased the yield of faba bean and improved the quality of the grains (Table 3).

Table 2. Response of faba bean to inoculation in 10 fields in Minia governorate (Saxena and Stewart 1983).

	Uninoculated		Inoculated	
	Range	Mean	Range	Mean
Nodule number/plant	122–285	173	212–479	367
Nodule dry weight (g/plant)	0.09–1.36	0.6	0.70–1.96	1.3
Seed yield (t/ha)	2.21–4.18	2.9	2.76–4.20	3.2
Straw yield (t/ha)	4.80–12.47	7.2	4.53–14.11	8.1

Table 3. Effects of inoculation with and without phosphorus and nitrogen fertilization on faba bean grain yield and nutrient uptake (Hamdi *et al.* 1966).

Treatment			Grain yield (t/ha)	Grain content (kg/ha)		
kg N/ha	kg P_2O_5/ha	Inoculation *		N	P	K
0	0	−	1.59	76.2	7.6	4.8
0	0	+	1.96	95.2	10.5	6.7
0	72	−	1.91	111.9	10.9	6.2
0	72	+	2.29	114.3	13.3	8.6
36	0	−	1.81	88.1	9.0	5.2
36	0	+	2.18	104.8	10.9	7.4
36	72	−	2.11	102.4	12.1	6.4
36	72	+	2.55	121.4	14.3	9.8

* (−) uninoculated; (+) inoculated.

Table 4. The response of faba bean to different types of inoculants (Abdel-Salam 1986).

	Control (uninoculated)	Inoculant		
		Local (Okadeen)	Nitragen	
			Powder	Granular
Nodule number/plant *	240	244	297	383
Nodule dry weight(mg/plant *)	570	677	834	1055
μmol C_2H_4/plant/h *	6.1	7.4	13.9	23.2
μmol C_2H_4/g dry nodule/h *	10.7	10.9	16.6	11.0
Seed yield, relative	100	121	126	137

* Plants 94 days old.

A field experiment was carried out at the Agricultural Experiment Station of Alexandria University during the winter season of 1981/82 to study the response of faba bean to different types of inoculum. The tested inoculants were (a) Okadeen (local), (b) powder, and (c) granular obtained from the Nitragin Company, Milwaukee, Wisconsin, USA. The results given in Table 4 show that both inoculated and uninoculated plants were well nodulated. However, better nodulation, higher nitrogenase activity, and increased yields were observed with the inoculated compared to the uninoculated treatments. The granular form of inoculum was found to be superior in all respects to the powder and the local inoculum (Table 4).

Nitrogen Fertilization

The effects of combined nitrogen on symbiotic nitrogen fixation depend on several factors including host cultivar, *Rhizobium* strain, stage of plant development when the nitrogen fertilizer is applied, form and level of supplied nitrogen, soil temperature, soil moisture, type of soil, day length and whether the legume is grown as monoculture or in association with a non-legume. In general, little attention is given to available native soil nitrogen in the study of the effect of nitrogen fertilizers, although this soil nitrogen may influence the overall effect obtained. This may explain the contradicting results on the effect of combined nitrogen on nodulation and nitrogen fixation.

Both beneficial and detrimental effects of mineral nitrogen have been observed on nodulation and nitrogen fixation by legumes under field conditions.

Faba beans require large quantities of nitrogen to attain full yield and protein potential. McEwen (1970) reported that 4.82 t/ha faba bean seeds contained 229 kg N/ha. Richards (1977) stated that the production of 2.0 t/ha faba bean seed containing 30% protein required 194 kg N/ha, 116 kg N/ha in the seeds and 78 kg N/ha in the straw and roots.

In Canada, Rogalsky (1972) and Sadler (1975) and in Czechoslovakia, Kralova and Mouchova (1974) reported that nitrogen fertilizer significantly increased faba bean seed yields, suggesting that symbiotic nitrogen fixation was not fully able to satisfy faba bean requirements. On the other hand, McEwen (1970), Rinno *et al.* (1973), and Richards and Soper (1979) proved that faba beans, when nodulated with effective strains of rhizobia, were capable of symbiotically fixing all of their nitrogen requirements.

In general, both formation of nodules and nitrogen fixing activity decrease as the availability of combined nitrogen to the root increases. In the presence of large amounts of mineral nitrogen, nodulation may be entirely depressed. This is thought to be due to the exhaustion of carbonaceous material within the plant (Mulder *et al.* 1969), reduction in the number of root hairs formed (Thornton 1936), abortion of infection threads (Munns 1968), and/or inactivation of indoleacetic acid which is supposed to be important for nodule formation (Tanner and Anderson 1964). In addition, combined nitrogen may depress nitrogen fixation when added to nodulated plants (Mulder *et al.* 1969).

Early nitrogen deficiency at nodule initiation may cause serious growth depression of the legume seedlings that cannot be recovered when symbiotically fixed nitrogen is available. The addition of small amounts of combined nitrogen at sowing or during the early stages of plant growth prevents nitrogen deficiency of the seedlings. Nodulation may be retarded but in the long run there are benefits from the added mineral nitrogen due to a more developed root system providing more sites for nodule formation (Mulder *et al.* 1969).

In field trials in Egypt, Taha *et al.* (1967b) showed that application of up to 18 kg calcium nitrate per hectare significantly increased the seed yield of faba bean. Increasing nitrogen fertilization above this level reduced seed yield but increased the dry weight of straw and deleteriously affected nodulation. Also, Hamissa (1973) and Hamissa *et al.* (1975) reported that seed yields of faba beans grown in the major faba bean growing areas in Egypt increased by 9–12% over the control with the application of 18 kg N/ha. Ibrahim *et al.*

Table 5. Effect of nitrogen fertilization on faba bean nodulation, seed yield, and amount of fixed nitrogen (Hamissa *et al.* 1982).

	Growth stage *	Nitrogen fertilizer (kg/ha)		
		0	20	100
Nodule number/plant	a	65	71	50
	b	42	45	39
Nodule dry weight (mg/plant)	a	203	141	129
	b	270	290	210
Symbiotic nitrogen fixed (kg/ha)	c		118.9	111.6
Seed yield, relative	c	100	109	130

* (a) flowering, (b) pod filling, (c) maturity.

(1982) reported that the highest faba bean seed yield was obtained with the application of 18 kg N/ha at Gemmeza Research Station (Lower Egypt) and with 36 kg N/ha in farmers' fields in Minia (Middle Egypt) in addition to inoculation and P fertilization with 72 kg P_2O_5/ha. These high yields were 1.25 t/ha for Gemmeza and 3.74 t/ha for Minia.

Using ^{15}N-labelled ammonium sulphate and non-leguminous crops as reference, Hamissa *et al.* (1982) conducted a field trial to measure the amount of N_2 fixed by faba beans. The results (Table 5) showed that: (a) the average amount of fixed nitrogen was 119 kg/ha with 20 kg N/ha, 70 kg P_2O_5/ha, and inoculation, (b) the higher the rate of nitrogen fertilizer applied, the lower the number and dry weight of nodules and the amount of fixed nitrogen, (c) the increase in seed yield due to nitrogen application at the rate of 20 and 100 kg N/ha was 9 and 30%, respectively, over the control treatment, and (d) most of the nitrogen required by faba bean is taken from the soil and the atmosphere.

Salem (1983) conducted field experiments at Alexandria University Farm for two seasons (1979-81) to study the effect of nitrogen fertilization on seed yield and some characters of faba beans. Seed yield per plant and other plant characters increased with the addition of 37.5 kg N/ha as calcium ammonium nitrate, then decreased with further applications of nitrogen. The increase in seed production per hectare due to nitrogen application was attributed to the increase in seed yield of individual plants, number of pods per plant, number of branches per plant and 100-seed weight. Salem (1983) stated that the response of faba bean seed yield to the amount of added nitrogen was highly and significantly quadratic between 0 and 113.5 kg N/ha.

Abdel-Ghaffar *et al.* (1982) indicated that nitrogen fertilization (180 kg N/ha) reduced nodule number per plant and supressed N_2 fixation (ARA), but grain yield was not significantly affected. Two field experiments were conducted, one at the agricultural experimental station of Alexandria University and the second in a farmer's field in El-Deshoudy area (El-Behera Governorate) during the winter seasons 1980/81 and 1981/82, respectively. At both sites, inoculated seeds were used and P was added preplanting at the rate of 72 kg P_2O_5/ha.

Table 6. Effect of nitrogen fertilization on nodulation, N^2 fixation (ARA), and seed yield of faba bean at two different sites (Abdel-Salam 1986).

	Site 1 **			Site 2 **		
	kg N/ha		Changes (%)	kg N/ha		Changes (%)
	0	72		0	119	
Nodule number/plant *	140	121	−13.6	205	191	−6.8
Nodule dry weight (mg/plant)	358	267	−25.4	729	674	−7.5
μmol C_2H_4/plant/h	10.8	4.4	−59.3	24.8	15.6	−37.1
μmol C_2H_4/g dry nodule/h	30.2	16.4	−45.7	33.8	23.2	−31.4
Seed yield, relative	100	102.3	+2.3	100	90.1	−9.9

* Plants examined at 85 days.
** Site 1; Agricultural Experimental Station of AlexandriaUniversity, Site 2; Farmer's field at El-Deshoudy area, El-Behera Governorate.

The data (Table 6) show that addition of nitrogen fertilizer (ammonium nitrate) at the rate of 72 and 119 kg N/ha reduced nodule number and dry weight per plant and this was accompanied by a large decrease in total and specific nitrogenase activities of the nodulated roots. With respect to grain yield, application of 72 kg N/ha at site 1 increased grain yield by only 2.3%, while the addition of 119 kg N/ha at site 2 decreased grain yield by 9.9% over the control. On the contrary, increases in nodulation and N_2 fixation (ARA) were observed with the application of only 48 kg N/ha (Table 6).

Phosphorus Fertilization

Legumes usually require large amounts of phosphorus, and their ability to utilize soil phosphate is often less pronounced than that of cereals and grasses (Mulder *et al.* 1969). Phosphate fertilization is very often a necessary prerequisite for the success of inoculation. A deficiency of this element can reduce nodulation and suppress N_2 fixation (Vincent 1965).

Field experiments carried out at Maatana (Upper Egypt) by Taha *et al.* (1967a) and at the experimental farm of the Faculty of Agriculture of Ain Shams University by Hamdi *et al.* (1966) showed that phosphorus application, even without inoculation, markedly increased the grain yield of faba bean and improved seed quality (Table 3). Hamissa (1973) and Hamissa *et al.* (1975) conducted field trials in the major faba bean growing areas. They reported that grain yields increased gradually and significantly as the rate of applied P was increased to 72 kg P_2O_5/ha. On average, seed yields increased by 9.8% and 15.7% over the control with 36 and 72 kg P_2O_5/ha, respectively. They also concluded that the highest yield was obtained from a combination of 36 kg N/ha and 72 kg P_2O_5/ha

Abdel-Salam (1986) conducted field experiments on the response of faba beans to increasing amounts of phosphate fertilizers at the Agricultural

Experiment Station (clay soil) and in a farmer's field at El-Deshoudy (sandy clay loam soil), north of the Delta near Alexandria. The results indicated that application of P fertilizer preplanting was much better than at 74 days from sowing and that triple superphosphate was superior to superphosphate. With regard to rate of phosphorus fertilization, the number and dry weight of nodules, N_2 fixation (ARA), plant dry matter, and grain yields increased with increasing P fertilizer up to 108 kg P_2O_5/ha. Decreases were observed with higher P levels (Table 7). The response of nodule dry weight, total and specific nitrogenase activities, plant dry weight, and plant uptake of N and P to the amount of added P was significantly quadratic between 0 and 180 kg P_2O_5/ha. The coefficient of determinations (R^2) indicated that more than 90% of the sum of squares at the different levels of phosphorus are a function of the variation in the amount of added phosphorus. The data presented in Table 7 are for plots without nitrogen fertilization. A similar trend was observed with 119 kg N/ha. Also, these data indicate that high levels of phosphorus not only partially inhibited nodulation and nitrogen fixation, but also depressed plant growth and grain yield. Shoot dry weight and nitrogen uptake with 180 kg P_2O_5 were even lower than for the control.

A similar depressive effect of excess phosphorus on nodulation, nitrogen fixation, and plant growth was reported by Demeterio *et al* (1972) and Shukla and Yadav (1982) for soybean and chickpea, respectively.

They attributed this effect to the interference of P with the Zn nutrition of nodules and host. It seems that a certain optimum level of P and Zn is essential not only for plant growth but also for maximum nodulation, production of leghaemoglobin, and absorption of essential nutrients such as Fe and K by nodules fixing atmospheric nitrogen. Plant utilization of P and Zn depends on the levels in the soil solution. Amer (1978) observed that Zn uptake by faba bean plants decreased with increasing P.

Table 7. Response of faba bean grown at El-Deshoudy farm to increasing amounts of superphosphate fertilizer without nitrogen fertilization, 1981/82 (Abdel-Salam 1986).

	kg P_2O_5/ha					
	0	36	72	108	144	180
Nodule number/plant *	188	203	205	242	223	197
Nodule dry weight(mg/plant)	618	665	729	775	697	642
μmol C_2H_4/plant/h	15	22	25	29	23	18
μmol C_2H_4/g dry nodule/h	25	33	34	37	34	28
Shoot dry weight (g/plant) *	13	16	16	17	14	9
N uptake (mg/plant)	666	841	841	903	778	523
P uptake (mg/plant)	35	55	57	63	57	42
Grain yield, relative	100	105	116	132	128	116

* Plants 86 days old.

Trace Elements

Of the trace elements required for plant growth, boron and molybdenum are of particular importance for symbiotic nitrogen fixation (Mulder *et al.* 1969). Boron is important in controlling the development of vascular tissue in the nodule, calcium intake, and carbohydrate translocation. Insufficient boron causes the formation of necrotic nodules that do not fix nitrogen. In molybdenum-deficient plants, nodulation proceeds normally but the nodules are more numerous, their color is yellow instead of the normal pink color, and they do not fix nitrogen. In addition, many workers found that legumes depending on symbiotic nitrogen fixation seem to require cobalt in trace amounts for the proper functioning of the nodules (Abdel-Ghaffar 1982).

Hakim (1974) reported that the application of 0.5 ppm molybdenum to clay loam and sandy loam soils, or pre-soaking faba bean seeds in 0.1% molybdic acid solution increased the nodulation, dry matter yield, and nitrogen content of plants. He also observed that soil application of boron at 5 ppm inhibited nitrogen fixation and attributed this effect to the abundance of this element in the soil. In field studies near Alexandria, El-Shiekh (1981) found that spraying faba bean plants with Zn, Mn, and Mo increased faba bean seed yields, while Fe had no effect.

The effect of foliar application of zinc on nodulation and N_2 fixation was studied in a field trial at the Agricultural Research Station of Alexandria University during the winter 1981/82. The results indicated that foliar application of Zn to inoculated faba bean plants with and without nitrogen fertilization increased nodule number and dry weight, nitrogenase activity, plant dry weight, and uptake of N and P (Table 8).

Soil Moisture

Several studies conducted in Egypt and elsewhere have shown that in nodulated faba bean plants there is a high degree of correlation between nitrogen fixation and soil water content. Sprent (1972) obtained a quadratic relationship between N_2 fixation and soil moisture during the vegetative growth period of faba bean, indicating that both water stress and waterlogging were inhibitory. Gallacher and Sprent (1978) reported that water stress could retard the development of nodules which were initiated under conditions of adequate water supply. This situation may occur quite frequently in the field since, under natural conditions, water stress usually occurs gradually and progressively as moisture is lost from the soil. Faba bean plants exposed to excess water grow better and fix more nitrogen than those under drought stress.

Abdel-Wahab and Zahran (1979) reported the depressing effects of moisture stress on faba beans. In this study, inoculated plants were subjected to water stress at 66, 33, or 16.5% field capacity for 3 weeks, after which water

Table 8. Response of faba bean to foliar application of zinc and nitrogen fertilization *, Agricultural Experiment Station, Alexandria University (Abdel-Salam 1986).

	N	Zn treatment		Increase (%)
		–	+	
Nodule number/plant **	–	163	214	31.3
	+	202	316	56.4
Nodule dry weight (mg/plant)	–	829	1019	22.9
	+	964	1134	17.6
μmol C_2H_4/plant/h	–	11.6	18.7	61.2
	+	15.1	23.5	55.6
μmol C_2H_4/g dry nodule/h	–	14.0	18.3	30.7
	+	15.7	21.4	36.3
Dry shoots (g/plant)	–	64.1	73.3	14.4
	+	66.7	77.5	16.2
N uptake (mg/plant)	–	3269	3857	18.0
	+	3415	4118	20.6
P uptake (mg/plant)	–	186	254	36.6
	+	223	280	25.6

* Zinc sulfate solution containing 0.3% Zn, N applied to soil at 48 kg N/ha.
** Plants 106 days old.

was applied to field capacity. After 1 week, plants growing at less than one-third of field capacity showed a marked reduction in nitrogen fixation (ARA). A progressive decline occurred and nitrogen fixation was completely inhibited on the third week of stress. At two-thirds of field capacity, only a slight reduction in nodule activity was observed. However, removing the stress by irrigation resulted in partial and progressive recovery. After 2 weeks of irrigation at field capacity, activity was normal in all treatments except the severest stress (16.5% field capacity) in which only partial recovery occurred.

Said (1981) conducted field experiments at Kafr El-Sheikh Governorate to investigate the effect of water stress on nodulation and nitrogen fixation by faba bean. The soil was irrigated after 25, 50, or 75% depletion of available soil moisture. The data presented in Table 9 indicate that nodule number and dry weight, total and specific nitrogenase activity in 107-day old plants, as well as straw and grain yields were higher when irrigation was at 50% depletion of available soil moisture than at either 25 or 75% depletion levels. In other words, four irrigations were better than either five or three irrigations.

Tadrous (1981) reported that the optimum irrigation regime for *Vicia faba* was 40-60% depletion, and that irrigation at 80% depletion decreased seed yields by 17–20%. Salem *et al.* (1983) conducted field experiments at the Agricultural Experiment Station of Alexandria University and obtained maximum seed yield with only two irrigations at an interval of about 4 weeks from sowing. The first irrigation was before flowering and the second during pod

Table 9. Effect of water stress on nodulation, N_2 fixation, (ARA) and yield of faba bean (Said 1981).

	Moisture depletion level at irrigation (%) **		
	25	50	75
Nodule number/plant *	18	54	53
Nodule dry weight (mg/plant)	220	856	823
μmol C_2H_4/plant/h	5.9	9.6	4.7
μmol C_2H_4/g dry nodule/h	2.7	11.3	5.7
Straw yield, relative	91.5	100	77.6
Grain yield, relative	92.4	100	80.3
Seed N (%)	3.04	3.40	3.74

* Plants 107 days old.
** Number of irrigations: 5 (25%), 4 (50%), and 3 (75%).

formation. This might be true for the northern part of Egypt because of the higher rainfall of this region. Metwally (1973) at Giza and El-Maghraby (1980) at Moshtohor in South Delta showed that five irrigations gave the best faba bean yields, and stated that increasing the depletion of available soil moisture before irrigation decreased seed yield.

El-Shakweer and El-Samanoudy (1985) conducted a field experiment at the farm of El-Fayoum Faculty of Agriculture to study the effect of water stress on faba bean growth from sowing to maximum flowering and from maximum flowering to harvest. They obtained highest grain yield by irrigating at 70% depletion of available water from sowing to maximum flowering followed by irrigation at 40% depletion.

Soil Salinity

Saline soils not previously utilized in agriculture are becoming more important as new land is brought into production to meet the demands of an increasing population or to replace land lost to urban development. Before saline soils can be cropped, they must be reclaimed by applying water to leach out excess salts. A high concentration of soluble salts in the root zone adversely affects the growth and yield of faba beans. This effect is attributed to: (a) direct toxicity of the salt, (b) reduction in availability of soil water by raising the osmotic pressure of the soil solution, (c) changes in availability of nutrients due to ion antagonism, and (d) changes in physical properties of the soil restricting water movement or reducing root penetration (Caesar and Rusitzka 1982). Subba Rao (1977) reported that high salinity or alkalinity decreased the formation of root hairs and the deposition of mucilage around the roots which are necessary pre-requisites for successful entry of rhizobia into roots.

Kamel *et al.* (1971) and Ibrahim Khadr (1974) indicated that leaching of

Table 10. Effect of reclamation of saline soil on nodulation and yield of faba bean grown in lysimeters (Kamel *et al.* 1971).

	Number of leaching			
	0	1	2	3
Soil T.S.S., %	5.20	0.98	0.60	0.23
Soil pH	7.80	8.00	8.00	8.20
Nodule number/plant *	0.00	63.00	104.00	204.00
Dry plant weight (g/plant)	7.90	21.10	21.80	24.50
Seed weight (g/plant)	4.60	12.30	12.20	13.90
N uptake (mg/plant)	237.40	522.50	533.40	649.50
P uptake (mg/plant	24.60	61.80	60.60	56.30

* Plants 60 days old.

toxic salts from saline soils resulted in significant increases in nodule number per plant, dry matter, seed yield, and nitrogen and phosphorus uptake by faba bean plants (Table 10).

Studies on the effect of salinity on nodulation and nitrogen fixation by faba beans under field conditions are rather limited. Lockerman *et al.* (1983) conducted a field experiment at the Montana Agricultural Experiment Station, USA, to study the effects of soil salinity on nodulation, N_2 fixation (ARA), and yields of grain and dry matter of faba bean cv Ackerperle. The results indicated that dry weight of roots, shoots, and seeds decreased as soil salinity increased from EC 2 to 14 mmhos/cm. Acetylene reduction activity was proportional to nodulation and reached a maximum at approximately EC 5 mmhos/cm. The depressive effect of salinity on nitrogenase activity was highly pronounced at 40 days after planting. As the plants aged (> 60 days), the differences between the different electrical conductivities, with respect to rate of acetylene reduction, gradually diminished. They stated that faba bean is moderately salt tolerant, decreasing to 50% of maximum yield at about EC 11 mmhos/cm, while lentil (*Lens culinaris*), green pea (*Pisum sativum*), and dry bean (*Phaseolus vulgaris*) yields decrease to 50% at EC 4 mmhos/cm. In Sudan, El-Karouri (1979) reported that the salinity at which 50% reduction in faba bean yield occurred was 9 mmhos/cm.

The detrimental effect of salinity not only depends on the concentration of soluble salts (EC) but also on the nature of the salts. Ibrahim *et al.* (1970) showed that nodule formation, dry weight, and nitrogen content of faba bean plants decreased with increasing concentration of the different sodium salts. Sodium chloride and sodium carbonate had a greater effect on nodule number than sodium sulfate.

At Sakha Experimental Station, Zaghloul (1981) conducted pot experiments to study the effect of soil salinity on the nodulation and growth of faba bean. The soil was salinized with a mixture of NaCl and $CaCl_2$ (1:1) before potting. As shown in Table 11, the numbers and dry weights of nodules were not affected by increasing the salinity from 2.2 to 10.0 mmhos/cm, although acetylene reduction activity increased. At a salinity of 15 mmhos/cm a

Table 11. Effect of soil salinity on nodulation, nitrogen fixation, and yield of faba bean (Zaghloul 1981).

	Soil salinity (EC_e) mmhos/cm			
	2.2	5.0	10.0	15.0
Nodule number/plant *	116	136	137	54
Nodule dry weight (mg/plant)	250	265	225	105
μmol C_2H_4/plant/h	13.5	14.5	15.4	5.9
μmol C_2H_4/g dry nodule/h	51.0	55.3	64.4	53.5
Dry matter (g/plant)	11.0	9.5	6.3	3.0
Total N (mg/plant)	288	288	158	55

* Plants 82 days old.

reduction in nodulation and nitrogenase activity was observed. Plant dry matter and nitrogen content were greatly reduced as salinity increased from 5 to 10 mmhos/cm. At high soil salinity (10-15 mmhos/cm), which greatly depressed plant growth, there was good nodulation and nitrogenase activity indicating that the rhizobia (*R. leguminosarum*) can tolerate higher levels of soil salinity than the host plant. These findings reflect the importance of selecting varieties or cultivars of leguminous crops capable of withstanding salinity rather than attempting to introduce salt-tolerant rhizobial strains into soil.

In addition, Abdel-Ghaffar *et al.* (1982) showed that irrigation with water salinized with a mixture of NaCl and $CaCl_2$ (1:1) to more than 2 mmhos/cm reduced nodulation, N_2 fixation (ARA), and yields of faba bean (Table 12). However, fertilization with 180 kg N/ha counteracted the adverse effect of salinity on nodulation and acetylene reduction but not on yield.

Summary

Faba bean (*Vicia faba* L.) is the most important grain legume in Egypt. It is produced on more than 50% of the total area under food legumes including soybeans. In 1984, it occupied about 113760 ha. Faba beans are the main source of protein for a majority of the population.

Faba bean, when nodulated with an effective strain of *Rhizobium* (*R. leguminosarum*), is an efficient nitrogen fixing legume. On average, it fixes about 135 kg N/ha. Roots of uninoculated plants from faba bean growing areas are well nodulated and the crop shows no signs of nitrogen deficiency. However, inoculation with effective strains of rhizobia capable of competing with the indigenous rhizobia has proved to be a practical means of increasing yield and improving grain quality.

Depending on the rate of nitrogen fertilization, both beneficial and detrimental effects are observed on nodulation and nitrogen fixation processes. In general, addition of small amounts of nitrogen (18-48 kg N/ha) at sowing or

Table 12. Effect of salinity of irrigation water and nitrogen fertilization (180 kg N/ha) on nodulation, acetylene reduction, and yield of faba bean in a pot experiment (Abdel-Ghaffar *et al.* 1982).

	N fertil-izer	EC of irrigation water (mmhos/cm)			
		0	2	5	10
		EC of soil extract at harvest (mmhos/cm)			
		1.5	8.8	15.5	26.0
Nodule number/plant *	−	382	97	79	4
	+	132	274	321	51
μmole C_2H_4/plant/h	−	12.8	7.2	7.6	0.9
	+	4.6	21.3	28.5	2.6
μmol C_2H_4/g dry nodule/h	−	113	171	120	3
	+	97	184	193	37
Dry matter yield (g/plant) **	−	14.12	12.72	11.56	3.05
	+	16.37	14.18	11.14	2.99
Seed yield (g/plant)	−	4.63	4.02	3.35	0.25
	+	4.45	4.21	3.62	0.33

* Plants 70 days old; ** plants 120 days old.

during the seedling stage usually stimulates nodulation and enhances symbiotic nitrogen fixation.

The responses of faba bean to phosphorus fertilization, in terms of number and weight of nodules, acetylene reduction, and grain yield is quadratic, between 0 and 180 kg P_2O_5/ha. Highest values are obtained at 72-108 kg P_2O_5/ha.

Presoaking seeds in, or foliar application of, solutions of the trace elements Mo, Zn, and perhaps Mn usually results in increased nodulation, nitrogen fixation, dry matter yields, and N and P contents of faba bean plants.

Water stress causes a marked reduction in nodulation, nitrogen fixation, growth, and yield. However, removing the stress by irrigation restores these activities. Usually, high yields are obtained when plants are irrigated at 50% depletion of available water.

Irrigation with saline water (EC > 5 mmhos/cm) negatively affects nodulation and growth. However, though soil salinity of more than 5 mmhos/cm markedly depresses the growth of faba beans, it may stimulate nodulation and nitrogen-fixing activity.

References

Abdel-Ghaffar, A.S. 1982. Nodulation problems and response to inoculation. First OAU/STRC International African Conference on Bio-Fertilizers. Cairo, Egypt, 22-26 March 1982.

Abdel-Ghaffar, A.S., El-Attar, H.A., El-Halfawi, M.H. and Abdel-Salam, A.A. 1982. Effects of inoculation, nitrogen fertilizer, salinity, and water stress on symbiotic N_2 fixation by *Vicia faba* and *Phaseolus vulgaris*. Pages 153-159 *in* Biological Nitrogen Fixation Technology for Tropical Agriculture (Graham, P.H. and Harris, S.C., eds.). Centro Internacional de Agriculture Tropical, Cali, Colombia.

Abdel-Salam, A.A. 1986. Effects of different fertilizers on yields of faba bean (*Vicia faba*) and dry beans (*Phaseolus vulgaris*). Ph.D. thesis. Faculty of Agriculture, Alexandria University, Egypt.

Abdel-Wahab, A.M. and Zahran, H.M. 1979. The effect of water stress on $N_2(C_2H_2)$ fixation and growth of four legumes. Pages 175-261 *in* Proceedings of the Society of Applied Microbiology, Annual Meeting, Cairo, Egypt.

Amer, M.A.O. 1978. Phosphorus-zinc interaction in relation to plant growth and their absorption by corn and bean plants grown in calcareous soil. M.Sc. thesis. Faculty of Agriculture, Alexandria University, Egypt.

Caesar, K. and Rusitzka, G. 1982. Tolerance to salinity. Pages 185-189 *in* Faba Bean Improvement (Hawtin, G. and Webb, C., eds.). Martinus Nijhoff, The Hague, The Netherlands.

Dawson, R.C. 1970. Potential for increasing protein production by legume inoculation. Plant and Soil 32: 655-673.

Demeterio, J.L., Roscoe, E. Jr., and Poulsen, G.M. 1972. Nodulation and nitrogen fixation by two soybean varieties as affected by phosphorus and zinc nutrition. Agronomy Journal 64: 566-568.

El-Karouri, M.O.H. 1970. Effects of soil salinity on broad bean (*Vicia faba*). Sudan, Experimental Agriculture 19: 59-63.

El-Maghraby, S. 1980. Effect of water regime, nitrogen and phosphatic fertilizers on growth and yield of broad bean. M.Sc. thesis. Faculty of Agriculture, (Moshtohor), Zagazig University, Egypt.

El-Shakweer, M.H.A. and El-Samanoudy, I.M. 1985. Effect of ground water level on legume crops in Fayoum Governorate. 6. Optimum irrigation of faba bean. Alex. Sci. Exch. 6 (3): 65-77.

El-Sheikh, M.H. 1981. Studies of faba bean. Effect of major cultural practices and microelements on faba bean. Ph.D. thesis. Alexandria University, Egypt.

Gallacher, A.E. and Sprent, J.I. 1978. The effect of different water regimes on growth and nodule development of greenhouse-grown *Vicia faba*. Journal of Experimental Botany 29 (109): 413-423.

Hakim, M.A.A. 1974. Symbiotic nitrogen fixation and growth of leguminous plants as influenced by some minor elements. Ph.D. thesis. Al-Azhar University, Cairo, Egypt.

Hamdi, Y.A. 1976. Field and greenhouse experiments on the response of legumes in Egypt to inoculation and fertilizers. Pages 289-298 *in* Symbiotic Nitrogen Fixation in Plants (Nutman, P.S., ed.). Cambridge University Press, London, UK.

Hamdi, Y.A. 1982. Symbiotic nitrogen fixation in faba beans. Pages 127-138 *in* Faba Bean Improvement (Hawtinu, G. and Webb, C., eds.). Martinus Nijhoff, The Hague, The Netherlands.

Hamdi, H., Taha, S., El-Damaty, A., and El-Sherif, A. 1966. The effect of bacterial inoculation and phosphatic fertilization on some legumes in a sandy loam soil of the UAR. Journal of Soil Science (A.R.E.) 6(1): 1-16.

Hamissa, M.R. 1973. Fertilizer requirements for broad beans and lentils. First FAO/SIDA Seminar for Plant Scientists from Africa and Near East, Cairo, Egypt.

Hamissa, M.R., Abdel-Samie, M.E., El-Banna, E.A., Abdel-Aziz, M.S., Abdel-Bary, S., and Abdel-Moneim, M.I. 1975. Response of bean plant (*Vicia faba*) to N, P and K. Agricultural Research Review (Egypt) 53: 155-166.

Hamissa, M.R., Khadr, M.S., Moustafa, A.H., Hammam, A., Hashim, Y., Shalaby S., Abido, Y., and Wahhab, S.A. 1982. Biological dinitrogen fixation by faba beans and chickpeas using the tracer technique. First OAU/STRC Inter-African Conference on Bio-Fertilizers. Cairo, Egypt, 22-26 March 1982.

Ibrahim, A.A. 1982. The Egyptian national program. Pages 373-379 *in* Faba Bean Improvement (Hawtin, G. and Webb, C., eds.). Martinus Nijhoff, The Hague, The Netherlands.

Ibrahim, A.N., Kamel, M. and Khadr, M.S. 1970. Nodule formation and growth of legumes as influenced by certain sodium salts. Agrokemia es Talajtan 19: 164-172.

Ibrahim, A.N. and Khadr, M.S. 1974. The effect of soil improvement on the nodule formation and plant growth of some leguminous crops in saline areas of the Nile Delta. Agrokemia es Talajtan 23: 111-120.

Ibrahim, A.A., Nassib, A.M., and El-Sherbeeny, M.H. 1982. Faba bean agronomy in Egypt. Pages 109-116 *in* Faba Bean Improvement (Hawtin, G. and Webb, C., eds.). Martinus Nijhoff, The Hague, The Netherlands.

Kamel, M., Ibrahim, A.N., Hamissa, M.R., and Khadr, M. 1971. Effect of reclamation of saline soil on nodule formation and yield of broad bean. Agricultural Research Review 49(6):145-155.

Kralova, M. and Mouchova, H. 1974. Nutrient uptake by horsebean from ammonium sulfate and potassium nitrate labelled by the ^{15}N-isotope. Rostlinna Vyroba 20: 335-343.

Lockerman, R.H., Kisha, T.J., Sims, J.R., and Abdel-Ghaffar, A.S. 1983. The effect of soil salinity on dinitrogen fixation and yield of faba bean (*Vicia faba* L.). FABIS Newsletter 7: 24-25.

McEwen, J. 1970. Fertilizer nitrogen and growth regulators for field beans. Journal of Agricultural Science, Cambridge 74: 61-72.

Metwally, M.A. 1973. Study of the effect of irrigation and fertilization on yield and technological properties in field bean (*Vicia faba* L.). M.Sc. thesis. Faculty of Agriculture, Al-Azhar University, Cairo, Egypt.

Mulder, E.G., Lie, T.A., and Woldendrop, J.W. 1969. Biology and soil fertility. Pages 163-208 *in* Soil Biology, Reviews of Research. UNESCO, Paris, France.

Munns, D.N. 1968. Nodulation of *Medicago sativa* in solution culture. Plant and Soil 28: 246-257 and 29: 33-47, 257-262.

Richards, J.E. 1977. The effect of fertilizer nitrogen on yield, crude protein content, and symbiotic fixation in *Vicia faba* L. var *minor*. M.Sc. thesis. University of Manitoba, Canada.

Richards, J.E. and Soper, R.J. 1979. Effect of N fertilizer on yield, protein content, and symbiotic N fixation in faba beans. Agronomy Journal 71:807-811.

Rinne, G., Ebert, K., and Wildner, M. 1973. The effect of nitrogen fertilization on the amino acid content of legumes grown for grain. Tag Ver. Akad. Landwirtach. Wiss, DDR. 122: 67-75.

Rizk, S.G. 1966. Atmospheric nitrogen fixation by legumes under Egyptian conditions. II. Grain legumes. Journal of Microbiology (Egypt) 1(1): 33-45.

Rogalsky, J.R. 1972. Summary of faba bean fertility trials. Pages 169-170 *in* 16th Annual Manitoba Soil Science Meeting, University of Manitoba, Canada.

Sadler, J.M. 1975. Soil, fertilizers, and symbiotically-fixed nitrogen as sources of nitrogen for faba beans under prairie conditions. Pages 121-129 *in* 19th Annual Manitoba Soil Science Meeting, University of Manitoba, Canada.

Said, M.M.A. 1981. Effects of water stress and nitrogen fertilization on symbiotic nitrogen fixation by faba beans (*Vicia faba*) and soybeans (*Glycine max*). M.Sc. thesis. Faculty of Agriculture, Alexandria University, Egypt.

Salem, S.A. 1983. Effect of nitrogen level and cultivars on the seed yield and some plant characters of faba bean (*Vicia faba* L.). Beitraege zur Tropischen Landwirtschaft und Veterinaermedizin 21 (H.2):159-165.

Salem, S.A., Caesar, K. and Bartels, M. 1983. Field studies on the irrigation levels and cultivars in faba bean (*Vicia faba*) L. in Northern Egypt. Zeitschrift fuer Acker-und Pflanzenbau 152: 325-335.

Saxena, M.C. and Stewart, R.A. (eds.). 1983. Pages 95-104 *in* Faba Bean in the Nile Valley. Martinus Nijhoff, The Hague, The Netherlands.

Seleem, M. 1947. Root Nodule Bacteria and their Importance. University of Cairo Press (Monograph in Arabic).

Shukla, U.C. and Yadav, O.P. 1982. Effect of phosphorus and zinc on nodulation and nitrogen fixation in chickpea (*Cicer arietinum* L.). Plant and Soil 65: 239-248.

Sprent, J.I. 1972. The effects of water stress on nitrogen-fixing nodules. IV. Effects on whole plants of *Vicia faba* and *Glycine max*. New Phytologist 71: 603-611.

Subba Rao, N.S. 1977. Soil Microorganisms and Plant Growth. Oxford and IBH Publishing Company, New Delhi, India.

Tadrous, H.W. 1981. The effect of soil moisture stress on yields and water stress use by faba beans. FABIS Newsletter 3:43.

Taha, S.M., Mahmoud, S.Z., and Salem, S.H. 1967a. Effect of inoculation with rhizobia on some leguminous plants in UAR. I. Phosphorus manuring. Journal of Microbiology (Egypt) 2: 17-29.

Taha, S.M., Mahmoud, S.Z., and Salem, S.H. 1967b. Effect of inoculation with rhizobia on some leguminous plants in UAR. II. Nitrogen fertilization. Journal of Microbiology (Egypt) 2: 31-41.

Tanner, J.W. and Anderson, I.C. 1964. External effect of combined nitrogen on nodulation. Plant Physiology 39: 1039-1043.

Thornton, H.G. 1936. Proceedings of the Royal Society (London), 119: 474.

Vincent, J.M. 1965. Environmental factors in the fixation of nitrogen by the legume. Pages 384-435 *in* Soil Nitrogen (Bartholomew, W.V. and Clark, F.E., eds.). American Society of Agronomy, Madison, Wisconsin, USA.

Zaghloul, E.M.A. 1981. Effects of salinity and nitrogen fertilizer on symbiotic nitrogen fixation in beans (*Phaseolus vulgaris* and *Vicia faba*). M.Sc. thesis. Faculty of Agriculture, Alexandria University, Egypt.

Discussion

Wery

How do you explain the stimulating effect of a small mineral nitrogen supply on nitrogen fixation of faba bean? Did you measure plant growth throughout the crop cycle?

Abdel Ghaffar

Application of small amounts of nitrogen fertilizer prevents the occurrence of nitrogen deficiency in faba bean seedlings and results in a more vigorous root system.

Plant growth was measured during growth but not at the seedling stage.

D.P. Beck and L.A. Materon (editors), Nitrogen Fixation by Legumes in Mediterranean Agriculture
© 1988 ICARDA – Printed in the Netherlands. ISBN 90 247 3624 2

Biological Nitrogen Fixation by Faba Bean, Lentil, and Chickpea in the Traditional Versus the New Areas of the Sudan

N.O. MUKHTAR and SAYDA A. ABU NAIB
Gezira Research Station, Wad Medani, Sudan

Summary. Faba beans, lentils, and chickpeas were grown at Rubatab and Hudeiba, where they are traditionally grown, and in new areas at Wad Medani and Rahad. Performance in the two areas was compared in terms of plant dry matter (DM), yields, nodule mass, and shoot tissue N%. Each experiment included an inoculant treatment and a treatment of 120 kg N/ha. The traditional area was superior to the new area in plant DM by 22-50%; in grain, straw, and seed nitrogen yields by 30-80%; in nodule dry matter production by 35-80%; but was inferior in shoot tissue N% by 13-50%. These differences differed with different crops. Chickpea in the new area nodulated only where inoculant was applied, while faba bean and lentil did not respond to either seed inoculation with *Rhizobium* or added fertilizer nitrogen.

Nitrogen yields were 180 and 160 kg/ha for faba bean and lentil, respectively, in the traditional area and 125-470 kg/ha in the new area. In the latter area, the soil mineralization capacity was 60 kg N/ha/season which could supply a larger proportion of plant requirements, masking the influence of both fixed and/or added nitrogen. Fixation is higher in the traditional area where plant demand is high and the higher nitrogen yields are far from being met by soil mineralization. The exact amounts fixed in both areas, however, are still to be assessed, using ^{15}N methodology.

Climatic factors, especially temperature, rather than soil type or nitrogen availability are thought to be the reason for the low crop performance in the new area.

Introduction

The importance of biological nitrogen fixation (BNF) was recognized in the Sudan as early as 1925 when the Gezira scheme started. Therefore, *Dolichos lablab* was included in crop rotations in the Gezira Scheme, followed later by groundnut. Beneficial effects of legumes were reported in subsequent cotton crops (Burhan and Mansi 1967), as well as an increase in soil organic nitrogen (Musa and Burhan 1974) in rotation with legumes.

However at this time, cheap industrial nitrogenous fertilizers were available and research was concentrated on cash crops, with less attention to food crops such as grain legumes. Because of unilateral development of crop and animal systems of production microbiologists also tended not to work with forage legumes.

Hitherto, work on BNF has been fragmentory and has always fallen short of the adoption of an inoculation program. During the past decade, however, some serious work on BNF has started in collaboration with international institutions and organisations. This has become more important following the drought in large parts of Africa, including the Sudan, where grain and forage legumes are among the major quick solutions to providing adequate food and feed for humans and animals. Research has included grain and forage legumes with cooperation from USAID, ICARDA/IFAD, UNEP/AGFUND, and FAO/IAEA.

Next winter, some laboratory-prepared inoculants for haricot bean, chickpea, faba bean, clitoria, and phillipasara are expected to reach the fields of some smallholders, but large-scale inoculant production, distribution, and quality control measures are still to be introduced and await further funding.

This paper presents some of the problems facing the introduction of grain legumes into new areas south of Khartoum where there is sufficient land and water.

Materials and Methods

Faba bean (*Vicia faba* L.), lentil (*Lens culinaris* Med.), and chickpea (*Cicer arietinum* L.) were grown at Rubatab and Hudeiba (traditional areas), and at Wad Medani and Rahad (new areas). Planting was on 60 cm ridges in subplots 2.4 x 6m. Seeds were sown on 1 November for faba bean and chickpea and on 20 November for lentil. Plants were watered every 7-10 days. Three treatments were used:

(1) uninoculated, not fertilized (control),
(2) inoculated with the best available isolate in a Nile silt carrier, and
(3) fertilized with 120 kg N/ha as urea.

A complete block design with four replicates was used. For all crops in all areas, samples were taken at 4, 6, and 8 weeks for estimations of nodule mass, shoot dry matter, and shoot tissue N%. Fortnightly samples of faba beans were taken at Wad Medani for C_2H_2 reduction assay. At harvest, seed, straw, and seed nitrogen yields were estimated.

Results

Nodulation

In lentils, nodule formation was not influenced by inoculation, and was rarely influenced in faba beans. Nodulation was significant in uninoculated

Table 1. Nodule mass (mg/plant) of faba bean, lentil, and chickpea at different locations.

Treatment	Faba bean		Lentil		Chickpea	
	Rubatab	W/Medani	Rubatab	W/Medani	Hudieba	Rahad
Control	1130	820	48.0	10.0	36.0	0
Inoculated	1580	720	78.0	10.0	45.0	135 **
Fertilized (120 kg N/ha)	630	630	53.0	10.0	11.0 *	0
	*	NS	NS	NS	*	**
Mean	1113	723	59.7	10.0	27.3	
S.E. (±)	300	410	17	1.0	4.7	

* differences significant at P = 0.05;
** differences significant at P = 0.01.
NS = differences not significant at P = 0.05.

chickpeas grown in the traditional area, but in the new area nodulation only occurred where inoculant was applied.

Compared to the new areas, nodule mass per plant was 35 and 83% more in the traditional area for faba beans and lentils, respectively (Table 1). At Wad

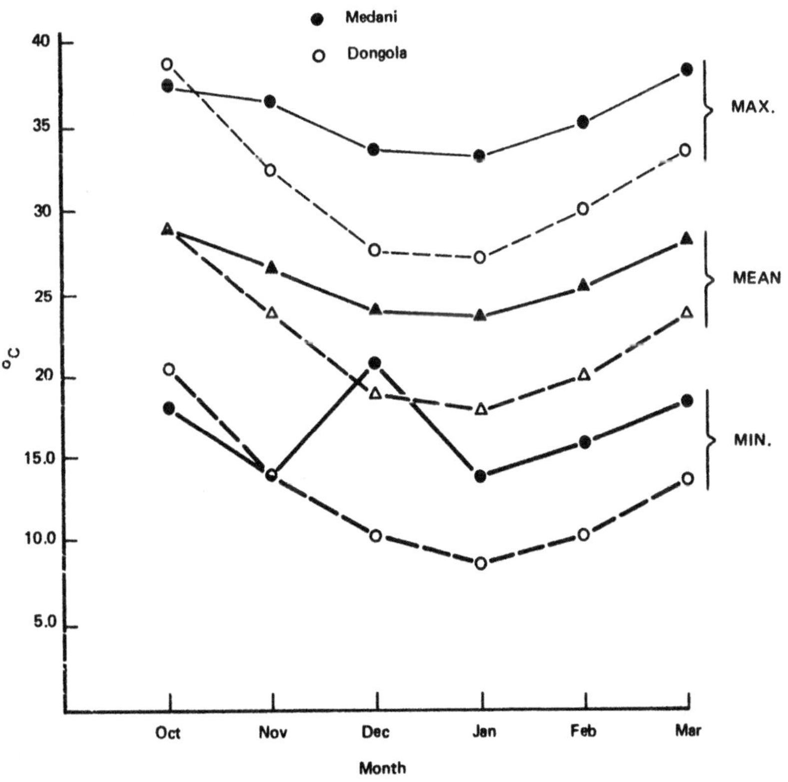

Fig. 1. Nitrogen fixed by faba bean with time as affected by starter nitrogen and phosphorus, Wad Medani.

Medani, acetylene reduction assays indicated high nitrogenase activity in faba beans which was enhanced by inoculation (Fig. 1).

Shoot dry matter

Shoot dry matter production was not affected by any treatment within the same location for all three crops. However, there were differences between locations, with the plants in the new area being much smaller. For example, 8 weeks after sowing, plant dry matter was 22, 50, and 15% more in the traditional area for faba bean, lentil, and chickpea, respectively (Table 2).

Shoot tissue N%

In the traditional areas, tissue N% in chickpea was significantly ($P = 0.05$) increased with added nitrogen, while in the new areas, fertilized faba bean plants accumulated significantly ($P = 0.05$) more nitrogen than the inoculated and control plants. Comparing locations however, the larger plants in the traditional area accumulated less nitrogen. The difference between locations

Table 2. Shoot dry matter (g/plant) for faba bean, lentil, and chickpea at different locations.

Treatment	Faba bean		Lentil		Chickpea	
	Rubatab	W/Medani	Rubatab	W/Medani	Hudeiba	Rahad
Control	14.40	11.78	4.10	2.40	1.30	3.00
Inoculated	13.54	10.52	3.80	2.50	3.30	1.80
Fertilized (N)	10.73	8.76	5.20	1.60	3.90	2.40
	NS	NS	NS	NS	NS	NS
Mean	12.89	10.34	4.36	2.17	2.83	2.40
S.E. (±)	1.92	1.77	0.75	4.7	0.75	0.16

NS = differences not significant at $P = 0.05$.

Table 3. Shoot tissue N% for faba bean, lentil, and chickpea at different locations.

Treatment	Faba bean		Lentil		Chickpea	
	Rubatab	W/Medani	Rubatab	W/Medani	Hudeiba	Rahad
Control	4.43	5.16	3.20	6.50	6.20	3.90
Inoculated	4.36	4.92	3.30	6.50	5.90	3.70
Fertilized (N)	4.63	5.50	3.30	6.60	6.50	3.80
	NS		*	NS		*
Mean	4.47	5.19	3.27	6.53	5.87	3.80
S.E. (±)	0.11	0.12	0.06	0.10	0.10	0.06

* Significant at $P = 0.05$.
NS = not significant.

Table 4. Seed, straw, and seed nitrogen yields (kg/ha x1000) for faba bean, lentil, and chickpea at different locations.

Treatment	Faba bean						Lentil						Chichpea			
	Seed yield		Straw yield		Nitrogen (seed)		Seed yield		Straw yield		Nitrogen (seed)		Seed yield		Nitrogen (seed)	
	Rub	W/M	Rub	W/M	Rub	W/M	Rub	W/M	Rub	W/M	Rub	W/M	Hud	Rahad	Hud	Rahad
Control	1.80	0.73	1.77	0.93			1.72	0.45	3.70	2.11			0.14	0.23		
Inoculated	1.95	0.73	1.65	1.55			2.17	0.37	4.10	1.93			0.20	0.19		
Fertilized (N)	1.60	0.49	1.54	1.08			2.18	0.47	4.30	2.18			0.40	0.13		
	NS	NS	NS	NS			NS	NS	NS	NS			*	NS		
Mean	1.78	0.65	1.65	1.19	176.7	125.3	2.02	0.43	4.00	2.27	158.9	73	0.25	0.18	10.3	7.5
S.E. (±)	0.17	0.09	0.16	0.26			0.34	0.08	0.38	0.26			0.06	0.01		

NS = not significant.
Rub = Rubatab, W/M = Wad Medani, Hud = Hudeiba.

was 14 and 50% for faba bean and lentil, respectively. In the case of chickpea, shoot tissue N% was greater by 35% in the traditional area compared to the new area (Table 3).

Seed, straw, and seed nitrogen yields

At Hudeiba, chickpea seed yield was significantly ($P = 0.05$) higher with nitrogen fertilization compared to inoculated and control crops. However, there were location differences. Seed, straw, and seed nitrogen yields were greater in the traditional areas compared with the new areas by 63, 28, and 29%, respectively for faba beans; 79, 43, and 54%, respectively for lentils; and 28% for chickpea grain (Table 4).

Discussion

On average, the differences in soil characteristics between the traditional grain legume-growing areas of northern Sudan and those of the new areas of central Sudan are not great (Table 5). On the other hand, mean temperatures in the two locations differ by about 5 °C throughout the growing season (Fig. 2). This difference could result in about a 30% reduction in faba bean pod setting and a 4% reduction in seed filling (Ageeb and Bashir 1984). Differences in environmental conditions, especially temperature, result in a field life in the south which is 20 and 30 days shorter for lentil and faba bean, respectively (El Sarrag and Nouai 1983).

In the traditional areas there are larger populations of *Rhizobium leguminosarum* as shown by the nodulation capacity (Table 1). In the new areas, populations are smaller for both lentil and faba bean. Rhizobia nodulating chickpea seem to be present in low numbers or absent in both areas.

The nitrogen yield for the crops grown in the traditional areas clearly indicated BNF activity, as the nitrogen yields of 180 and 160 kg N/ha for faba bean and lentil, respectively, are greater than could be obtained in a soil known to be low in nitrogen. In the new area, however, nitrogen yields are much lower, especially in the case of lentil and chickpea, and the nitrogen mineralized from soil (60 kg N/ha, Fig. 3) could meet all or part of the nitrogen demand.

Table 5. Some soil properties in the Gezira and Zeidab schemes.

Site	pH	Clay (%)	Carbon (%)	Total nitrogen (ppm)	Salts	Available P (ppm)
Gezira	8.5	60	0.4	400	low	4.0
Zeidab	8.0	50	0.75	600	low	4.0

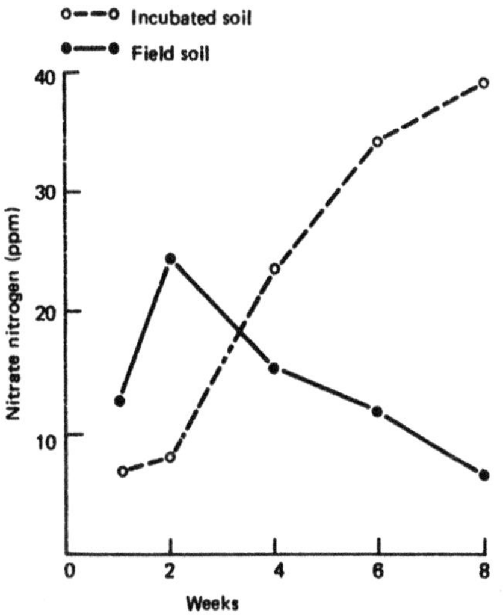

Fig. 2. Maximum, minimum, and mean temperature (30 days mean) at Dongola (traditional area) and Wad Medani (new area).

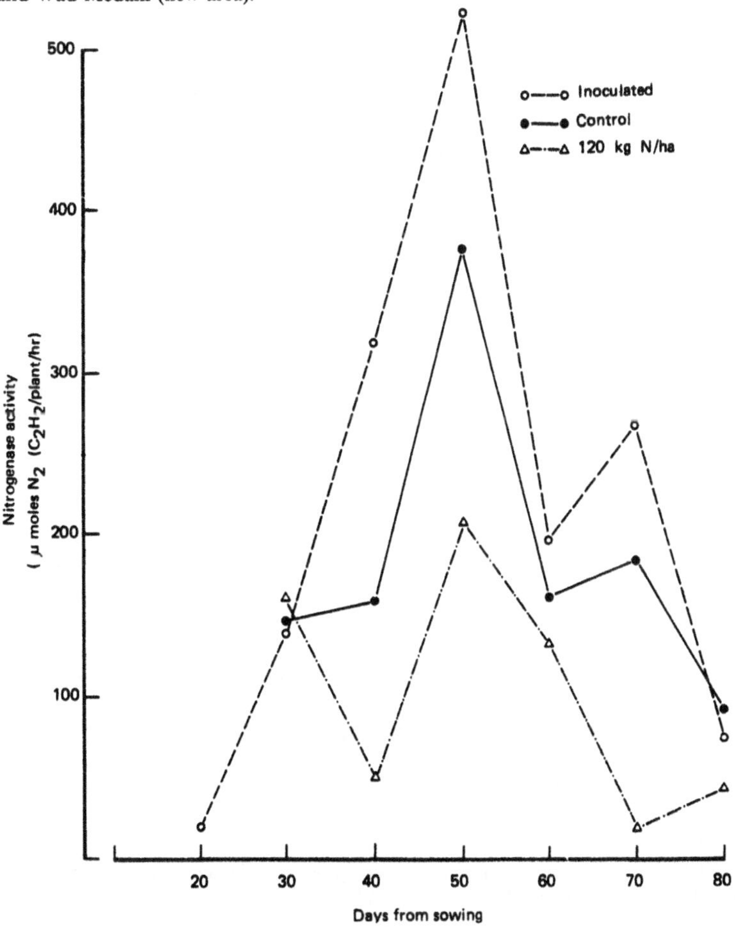

Fig. 3. Nitrate nitrogen mineralized (ppm) with time from field and incubated Gezira soil.

In a few cases, seed inoculation increased nodulation and plant dry matter production by faba bean. However this was not reflected in seed yield, except where the rhizobial population was low (Mukhtar and Nourai 1984; Mahdi *et al.* 1981).

Shoot N% was usually greater in the new area, due to less nitrogen dilution by the smaller plants and less nitrogen demand by the reproductive sink. It is possible that environmental factors govern legume production through the effect on plant size and nitrogen demand rather than through nitrogen availability. Hence, the supply of nitrogen may not be the first priority for high yields and emphasis should be placed on breeding for high yield, as well as for efficiency of BNF.

Conclusions

(1) Environmental factors, especially temperature, are perhaps the major constraints to high grain legume production in the new production areas of the Sudan. Longer, cooler winters resulted in increased yields (Mukhtar and Nourai 1984).

(2) Short, warm growing periods resulted in lower yields. Fluctuations in seed yields seem to be related to small changes in temperature.

(3) Short, low-yielding plants demand comparatively less nitrogen, most of which might be secured from soil and/or nitrogen fixation. Thus, the lack of effect due to added nitrogen is probably a result of a lower overall nitrogen requirement, rather than fixation being very efficient.

(4) Use of ^{15}N methodology to assess actual amounts of nitrogen fixed and a mutation breeding program seem feasible.

References

Ageeb, O.A.A. and Bashir, M. 1984. Paper presented at the Fifth Annual Coordination Meeting of ICARDA/IFAD Nile Valley Project, ICARDA/IFAD, 13-17 Sept 1984, Cairo, Egypt. ICARDA, Aleppo, Syria.

Burhan, H.O. and Mansi, M.G. 1967. Journal of Agricultural Science Cambridge 68: 255-261.

El Sarrag, G. and Nourai, A.H. 1983. A review of research on lentils (*Lens culinaris*) in the Sudan. LENS 10 (1): 1-12.

Mahdi, A.A., El Sarrag, G. and Mukhtar, N.O. 1981. Pages 101-104 *in* Faba Bean in the Nile Valley (Saxena, M.C. and Stewart, R.A., eds.). Martinus Nijhoff, The Hague, The Netherlands.

Mukhtar, N.O. and Nourai, A.H. 1984. Paper presented at the Fifth Annual Coordination Meeting of ICARDA/IFAD Nile Valley Project, ICARDA/IFAD, 13-17 Sept 1984, Cairo, Egypt. ICARDA, Aleppo, Syria.

Musa, M.M. and Burhan, H.O. 1974. The relative performance of forage legumes as rotational crops in the Gezira. Experimental Agriculture 10: 131-140.

Discussion

Weaver

Is it proper to conclude that treatment responses to fertilizer nitrogen or inoculation are not significant when experimental errors do not allow detection of yield changes of 50% or more?

Mukhtar

Such a difference was only encountered with faba bean straw yield, a parameter difficult to determine due to foliage losses and plant lodging at the end of the season. The discussion in addition considered nodule mass, shoot dry matter, and the seed yield which contained no such variation.

D.P. Beck and L.A. Materon (editors), Nitrogen Fixation by Legumes in Mediterranean Agriculture
 ISBN 90 247 3624 2

Measurement of Nitrogen Fixation by the Acetylene Reduction Assay; Myths and Mysteries

J.F. WITTY [1] and F.R. MINCHIN [2]

[1] *Rothamsted Experimental Station, Harpenden, Herts, UK*
[2] *Animal and Grassland Research Institute, Hurley, Berks., UK*

Summary. The acetylene reduction assay has been widely used in research on biological nitrogen fixation. However, evidence suggests that the assay is inaccurate. This paper discusses the problems of the assay and possible sources of inaccuracy.

Introduction

The acetylene reduction assay has had a profound effect on the progress of research on biological nitrogen fixation (Havelka and Hardy 1976). Since its development, almost every researcher in this field has used the method at one time or another. Such wide acceptance can be ascribed to the fact that it is inexpensive, rapid, sensitive, and apparently accurate, at least when used with free-living microorganisms and isolated enzyme systems. However, increasing evidence shows that there are several major sources of inaccuracy when the method is applied to the legume symbiosis. Some of these errors are associated with the specific adaptations of legume nodules for the regulation of oxygen supply to the bacteroids (Witty, Minchin, Skot and Sheehy 1986). This review examines the problems involved in obtaining quantitative or even comparative estimates of nitrogen fixation by field-grown legume crops using the acetylene reduction assay.

The "normal" assay (Hardy *et al.* 1973) involves the incubation of detached nodules, nodulated root pieces, or detopped root systems with 10% acetylene in a closed container of known volume. Gas phase samples are analysed by gas chromatography to measure the concentration of accumulated ethylene. This measurement does not relate directly to nitrogen fixation. It only measures the electron flux through nitrogenase in the sample material under the conditions which prevail during the assay period. To obtain useful practical information from such data, a number of assumptions or conversion factors must be made. The steps necessary to derive various meaningful parameters from basic acetylene reduction data and the related assumptions are outlined in Table 1. This example is based on measurements made in the "normal" way on nodulated root systems or detached nodules taken from field-grown plants. It is assumed that the experimental objective is to obtain either quantitative or,

Table 1. Assumptions or conversion factors used to obtain useful information from the 'normal' acetylene reduction assay.

Parameters measured and derived (typical units)	Assumptions or conversion factors used to derive parameter	Errors
1) Electron flux through nitrogenase over assay period per sample. (eg $\mu M\ C_2H_4$/sample/h)	Conversion factor normally taken as one on the assumption that the rate of activity during measurement is equal to the pre-assay rate. This is a myth because	
STEP 1	[A] The addition of acetylene reduces nitrogenase activity i.e., the assay procedure itself alters the rate of activity.	50%
	[B] Activity is also reduced by detopping the plant, shaking the root system, or removing nodules during sample preparation.	
2) Electron flux per sample prior to assay (eg $\mu M\ C_2H_4$/sample/h)		
STEP 2	[C] Conversion factor normally taken to be unity on the assumption that all nodules are recovered. This depends on soil structure, root depth, and nodule distribution.	60%
3) Electron flux per plant or area per hour (eg $\mu M\ C_2H_4$/plant/h)		
STEP 3	[D] Conversion factor normally taken as 4 C_2H_4: 1 N_2 on the assumption that the electron allocation coefficient to H_2 is 0.25. This is a theoretical minimum; measured values are normally larger.	30%
4) Rate of N_2^- fixation per plant or area per hour (eg kg N/ha)		

at least, comparative data relating to N_2-fixation by the crop. The following text discusses the validity of conversion factors or assumptions shown under headings [A] to [D] in this table.

The first assumption normally made (step 1, Table 1) is that the assay procedure does not, itself, affect the rate of nitrogenase activity, i.e., the rate of ethylene accumulation is constant and the values obtained relate to the normal operation of the nitrogen-fixing system prior to the assay. This is a myth (a prevalent but false belief (Fowler and Fowler 1942)). Both exposure to acetylene and disturbance of the plant during sample preparation have a large effect on activity.

The Acetylene-Induced Decline

When detached nodules or intact nodulated root systems are exposed to acetylene in an open flow-through gas system, a rapid decline in both nitrogenase activity and associated respiration have been observed (Minchin *et al.* 1983). The decline begins within a few minutes of exposure to acetylene and continues for about 30 minutes (Fig. 1). A similar decline in H_2 evolution and respiration is observed when N_2 in the gas phase is replaced with Ar or He (Layzell *et al.* 1984). The common causal factor appears to be the cessation of ammonia production within the nodule.

The open flow-through system used to obtain these results contrasts with the normal assay procedure in that it measures the rate of ethylene production rather than its accumulated concentration. Air containing 10% acetylene is passed continuously through the assay vessels and the effluent gas stream is analysed for CO_2 by infra-red gas analysis and ethylene by gas chromatography. Rates of production are calculated as the product of concentration and flow rate (Cooper *et al.* 1982; Minchin *et al.* 1983).

Three reasons can be suggested as to why this effect has not been previously noted. First, original work which validated the method was done with free-liv-

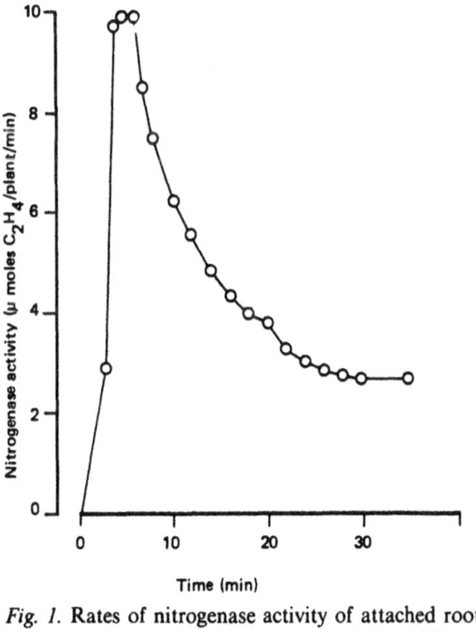

Fig. 1. Rates of nitrogenase activity of attached roots of white clover following exposure at time zero to a continuous gas stream containing 10% acetylene. (Minchin, Witty, Sheehy and Muller 1983).

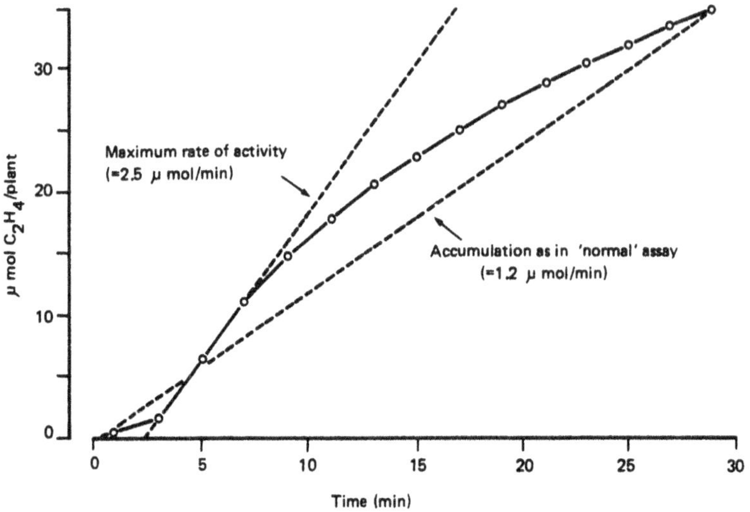

Fig. 2. Accumulated acetylene reduction data calculated for the first 30 minutes of the decline curve presented in Fig. 1. Also shown as hatched lines are the rate of accumulation based on the maximum value in Fig. 1, and the rate of accumulation as it would appear from measurements taken at zero and 30 minutes as in the "normal" assay procedure. The difference between these gradients illustrates the type of error associated with "normal" assay.

ing microorganisms (Dilworth 1966) which do not show the decline because they lack the complex O_2 regulation systems of root nodules. Secondly, the effect is masked by measurements of accumulated ethylene concentration in closed vessels. The data presented in Fig. 2, which shows results as usually presented for a "normal" assay, were produced by accumulating the rates of activity shown in Fig. 1. The decline in activity is much less clear and the acceptance of a linear relationship will depend upon the number of points and their experimental error. Thirdly, disturbance of the nodule system reduces both maximum activity and the acetylene effect, making it more difficult to detect non-linearity in ethylene accumulation (Fig.4).

The decline is not due to impurities in the acetylene, since cleaning this gas has no effect (Minchin *et al.* 1983), nor is it due to water stress, as suggested by de Visser and Poorter (1984), since it occurs in well-watered rooting medium. The effect persists when the input gas stream is amended with 1000 ul/1 CO_2 so that this gas does not appear to be involved, as suggested by

Table 2. The oxygen diffusion resistance of disturbed and undisturbed soyabean nodules (Minchin *et al.* 1986).

System measured	Diffusion resistance ($S/m \times 10^{-6}$)	
	Before C_2H_2	After C_2H_2
Undisturbed intact roots	0.71 + 0.054	1.62 + 0.179
Detopped shaken roots	3.05 + 0.312	4.60 + 0.362

Table 3. Comparison of alternative measurement techniques for nitrogenase activity of intact plants (Minchin *et al.* 1986).

Technique	umol/h/plant
^{15}N uptake	13.5+0.76
H_2 production	16.6+0.83
Equivalent C_2H_2 production	57.9+2.78
Measured C_2H_2 reduction rates	
a) From maximum in flow-through system	53.4+3.12
b) From 30 ethylene accumulation	36.8+3.07

* Calculated as $(3 \times {}^{15}N_2$ uptake$) + H_2$.

Bethonod *et al.* (1984). Measurements of nitrogenase activity and respiration at different external O_2 concentrations show that the true cause of the acetylene effect is O_2 limitation of bacteroid respiration. This is produced by an increase in resistance of the nodule to inward O_2 diffusion (Witty *et al.* 1984; Table 2).

Comparisons of the combined rate of $^{15}N_2$-fixation and H_2 production with the rate of acetylene reduction, determined in a flow-through system, indicate that the maximum (pre-decline) rate of acetylene reduction reflects the real rate of activity prior to the assay (Table 3). Calculations of nitrogenase activity based on accumulated ethylene production, where this maximum is obscured, produce an underestimate of up to 50% (Fig. 2).

Effects of Plant Disturbance

Prior to the 'normal' assay, plant tops are removed and roots are separated from the growth medium by washing or vigorous shaking. Washing is known to reduce nitrogenase activity (Mague and Burris 1972), as is nodule detachment (Hardy *et al.* 1973). Despite these problems, such treatments have been used in the preparation of samples for comparative assays.

The possibility that removing the shoot and shaking the root system has a detrimental effect on nitrogenase activity has been largely ignored, despite warnings during the development of the assay procedure (Hardy *et al.* 1973). This may be because of conflicting reports on the comparative activities of intact plants and detopped/shaken roots. A substantial decrease has been reported by some workers (Mague and Burris 1972; Hardy *et al.* 1977; Wych and Rains 1978) whilst in contrast, others have reported no effect (Trinick *et al.* 1976; Saito *et al.* 1980). Such anomalies may be due to complications arising from an acetylene-induced decline in the activity of intact control plants.

The effects of shoot removal and shaking of roots have been investigated recently using a flow-through gas system (Minchin, Sheehy and Witty 1986). A rapid decline was caused by shaking the detopped roots, which resulted in a

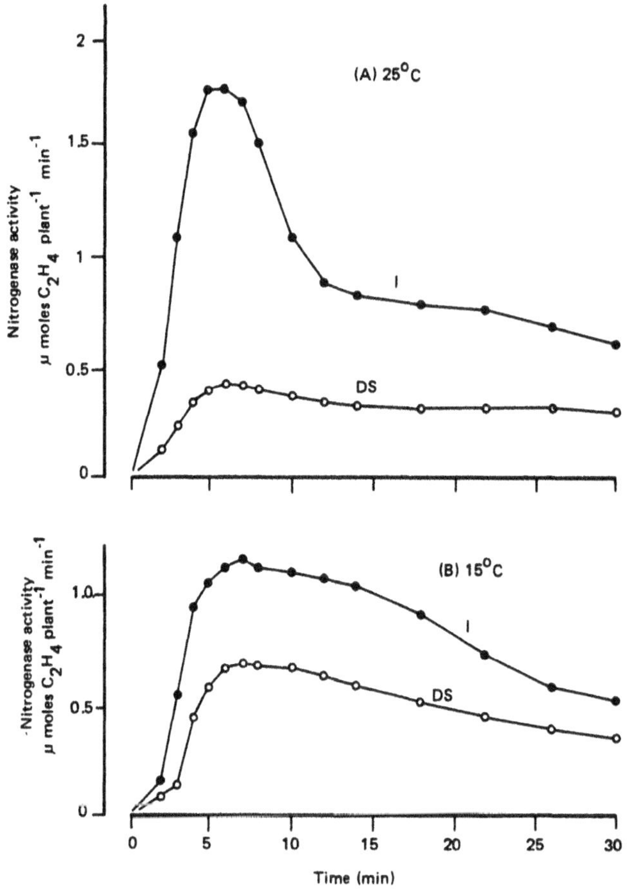

Fig. 3. Rate of acetylene reduction of soyabean roots following exposure to 10% acetylene at time zero. Data are shown for intact, undisturbed (I) and the topped/shaken roots (DS) at (A) 25 °C and (B) 15 °C. The activities prior to the addition of acetylene are equal to the maximum (pre-decline) rate of ethylene production. (Minchin, Sheehy and Witty 1986).

75% reduction in maximum activity (Fig. 3a). Evidence from responses to increased external O_2 concentration indicates that this effect, like that of acetylene, is caused by an increase in the O_2 diffusion resistance of the nodule (Table 2). Thus, if there is a large response to disturbance there will be only a small further response to acetylene (shaken roots, Fig. 3a). Conversely, where disturbance effects are small, the acetylene-induced decline will be larger. In either case, substantial quantitative errors will result when ethylene accumulation is measured in closed assay vessels.

Furthermore, because of treatment interactions, even comparative estimates using the "normal" procedure are invalid. Nodules which have already increased their diffusion resistance in response to adverse environmental factors, such as water stress or shading, show a reduced response to disturbance acetylene. Thus, reduction in activity caused by the assay procedure will be

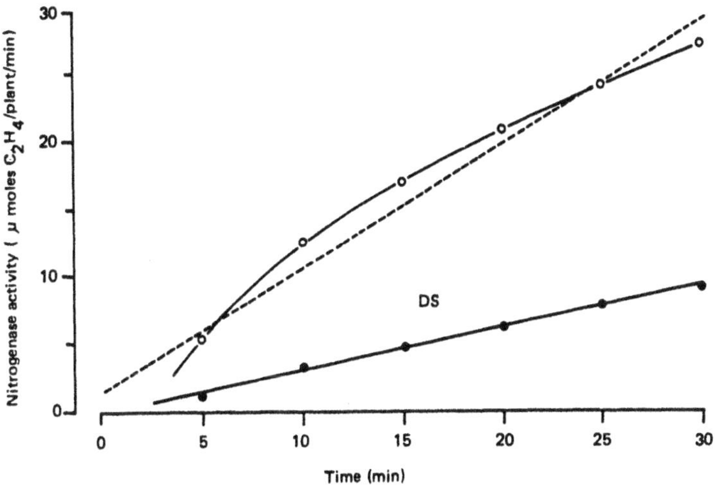

Fig. 4. Accumulated ethylene production calculated from the curves shown in Fig. 3A. Linear correlations are shown for intact (I) and de-topped/shaken (DS) roots. For intact roots the real accumulation curve is represented as a solid line.

less with stressed than unstressed symbioses, offsetting real treatment differences. With soyabean for example, measurement of maximum rates of activity of undisturbed plants in the flow-through system showed that the nitrogenase activity was higher at 25 °C than at 15 °C (Treatment I, Figs 3a and b). However, when the activity of detached shaken root systems was compared, this observation was reversed because the disturbance-induced decline was much larger with the unstressed nodules at the higher temperature (Treatment DS, Figs 3a and b).

The smaller acetylene-induced decline occurring in disturbed systems is almost completely masked in plots of accumulated ethylene concentration. The data shown in Fig. 4 were produced by accumulating the rate measurements shown in Fig. 3a. Although the plot for intact, undisturbed plants (Fig. 4, I) is obviously non-linear, a linear relationship would be fitted for the de-topped, shaken root system (Fig. 4, DS), despite a decline in activity of 33%, apparent in Fig. 3.

Recovery of Nodules

To obtain a value for nitrogenase activity per plant from the activity of the nodule population sampled (Step 2, Table 1) it must be assumed either that all nodules are recovered or a conversion factor describing nodule loss must be used. With shallow-rooted crops growing in soil of good texture the recovery of the entire nodule population is difficult. With deep-rooted legumes or where soil is dry and structurally difficult, the recovery of all nodules is almost impossible.

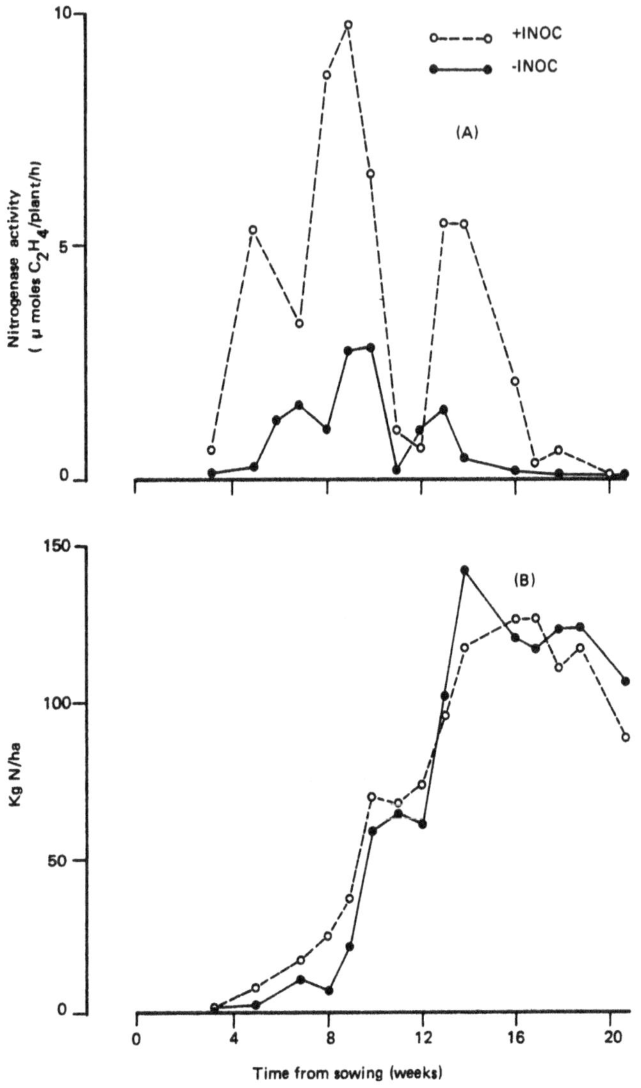

Fig. 5. Seasonal profiles of nitrogenase activity (A) and shoot nitrogen content (B) of uninoculated (●) and inoculated (○) plants of lupin grown as a friable, low nitrogen soil at Rothamsted.

The data presented in Fig. 5 were obtained in a field experiment designed to evaluate the benefits of inoculation to lupin (*Lupinus albus*). In initial glasshouse experiments using soil from the experimental site, the number of nodules for three replicate plants varied from 0 to 12, suggesting that indigenous *Rhizobium lupini* was scarce. Seeds with and without an inoculant coating were planted in mid-April and harvested at approximately weekly intervals thereafter by digging around groups of 3-5 plants with a spade to a width and depth of about 200 mm. Detached root systems were assayed in the 'normal' way and plant and plant tops were analysed for total nitrogen content. The results from the acetylene reduction assay appear to confirm the benefit of inoculation (Fig. 5a). However, such benefit was not reflected in the nitrogen accumulation data (Fig 5b); although inoculation gave a slight benefit early in the season, neither the maximum nor the final nitrogen content was altered. A careful excavation of root systems late in the season, when this anomaly had become apparent, showed that the majority of nodules on the inoculated plants were near the soil surface on primary and secondary roots. Equal numbers of nodules were present on the roots of uninoculated plants, but these were randomly distributed over the entire root system, to a depth and circumference of about 400mm and were largely excluded in the sampling procedure used.

In suitable soil and if time is available, this difficulty can be overcome by an extensive sampling procedure but, in many experiments, comparisons will be invalidated by interactions between treatments and nodule recovery. Particular problems will occur with treatments which modify soil structure, such as irrigation or the addition of organic matter, and in comparisons between sites of differing soil type. Similarly, treatments which affect nodule distribution, such as rate of fertilizer application, plant spacing, plant age, plant genotype, and *Rhizobium* strain, may render comparisons invalid.

Conversion Ratio of Acetylene to Nitrogen

To obtain values for the rate of N_2-fixation from rates of acetylene reduction (Table 1, Step 3) an appropriate conversion ratio for C_2H_2: N_2 is needed. In early studies a ratio of 3: 1 was used on the basis that the reduction of N_2 to $2NH_3$ required six electrons and the reduction of C_2H_2 to C_2H_4 required two (Hardy *et al.* 1973). As it became apparent that with each N_2 reduced at least two protons were also reduced to H_2 (Chatt 1980), a conversion ratio of 4: 1 was adopted; in reality the proportion of electrons allocated to H_2 production is variable. If the number of electrons per N_2 used in H_2 production is designated as n, then nitrogen fixation may be represented as

$$N_2 + 6e^- + 6H^+ \rightarrow 2NH_3 \quad (1)$$

$$ne^- + 2H^+ \rightarrow \tfrac{1}{2}H_2 \quad (2)$$

This gives the overall reaction

$$N_2 + (6+n)e^- + (6+n)H^+ \rightarrow 2NH_3 + \tfrac{1}{2}H_2 \quad (3)$$

From equation (3), the conversion ratio for C_2H_2: N_2 is $(6+n)/2$ i.e., $(6+n)$ e^- for each N_2 and $2e^-$ for C_2H_4. In experiments with a range of different *R. leguminosarum* strains nodulating the same host plant (*Pisum sativum*) grown under the same conditions, the value of n varied from 2 to 6, giving strain-dependent conversion ratios of 4–6 (Witty and Skot, unpublished). If such conversion ratios are not determined, then ranking of treatments (for example, *Rhizobium* strains) based on their ability to reduce acetylene is invalid. Those with the highest acetylene reduction activity may have a low electron allocation to N_2 and consequently fix less N_2 than other strains with a lower rate of acetylene reduction but a better electron allocation coefficient.

Other factors which apparently affect electron allocation to H_2 include temperature (Dart and Day 1971), irradiance (Bethlenfalvay and Phillips 1979), or treatments which modify the rate of electron flux through nitrogenase (Hagemen and Burris 1980).

Experimentally determined electron allocations to H_2 are normally higher than the theoretical minimum; proportions in the range 40–60% are not uncommon (Schubert and Evans 1976), giving values for n of 4–9 and C_2H_2: N_2 ratios of 5–7.5: 1. However, these data may themselves be suspect. Electron allocation coefficients are normally calculated as [H_2 production in air]/ [C_2H_4 production in C_2H_2] and these estimates are therefore subject to errors caused by the acetylene-induced decline. The extent of the resulting overestimate will depend upon the physiological state of the nodules prior to and during the assay. Paradoxically, it may be that values obtained using severely disturbed systems e.g., detached nodules (van Kessel and Burris 1983), where the acetylene-induced decline is almost absent, are a better estimate of *in vivo* electron allocation ratios than could be obtained using an undisturbed system. The validity of this approach depends upon the extent to which the reduced O_2 supply caused by disturbance modifies the electron allocation coefficient.

Conclusions

The preceeding sections indicate that the "normal" acetylene reduction assay produces large quantitative errors, even with pot-grown plants where all nodules can be recovered with little disturbance. These errors are compounded in the field and the repetition of such "spot" measurements to obtain seasonal rates of N_2-fixation (Fig. 5a), often without regard to diurnal or daily changes in environmental conditions, is meaningless.

Using the assay for comparative purposes is no more meaningful. Factors known to alter the magnitude of errors include soil type, soil moisture, plant genotype, plant variety, plant age, *Rhizobium* strain, fertilizer treatment, temperature, and irradiance (Table 4). Comparisons between plants differing in these treatments are invalid.

A number of other methods, more valid than the acetylene reduction assay, can be used to measure the amount of N_2 fixed by a crop over the growing season. These range from simple difference methods based on crop nitrogen content, to techniques using ^{15}N to distinguish between nitrogen derived from the soil and from the atmosphere. The flow-through system, as outlined in the introduction, could be modified for field use. This would make the acetylene

Table 4. Sources of error in comparative acetylene reduction assays.

Error in measurement of absolute rates	Causes of variation	Treatments responsible
The acetylene-induced decline The disturbance-induced decrease	Differences in the initial resistance of the nodule to O_2 diffusion alter the extent of the effect.	Plant species Plant cultivar Water stress Defoliation Temperature *Rhizobium* strain
Failure to recover all nodules	Differences in nodule distribution and soil structure.	Applied N Plant spacing Plant age Plant species Plant cultivar *Rhizobium* strain
C_2H_2: N_2 ratios greater than 4:1	Changes in electron allocation between N_2 and H_2	Temperature Irradiance Plant species Plant cultivar Rate of e^- flux

reduction assay a valuable tool for measuring instantaneous rates of nitrogenase activity in studies of crop physiology. However, for the reasons already described, the data could not be extrapolated to estimate seasonal fixation. The problem of the acetylene-induced decline could be overcome in such systems if the size of the *in situ* vessel and the flow rate of the gas stream allowed for a clear resolution of the maximum rate of activity.

Flow-through techniques have been proposed which involve open-ended assay chambers placed in the soil at the beginning of the season (Hardy *et al.* 1977; Denison *et al.* 1983). However, the restriction to growth imposed by such vessels may alter the morphology of plants allowed to grow continuously within them (Denison *et al.* 1983). The alternative of driving the assay cylinder into the ground around established plants can result in lateral shearing of the roots (Hardy *et al.* 1977) and soil compaction. In either case, the depth of the assay vessel may not be adequate to include the whole nodule population (Denison *et al.* 1983). As discussed above, the distribution of nodules is modified by inoculation and environment.

An alternative to measuring the maximum rate of activity is to use non-saturating concentrations of acetylene (0.2%) which do not induce the decline (Denison *et al.* 1983). However, ethylene production data obtained with this technique must be extrapolated to that which would have occurred under saturating acetylene concentrations. The necessary calibration of the kinetics of nitrogenase requires that the acetylene concentration be increased to 1% or more; for soyabeans we have found the decline to be induced by acetylene levels as low as 0.4%. Thus, errors are likely to occur in the calibration procedure.

When these difficulties are resolved, the flow-through system will be particularly useful since root respiration can also be measured to provide both a reliable estimate of nitrogenase activity and, using the regression technique proposed by Witty *et al.* (1983), an assessment of the amount of photosynthate used in fixation and the efficiency of its use.

The objective of studies on nitrogen fixation at the International Institutes is to increase the productivity of local agricultural systems. Whilst modified flow-through systems may help to understand basic crop physiology and ^{15}N labelled fertilizers can be used in an attempt to estimate seasonal fixation, neither can substitute for clear definition of problems and a good experimental approach. Despite progress in understanding many areas of biological nitrogen fixation, its accurate measurement in the field remains a mystery (a truth beyond human intelligence (Fowler and Fowler 1942)). Elaborate techniques undoubtedly have their place but direct methods should not be ignored merely because they lack the allure of high technology. The best way forward may often be by the application of techniques as simple as the measurement of crop yield.

References

Bethlenfalvay, G.J. and Phillips, D.A. 1979. Variation in nitrogenase and hydrogenase of Alaska pea root nodules. Plant Physiology 63: 816-820.

Bethonod, O, Prioul, J.L. and Deroche, M.E. 1984. Short-term inhibition of acetylene reduction activity by low CO_2 concentration around the nodulated roots of various legumes. Physiologie Vegetale 25: 565-570.

Chatt, J. 1980. Chemistry relevant to biological nitrogen fixation. Page 1018 *in* Nitrogen Fixation (Stewart, W.P.D. and Gallon, J.R., eds.). Academic Press, London, UK.

Cooper R.E., Sheehy, J.E. and Minchin, F.R. 1982. An open continuous flow shoot/root enclosure apparatus for carbon and nitrogen assimilation studies. Pages 119-222 *in* Grassland Research Institute, Annual Report 1981. Grassland Research Institute, Hurley, UK.

Dart, P.J. and Day, J.M. 1971. The effect of incubation temperature and oxygen tension on nitrogenase activity of legume root nodules. Plant and Soil, Special Volume: 167-184.

Denison, R.F., Sinclair, T.R., Zobel, R.W., Johnson, M.M. and Drake, G.M. 1983. A non-destructive field assay for soybean nitrogen fixation by acetylene reduction. Plant and Soil 70: 173-182.

de Visser, R. and Poorter, H. 1984. Growth and root nodule nitrogenase activity of *Pisum sativum* as influenced by transpiration. Physiologia Plantarum 61: 637-642.

Dilworth, M.J. 1966. Acetylene reduction by nitrogen fixing preparations from *Clostridium pasteurianum*. Biochimica Biophysica Acta 127: 285-294.

Fowler, F.G. and Fowler, H.W. 1942. The Pocket Oxford Dictionary. Clarenden Press, Oxford, UK.

Hageman, R.V. and Burris, R.G. 1980. Electron allocation to alternative substrates of *Azotobacter* nitrogenase is controlled by electron flux through dinitrogenase. Biochimica Biophysica Acta 591: 63-77.

Hardy, R.W.F., Burns, R.C. and Holsten, R.D. 1973. Applications of the acetylene reduction assay for measurement of nitrogen fixation. Soil Biology and Biochemistry 5:47-81.

Hardy, R.W.F., Criswell, J.G. and Havelka, U.D. 1977. Investigation of possible limitations of nitrogen fixation by legumes: (1) methodology, (2) identification and (3) assessment of significance. Pages 451-467 *in* Recent Developments in Nitrogen Fixation (Newton, W., Postgate, J.R. and Rodriguez-Barrueco, C., eds.). Academic Press, London, UK.

Havelka, U.D. and Hardy, R.W.F. 1976. Legume N_2 fixation as a problem in carbon nutrition. Pages 456-475 *in* Proceedings of the First International Symposium on Nitrogen Fixation. Volume 2. (Newton, W.E. and Nyman, C.J., eds.). Washington State University Press, Pullman, Washington, USA.

Layzell, D.B., Weagle, G.E. and Canvin, D.T. 1984. A highly sensitive flow through H_2 analyser for use in nitrogen fixation studies. Plant Physiology 75: 111-115.

Mague, T.H. and Burris, R.H. 1972. Reduction of acetylene and nitrogen by field grown legumes. New Phytologist 71: 275-286.

Minchin, F.R., Sheehy, J.E. and Witty, J.F. 1985. Factors limiting N_2-fixation by the legume-*Rhizobium* symbiosis. Pages 285-291 *in* Nitrogen Fixation Research Progress (Evans, H.J., Thornley, P.J. and Newton, W.E., eds.). Martinus Nijhoff, The Hague, The Netherlands.

Minchin, F.R., Sheehy, J.E. and Witty, J.F. 1986. Further errors in the acetylene reduction assay: effects of plant disturbance. Journal of Experimental Botany 37: 1581-1591.

Minchin, F.R., Witty, J.F., Sheehy, J.F. and Muller, M. 1983. A major error in the acetylene reduction assay: decreases in nodular nitrogenase activity under assay conditions. Journal of Experimental Botany 34: 641-649.

Mortenson, L.E. and Thornley, R.N.F. 1979. Structure and function of nitrogenase. Annual Reviews of Biochemistry 48: 387-418.

Mortenson, L.E. and Upchurch, R.G. 1981. Effect of adenylates on electron flow and efficiency of nitrogenase. Pages 75-78 *in* Current Perspectives in Nitrogen Fixation (Gibson, A.H. and Newton, W.E., eds.). Australian Academy of Science, Canberra, Australia.

Saito, S.M.T., Matsui, E. and Salati, E. 1980. $^{15}N_2$ fixation, H_2 evolution, and C_2H_2 reduction relationships in *Phaseolus vulgaris*. Physiologia Plantarum 49: 37-42.

Schubert, K.R. and Evans, H.J. 1976. Hydrogen evolution: A major factor affecting the efficiency of nitrogen fixation in nodulated symbionts. Proceedings of the National Academy of Science, USA 73: 1207-1211.

Sprent, J.I. and Bradford, A.M. 1977. Nitrogen fixation by field beans (*Vicia faba*) as affected by population density, shading and its relationship with soil moisture. Journal of Agricultural Science 88: 303-310.

Trinick, M.J., Dilworth, M.J. and Grounds, M. 1976. Factors affecting the reduction of acetylene by root nodules of *Lupinus* species. New Phytologist 77: 359-370.

van Kessel, C. and Burris, R.H. 1983. Effect of H_2 evolution on $^{15}N_2$ fixation, C_2H_2 reduction and relative efficiency of leguminous symbionts. Physiologia Plantarum 59: 329-334.

Witty, J.F., Minchin, F.R., Sheehy, J.E. and Miguez, M.I. 1984. Acetylene-induced changes in the oxygen diffusion resistance of legume root nodules. Annals of Botany 53: 13-20.

Witty, J.F., Minchin, F.R. and Sheehy, J.E. 1983. Carbon costs of nitrogenase activity in legume root nodules determined using acetylene and oxygen. Journal of Experimental Botany 34: 951-963.

Witty, J.F., Minchin, F.R., Skot, L. and Sheehy, J.E. 1986. Nitrogen fixation and oxygen in legume root nodules. Oxford Surveys of Plant Molecular and Cell Biology. Oxford University Press. (In press).

Witty, J.F., Skot, L. and Revsbech, N.P. 1986. Direct evidence for changes in the resistance of legume root nodules to O_2 diffusion. Journal of Experimental Botany (submitted).

Wych, R.D. and Rains, D.W. 1978. Simultaneous measurements of nitrogen fixation estimated by the acetylene reduction assay and nitrate absorption by soybeans. Plant Physiology 62: 443-448..lm 10;

Discussion

Ismaili

Do you think that the ^{15}N method also suffers from imprecision as does the acetylene reduction method?

Witty

The ^{15}N method certainly is subject to imprecisions. These are, however, much less than those in the acetylene reduction assay and many of these can be overcome by the choice of an appropriate control and method of labelling soil.

Ruiz-Argueso

You mentioned O_2 diffusion deficiencies as one of the main causes of O_2H_2 reduction decline. Do you find significant variations in O_2 diffusion resistance among nodules from different legume species?

Witty

Yes. To expand on this, the diffusion resistance of a nodule correlates with its activity. The more active the nitrogenase the more O_2 is used in oxidative ATP production. Thus the nodules of more active species have a lower resistance to O_2 diffusion.

Cocks

When measuring nitrogen fixation using acetylene reduction we have noticed that the rate peaks in winter, declining sharply in spring, whereas nitrogen accumulation is most rapid in spring. Can you explain this?

Witty

I find this very interesting and can only speculate as to the answer. My guess would be that it is due to poor recovery of nodules at the later harvest, because a second generation of nodules which formed did so at some depth, and away from the area sampled.

D.P. Beck and L.A. Materon (editors), Nitrogen Fixation by Legumes in Mediterranean Agriculture
 ISBN 90 247 3624 2

The Use of ^{15}N Enriched Fertilizers for Estimating Nitrogen Fixation in Grain and Pasture Legumes

S.K.A. DANSO

Joint FAO/IAEA Division Vienna, Austria

Summary. The optimization of the potential of many legumes to fix atmospheric nitrogen (N_2) in different environments requires suitable methodologies to distinguish between the contribution from this process and from soil and fertilizer nitrogen uptake. Most of the available methods suffer from serious drawbacks, and therefore cannot give accurate quantitative estimates of the amount of N_2 fixed.

The ^{15}N isotope labelling techniques involving the application of ^{15}N-enriched fertilizers to soil provide reliable integrated estimates of the proportions and amounts of N_2 fixed by several grain and pasture legumes. Such studies can be performed directly in the field. Two approaches, the isotope-dilution and A-value methodologies, have been used and some of the underlying concepts in these techniques are discussed in this paper. Currently, the isotope-dilution approach seems to be more popular than the A-value approach.

The precision and accuracy of N_2-fixation estimates made with these techniques are strongly influenced by the reference crop used to assess the $^{15}N/^{14}N$ ratio or the available nitrogen in the soil (A-soil). The use of an appropriate reference crop is therefore crucial. The criteria for the selection of the reference crop are discussed, as well as practices that reduce errors in the estimates of N_2 fixed. While these errors may be serious at low levels of fixation, their influence at high fixation levels is small.

A major disadvantage of ^{15}N techniques is their high cost. Practices such as the use of residual ^{15}N or ^{15}N-labelled plant residues from previous experiments result in reduced costs, and the estimates of N_2-fixation obtained have been shown to have high precision.

Introduction

When grown under suitable conditions, many legumes derive substantial amounts of their nitrogen from atmospheric N_2-fixation. Several factors have been shown to influence the amount or proportion of N_2 fixed in plants, including inoculation with *Rhizobium* (Agboola and Fayemi 1972; Habish and Ishag 1974; Hardarson *et al.* 1984a), soil phosphorus status (Hernandez and Focht 1985; Jakobsen 1985; Olsen and Moe 1971; Zaroug and Munns 1979), and soil inorganic nitrogen status (Allos and Bartholomew 1959; Hardarson *et al.* 1984b; Richards and Soper 1979; Weber 1966). To optimise the amount of N_2 fixed in any system it is therefore necessary to be able to accurately estimate the magnitude of fixation under different conditions, and this requires appropriate techniques.

Although several methods, such as dry matter yield (Haydock *et al.* 1980; Edwards *et al.* 1981; Sekhon *et al.* 1978), nodule number or mass (Graham 1981; Herridge 1980), total nitrogen difference (Broadbent *et al.* 1982; Hardarson *et al.* 1984b; Talbott *et al.* 1982; Rennie 1984; Vasilas and Ham 1984), and the reduction of acetylene to ethylene (Edwards *et al.* 1981; Goh *et al.* 1978; Hardy *et al.* 1968; Hudd *et al.* 1980) have been used to assess N_2 fixed, most of these techniques are based on indirect criteria, and cannot distinguish between the various sources of nitrogen in a fixing plant. The advantages and limitations of these techniques have been pointed out in several reviews (Bremner 1977; Danso 1985; Danso *et al.* 1986; Fried *et al.* 1983; Hardy *et al.* 1968; Knowles 1981; Rennie and Rennie 1983).

In his report at this workshop, Dr. Witty provided further evidence why the use of the once popular acetylene reduction technique could give unreliable estimates of N_2 fixed. It therefore seems that at this moment the use of ^{15}N-labelled fertilizers to quantify N_2 fixed, distinct from soil and fertilizer-derived nitrogen, is the most reliable method.

Two common approaches have been adopted with the ^{15}N method. These have been termed the isotope-dilution technique (Broadbent *et al.* 1982; Rennie and Rennie 1983; Ruschel *et al.* 1979; Vasilas and Ham 1984) and the A-value technique (Fried and Broeshart 1975; Vasilas and Ham 1984; Wagner and Zapata 1982). Each of these methods is based on two concepts, i.e., (i) the isotopic dilution of the fertilizer- or soil-derived nitrogen of higher $^{15}N/^{14}N$ ratio by the lower $^{15}N/^{14}N$ ratio of atmospheric N_2, and (ii) the assimilation by plants of nitrogen from soil, fertilizer, and atmosphere, in direct proportion to the amounts available from each of these sources (the principle upon which the A-value concept is based).

This paper will attempt to elucidate the principles and procedures to be followed in using ^{15}N-labelled fertilizers to estimate N_2 fixed in grain and forage legumes. Suitable reviews are available elsewhere (Bremner 1977; Chalk 1985; Danso *et al.* 1986; Fried *et al.* 1983; Rennie *et al.* 1978; Rennie and Rennie 1983).

The Isotope Dilution Approach

The isotope dilution principle and the equations involved have been outlined by Danso *et al.* (1986), Rennie *et al.* (1978), and Rennie and Rennie (1983). Its use for estimating N_2-fixation requires the application of equal nitrogen rates and ^{15}N enrichment of a given fertilizer to both the legume and a suitable non-fixing reference or control crop. In principle, since the legume and the reference crop are absorbing nitrogen from a similar zone, the $^{15}N/^{14}N$ ratio of soil-derived nitrogen is expected to be the same for both crops. However, in addition to the nitrogen absorbed from soil, the N_2-fixing legume also assimilates atmospheric N_2 of lower $^{15}N/^{14}N$ ratio than the soil nitrogen. This results in a dilution of the $^{15}N/^{14}N$ ratio of soil nitrogen assessed by the reference crop. The extent to which this soil $^{15}N/^{14}N$ ratio is diluted is an indication of the magnitude or the efficiency of N_2-fixation. For example, if it is assumed that upon analysis, the nitrogen in a reference crop contained 1 atom % ^{15}N excess while the legume contained 0.5 atom %, this indicates that the nitrogen uptake from soil sources and fixation was in the ratio of 1:1, or 50% of the nitrogen in the plant was derived from atmospheric N_2-fixation. If, in the same study, another legume treatment was found to contain 0.3 atom % ^{15}N excess, this implies a greater dilution of the 1 atom % ^{15}N excess of soil nitrogen or higher N_2-fixation capacity than the first legume treatment. To calculate N_2 fixed using the isotope dilution technique, Fried and Middelboe (1977) used the following equations.

$$\%\ \text{Ndfa} = \left(1 - \frac{\text{atom}\ \%\ ^{15}\text{N excess in legume}}{\text{atom}\%\ ^{15}\text{N excess in reference}}\right)100 \qquad \text{(i)}$$

$$\text{Ndfa (kg N/ha)} = \frac{\%\ \text{Ndfa}}{100} \times \text{Total N in legume} \qquad \text{(ii)}$$

where Ndfa is the nitrogen derived from atmospheric N_2-fixation.

The A-Value Approach

On soils of low nitrogen status, non-fixing crops may grow poorly, while N_2-fixing plants grow well. This could cause a mismatch in growth and nitrogen uptake patterns between the reference and the fixing crop. But the addition of large amounts of fertilizer nitrogen (which may be needed to promote the healthy growth of the reference crop) to legumes has been shown to inhibit N_2-fixation (Allos and Bartholomew 1955; 1959; Deibert *et al.* 1979; Hardarson *et al.* 1984b; Wagner and Zapata 1982). The use of the A-value approach however, allows the application of different nitrogen rates and ^{15}N enrichments of a given fertilizer to legume and reference crop plots to estimate N_2 fixed (Fried and Broeshart 1975).

The principle behind this approach is that a plant with more than one source of available nitrogen will absorb from each of these in proportion to their relative amounts, i.e., there will be no preference for one source of nitrogen over another. This is based on the A-value concept first elucidated by Fried and Dean (1952), which has been used to estimate N_2 fixed in field-grown legumes (Fried and Broeshart 1975). Thus, the amount of plant-available nitrogen in soil (A-soil), expressed in kg N/ha of the applied fertilizer, remained constant at different fertilizer nitrogen rates (Aleksic *et al.* 1968;

Legg and Allison 1959; Legg and Stanford 1967). This is also the amount that is available for both fixing legume and reference crop which absorb nitrogen from a similar soil zone. The A-soil is assessed by the reference non-fixing crop, and is expressed mathematically as

$$\text{A-soil} = \frac{\%\ \text{Ndfs}}{\%\ \text{Ndff}} \times \text{R} \tag{iii}$$

$$= \frac{100 - \%\ \text{Ndff}}{\%\ \text{Ndff}} \times \text{R} \quad \text{(Danso 1985)} \tag{iv}$$

where % Ndfs and % Ndff are the percentages of nitrogen in the plant derived from soil and fertilizer, respectively, and R is the amount of fertilizer nitrogen added. For a N_2-fixing legume however, the estimated A-value includes nitrogen derived from fixation i.e., A-soil + A-fixed.

$$\text{Thus, A-soilA-fixed} = \frac{100 - \%\ \text{Ndff}}{\%\ \text{Ndff}} \times \text{R(v)}$$

Since A-soil is the same for both fixing and non-fixing crops,

$$\text{A-soil} + \text{A-fixed-A-soil} = \text{A-fixed} \tag{vi}$$

(Legume A-value) (Reference crop A-value)

Based on the A-value concept, proportional uptake may be expressed as

$$\frac{\%\ \text{Ndfa}}{\text{A-fixed}} = \frac{\%\ \text{Ndff}}{\text{R}} = \frac{\%\ \text{Ndfs}}{\text{A-soil}} \tag{vii}$$

$$\text{Therefore, } \%\ \text{Ndfa} = \frac{\%\ \text{Ndff}}{\text{R}} \times \text{A-fixed} \tag{viii}$$

From this, Ndfa (kg/ha) can be calculated as in equation (ii). The advantage in using equation (viii) over that used by Fried and Broeshart (1975), is that it avoids the use of yield-dependent parameters.

Although the A-value approach has been a subject of controversy (Broadbent 1970; Chalk 1985; Deibert *et al.* 1979; Vasilas and Ham 1984), several studies using this concept have reported satisfactory estimates of N_2 fixed (Heichel *et al.* 1981; Phillips and Bennett 1978; Wagner and Zapata 1982).

Reference Crops

For each of the two approaches described above, a suitable reference crop is needed to assess either the integrated $^{15}N/^{14}N$ ratio of available soil nitrogen or the A-value of soil nitrogen. The selection of an appropriate reference crop is therefore crucial for the accurate estimation of N_2 fixed (Fried *et al.* 1983; Rennie and Rennie 1983; Witty 1983). The criteria used in the selection of suitable reference crops have been discussed by Fried *et al.* (1983) and Danso (1985), and include:

1) The absence of N_2-fixing ability. This may be ascertained simply by testing

for the inability of the test plant to reduce acetylene to ethylene (Phillipps and Bennett 1978).

2) Similarity in the rooting zone of the reference and fixing crops. This can be examined by assessing root lengths/habits, or indirectly through comparisons of ^{15}N enrichments in different potential reference crops, particularly if a non-nodulating isoline of the fixing legume (or uninoculated legume if native rhizobia are absent from the soil) can be included (Fried *et al.* 1983; Rennie 1986; Rennie *et al.* 1982; Wagner and Zapata 1982).

3) Relative nitrogen uptake patterns of reference and fixing crops. As pointed out by Fried *et al.* (1983) and Witty (1983), these should match otherwise there could be serious errors in the estimation. Witty (1983) described how to examine this, and the conditions under which such errors can be minimized. Ledgard *et al.* (1985a; 1985b) devised a mathematical formula that allows the assessment of the relative uptake of added and indigenous soil nitrogen by nodulated legume and reference crops.

4) Duration of growth. Since the $^{15}N/^{14}N$ ratio of a soil to which ^{15}N-enriched fertilizer has been added has been shown to decline sharply with time (especially during the initial periods following ^{15}N application), it is essential that both crops absorb similar ratios of $^{15}N/^{14}N$ during the entire growing season of the fixing crop (Fried *et al.* 1983) otherwise the soil $^{15}N/^{14}N$ ratio assessed by the reference crop could differ from that of the fixing crop. This implies that both crops must be planted and harvested at the same time.

5) The cropping system used. In an intimate mixture of N_2-fixing and non-fixing species, the possible transfer of nitrogen from the former to the latter has been controversial. Some reports have demonstrated that this occurs (Agboola and Fayemi 1972; Broadbent *et al.* 1982), while others (Ezedinma 1964; Henzell *et al.* 1968; Vallis *et al.* 1967) could not demonstrate such transfer. In any case, for the non-fixing crop in a legume- non-legume mixture to be used as a valid reference crop, there should be no significant transfer of nitrogen (Fried *et al.* 1983; West and Wedin 1985).

The following have been used as reference crops to estimate N_2 fixed in various legumes.

1) A non-nodulating isoline of the N_2-fixing legume in question (Deibert *et al.* 1979; Ham and Caldwell 1978; Talbott *et al.* 1982; Wagner and Zapata 1982) or another legume (Eaglesham *et al.* 1982; Patterson and La Rue 1983; Talbott *et al.* 1982; Vasilas and Ham 1984). According to Fried *et al.* (1983), the non-nodulating legume would closely resemble its nodulating isoline in most respects and thus possibly in patterns of growth and nitrogen uptake and would therefore be ideal as a reference crop. This is supported by the reported similarity in nitrogen uptake patterns for nodulating and non-nodulating soybean isolines (Vasilas and Ham 1984). However, differences in fertilizer or soil nitrogen uptake between non-nodulating and nodulating soybean isolines have been reported (Boddey *et al.* 1984; Deibert *et al.* 1979; Harper 1974; Ruschel *et al.* 1979).

2) An uninoculated N_2-fixing legume (Fried *et al.* 1983; Rennie *et al.* 1982; Wagner and Zapata 1982). Where no indigenous rhizobia exist, the uninoculated nodulating line has been found to be an excellent control (Fried *et al.* 1983; Rennie 1986; Rennie *et al.* 1982). However, care must be taken to thoroughly examine roots of such uninoculated legumes to ensure that they are not nodulated and that their use is therefore valid. If possible, they should be compared with other reference crops (Fried *et al.* 1983; Rennie *et al.* 1982). One problem in using an uninoculated control is that cross-contamination from the inoculated treatment may be difficult to avoid, unless special precautions are taken against this.

3) A fixing legume inoculated with an ineffective rhizobial strain (Fried *et al.* 1983; Rennie *et al.* 1982). Again, this is valid in soils devoid of effective native soil rhizobia. Even where these are few and/or not highly competitive, it is uncertain if 100% nodulation could be attributed to only the ineffective strains.

4) A fixing legume fertilized with a high rate of nitrogen fertilizer to completely inhibit nodulation and N_2-fixation (Wagner and Zapata 1982). This is difficult to ensure, since although large amounts of nitrogen may severely inhibit nodulation and N_2-fixation (Allos and Bartholomew 1959; Deibert *et al.* 1979; Eaglesham *et al.* 1982; Hardarson *et al.* 1984b), the process may not be completely suppressed (Richards and Soper 1979; Wagner and Zapata 1982).

5) Non-legume crops, e.g., cereals (Broadbent *et al.* 1982; Fried *et al.* 1983; Phillips and Bennett 1978; Rennie 1984; Wagner and Zapata 1982). The important prerequisites are to ensure that the conditions outlined above for an appropriate reference crop (such as the lack of N_2-fixation and similar zones for nitrogen uptake) are fulfilled. The validity of using such non-legumes can be established by comparing the isotopic composition of nitrogen in these plants with that of a non-nodulating isoline of the fixing legume if this is available, or the uninoculated legume if no native rhizobia are present in the soil (Fried *et al.* 1983; Rennie *et al.* 1982).

Greenhouse and Field Application in Grain and Pasture/Forage Legumes

Both the isotope dilution and A-value approaches have been used extensively to estimate N_2 fixed in several grain legumes. However, because of the controversy surrounding the A-value of soil remaining constant at different fertilizer nitrogen rates (Broadbent 1970; Deibert *et al.* 1979; Vasilas and Ham 1984), the A-value approach now appears to be used less frequently than the isotope dilution approach. It is worth noting however that under severe nitrogen deficiency, the growth of a non-fixing crop may be so poor that it may exhibit nitrogen uptake or growth symptoms uncharacteristic of the healthy plant (Harper 1974; Vasilas and Ham 1984); the addition of a higher nitrogen rate to the control in such a case (and thus use of the A-value approach) may be a better alternative. For pasture legumes, predominant use of the isotope dilution method can be attributed largely to the practical need to add the same amount of nitrogen to legumes and non-legumes growing in mixed pastures, or even for pasture/forage legumes alone to the difficulty in assessing the A-value of a soil to which different fertilizer nitrogen rates have been applied not only once, but in several split applications (since many harvests over a long period may necessitate repeated applications).

Types of ^{15}N-Enriched Fertilizers

The different types of ^{15}N-enriched fertilizers used can be classified broadly into a) inorganic nitrogen fertilizers, and b) immobilized ^{15}N.

The inorganic fertilizers can be further sub-divided into the fast-release or normal fertilizers and the slow-release types. Almost all types of the usual agricultural nitrogen fertilizers, such as ammonium sulphate, urea, potassium nitrate, calcium nitrate, and sodium nitrate have been used (Chalk 1985). Although the type of nitrogen fertilizer used has been found to influence N_2-fixation estimates (Steele and Littler 1986; Witty and Ritz 1984), more research is needed to ascertain why one ionic form of nitrogen should result in a different estimate of nitrogen than another, and also, under such circumstances, which estimate is more accurate. Several factors could contribute to these discrepancies. One of these relates to the greater mobility of NO_{3-} in soil than NH_4^+. Thus, where root length differences exist, much more of the nitrogen in NO_{3-} than NH_4^+ could move out of the zone of nutrient uptake of a shallow-rooted crop. The relative availability of NO_{3-} and NH_4^+ for such reference and fixing crops would thus be different and could therefore result in different estimates of N_2 fixed (Steele and Littler 1986).

Also, as reported by Fried *et al.* (1983) and Witty (1983), the decline in the $^{15}N/^{14}N$ ratio in soil influences the accuracy of N_2-fixation estimates; the lower the rate of decline, the fewer the errors associated with any mismatch between nitrogen uptake of reference and fixing crops (Witty 1983). Differences in N_2-fixation estimates using different fertilizer forms could therefore be at least partially attributed to differences in the $^{15}N/^{14}N$ decline rates associated with different fertilizer types. The slow-release, inorganic nitrogen types, such as oxamide and gypsum-pelleted KNO_3, exhibit lower $^{15}N/^{14}N$ decline rates. They have been found by Witty and Ritz (1984) to give better estimates of N_2 fixed, and smaller errors were observed when there was a mismatch between the reference and fixing crops. Also, Rennie (1986) reported good estimates of N_2 fixed when ^{15}N-enriched ammonium sulphate was used together with "N-serve" compound to slow down the rate of nitrification of the applied fertilizer.

Immobilization of nitrogen for the isotope dilution approach of estimating N_2-fixation has been achieved by:

1) Addition of ^{15}N-labelled plant material (Boddey *et al.* 1984; Broadbent *et*

al. 1982; Butler and Ladd 1985; De Freitas *et al.* 1984; Henzell *et al.* 1968; Rennie 1986).

2) Addition of sucrose or a high energy source to hasten microbiological immobilization of the added ^{15}N fertilizer (Fried *et al.* 1983; Wagner and Zapata 1982).
3) A combination of (1) and (2) (Chalk *et al.* 1983; Legg and Sloger 1975; Talbott *et al.* 1982).
4) Use of residual ^{15}N from previous ^{15}N application (Broadbent *et al.* 1982; Fried *et al.* 1983; Kohl and Shearer 1981; Rennie *et al.* 1982).

The idea behind immobilizing soil nitrogen is to ensure a gradual release of nitrogen from the immobilized ^{15}N and thereby prevent the rapid decline of $^{15}N/^{14}N$ ratio in soil (Fried *et al.* 1983; Witty and Ritz 1984).

Nitrogen Application Rates

Application rates have varied for different studies, crops, and locations (Chalk 1985). However, decisions on the rate of ^{15}N-labelled fertilizer to use must often take into consideration the nitrogen status of the soil, the influence of the added nitrogen on N_2-fixation, the length of the growing season, how many ^{15}N applications will be made, the level of precision needed (and/or the type of ^{15}N measuring equipment evailable), and the cost of the experiment. In general, single rates higher than 20 kg N/ha of 5 atom % ^{15}N have seldom been applied to legume crops. This is the rate that was first used in the A-value approach to estimate N_2 fixed in field-grown legumes (Fried and Broeshart 1975). This rate and ^{15}N-enrichment are equivalent to 0.1g $^{15}N/m^2$ which, from the experience gained from several FAO/IAEA-Coordinated Research Programmes, is sufficient for the precision needed in most N_2-fixation studies (personal communication). For a reference crop and using the A-value approach, this rate was achieved by adding 100 kg N/ha of 1 atom % ^{15}N excess ammonium sulphate (Fried and Broeshart 1975).

For forage and pasture legumes, many of which are either perennial and/or several harvests are made at relatively short time intervals, it has been usual to make repeated additions of small quantities of ^{15}N-labelled fertilizer at frequent or short intervals (Goh *et al.* 1978; Vallis *et al.* 1967; 1977; Williams *et al.* 1977). Thus, in the IAEA Seibersdorf Laboratory and the FAO/IAEA Coordinated Research Programme on Pasture Management, 5 kg N/ha of 5 atom % ^{15}N excess nitrogen fertilizers (equivalent to 0.025 g ^{15}N harvest) have been used to satisfactorily estimate N_2 fixed in several legume-grass mixtures (personal communication). Vallis *et al.* (1977) used a lower ^{15}N rate (0.009 g ^{15}N/application or 0.018 g ^{15}N/harvest) to estimate N_2 fixed by several legumes in mixed swards. A single application of a large quantity of ^{15}N-labelled fertilizer to pastures at planting was found to be inferior to repeated small applications, and the former method led to significant inhibition of N_2-fixation (Danso *et al.* 1986).

Methods of Fertilizer Application

No standard approach has been adopted for the application of ^{15}N-labelled fertilizers for plant uptake to estimate N_2 fixed (Chalk 1985 and Danso *et al.* 1986). The methods that have been used so far include foliar application (Fried and Broeshart 1981), broadcasting (Ham 1978; Fried and Broeshart 1981; Fried *et al.* 1983), banding (Fried and Broeshart 1981; Fried *et al.* 1983), and solution spraying onto or injecting into soil (Fried *et al.* 1983; Rennie 1986; Rennie *et al.* 1982; West and Wedin 1985; Wagner and Zapata 1982). The major advantage of the application of solutions is that a small quantity of a ^{15}N-labelled fertilizer can be applied in a large volume of water (about 200 ml/m^2, depending on soil type), ensuring a more uniform distribution than would otherwise be possible. It also allows point placements to be made to study the effect of, for example, placement on the relative availabilities of different fertilizers to various crops. For uniform incorporation of the applied ^{15}N in soil, irrigation after nitrogen application has been practiced (Vallis *et al.* 1967 and Vasilas and Ham 1984).

Advantages of the ^{15}N Methodology

The advantages of the ^{15}N methodology are that:

a) It gives a truly integrated value for N_2-fixation for whole growing seasons.
b) It can be used directly in field situations.
c) The nitrogen derived from fixation can be distinguished from that derived from the soil or fertilizer. This is very important especially in comparing varieties/species/treatments with different abilities to absorb soil or fertilizer nitrogen.
d) The proportion of N_2 fixed by a crop can be determined, even if the yield cannot be measured because of damage due to disease, animals, soil, etc.
e) Unlike the total nitrogen difference method, it is not necessary to grow the reference crop without nitrogen fertilizer or at a very low level of nitrogen.
f) The isotope dilution approach allows comparisons of N_2 fixing abilities of, for example, different legume varieties, to be made simply by ranking them on the basis of $^{15}N/^{14}N$ ratios. The lower this ratio, the higher the amount of atmospheric N_2 incorporated (greater N_2 fixing capability). Thus, the need for a reference crop can be avoided in such cases.
g) The ^{15}N methods are more sensitive and therefore accurate than most other methods for estimating N_2-fixation (Hardarson *et al.* 1984a; Phillips and Bennett 1978; Williams *et al.* 1977). Also, errors in ^{15}N-aided estimates of N_2-fixation tend to be small at high values of fixation (Danso 1986; Danso *et al.* 1986) and with practices that result in a less drastic decline of the $^{15}N/^{14}N$ ratio in soil (Witty and Ritz 1984).
h) In addition to the estimates of N_2 fixed, additional information, such as the nitrogen fertilizer use efficiency of the legume or reference crop (particu-

larly if it happens to be an economic crop) can be obtained at no extra cost, while the unused ^{15}N-labelled straw can be incorporated into soil for organic matter decomposition studies. Also, the nitrogen distribution pattern in plants can be followed by splitting the crop into different plant parts and analysing the ^{15}N enrichments in each separately, a procedure which is already recommended in N_2-fixation estimation studies (Fried *et al.* 1983).

Disadvantages of the ^{15}N Methodology

The ^{15}N methodology has a few disadvantages:

1) Cost has been the major disadvantage most often cited with the ^{15}N methodologies. This however, has often been exaggerated or can be substantially reduced without sacrificing much precision (Khol and Shearer 1981). For example, ^{15}N emission spectrometers (e.g., NOI-5e and Jasco) are considerably cheaper than the least expensive mass spectrometer available. Also, ^{15}N-labelled urea costs far more than ammonium sulphate (on a per gram ^{15}N basis). This is due to the more laborious procedure for preparing the former. The ^{15}N content of soil can also be increased either by using higher ^{15}N enrichments or through increased fertilizer nitrogen rates. However, ^{15}N enrichments above 10 atom % ^{15}N are usually more expensive than lower enrichments (on a per gram ^{15}N basis). Excessively high ^{15}N enrichments may therefore increase the cost of experiments unnecessarily. The addition of ^{15}N-labelled organic materials from previous studies (Legg and Sloger 1975; Rennie 1986) or residual ^{15}N from previous experiments (Khol and Shearer 1981) avoids the need for further ^{15}N purchases. Furthermore, the use of small isotope sub-plots (incorporated in larger yield plots to which ordinary nitrogen fertilizers are applied) significantly reduces the cost of field experiments.2) The accuracy of determinations of N_2 fixed are largely influenced by the reference crop selected. However, this problem can be minimized either by careful selection of the reference crop, or by creating conditions that result in a less drastic decline in the $^{15}N/^{14}N$ ratio of the soil (Witty 1983; Witty and Ritz 1984). In addition, these errors are critical at low values of fixation and are of little consequence at high levels of fixation (Danso 1986; Danso *et al.* 1986).

References

Agboola, A.A. and Fayemi, A.A.A. 1972. Fixation and excretion of nitrogen by tropical legumes. Agronomy Journal 64: 409-412.

Aleksic, Z., Broeshart, H. and Middelboe, V. 1968. The effect of nitrogen fertilization on the release of soil nitrogen. Plant and Soil 29: 474-477.

Allos, H.F. and Bartholomew, W.V. 1955. Effect of available nitrogen on symbiotic fixation. Soil Science Society of America Proceedings 19: 182-184.

Allos, H.F. and Bartholomew, W.V. 1959. Replacement of symbiotic fixation by available nitrogen. Soil Science 87: 61-66.

Boddey, R.M., Chalk, P.M., Victoria, R., and Matsui, E. 1983. The ^{15}N-isotope dilution technique applied to the estimation of biological nitrogen fixation associated with *Paspalum notatum* cv. Batatatais in the field. Soil Biology and Biochemistry 15: 25-32.

Boddey, R.M., Chalk, P.M., Victoria, R., and Matsui, E. 1984. Nitrogen fixation by nodulated soybean under tropical field conditions estimated by the ^{15}N isotope dilution technique. Soil Biology and Biochemistry 16: 583-588.

Bremner, J.M. 1977. Use of nitrogen-tracer techniques for research on nitrogen fixation. Pages 335-352 *in* Biological Nitrogen Fixation in Farming Systems of the Tropics (Ayanaba, A. and Dart, P., eds.).

Broadbent, F.E. 1970. Variables affecting A values as a measure of soil nitrogen availability. Soil Science 110: 19-23.

Broadbent, F.E., Naskashima, T. and Chang, G.Y. 1982. Estimation of nitrogen fixation by isotope dilution in field and greenhouse experiments. Agronomy Journal 74: 625-628.

Broeshart, H. 1974. Quantitative measurement of fertilizer uptake by crops. Netherlands Journal of Agricultural Science 22: 245-254.

Butler, J.H.A. and Ladd, J.N. 1985. Symbiotically fixed and soil-derived nitrogen in legumes grown in pots in soils with different amounts of available nitrate. Soil Biology and Biochemistry 17: 47-55.

Chalk, P.M. 1985. Review. Estimation of N_2 fixation by isotope dilution: An appraisal of techniques involving ^{15}N enrichment and their application. Soil Biology and Biochemistry 17: 389-410.

Chalk, P.M., Douglas, L.A. and Buchanan, S.A. 1983. Use of ^{15}N enrichment of soil mineralizable N as a reference for isotope dilution measurements of biological fixed nitrogen. Canadian Journal of Microbiology 29: 1046-1052.

Danso, S.K.A. 1985. Methods for estimating biological nitrogen fixation. Pages 213-244 *in* Biological Nitrogen Fixation in Africa (Ssali, H. and Keya, S.O., eds.).

Danso, S.K.A. 1986. Review. Estimation of N_2-fixation by isotope dilution: An appraisal of techniques involving ^{15}N enrichment and their application-Comments. Soil Biology and Biochemistry. (In press.)

Danso, S.K.A., Hardarson, G. and Zapata, F. 1986. Assessment of dinitrogen fixation potentials of forage legumes with ^{15}N technique. Pages 26-57 *in* Proceedings of a Workshop on Potentials of Forage Legumes in Farming Systems of Sub-Saharan Africa, ILCA, Addis Ababa, 16-19 Sept 1985 (Haque, I., Jutzi, S. and Neate, P.J.H., eds.).

Deibert, E.J., Bijeriego, M. and Olson, R.A. 1979. Utilization of ^{15}N fertilizer by nodulating and non-nodulating soybean isolines. Agronomy Journal 71: 717-723.

Eaglesham, A.R.J., Ayanaba, A., Ranga Rao, V. and Eskew, D.L. 1982. Mineral N effects on cowpea and soybean crops in a Nigerian soil. II. Amounts of N fixed and accrual to the soil. Plant and Soil 68: 183-192.

Edwards, D.G., Kang, B.T. and Danso, S.K.A. 1981. Differential response of six cowpea (*Vigna unguiculata* (L.) (Walp) cultivars to liming in an ultisol. Plant and Soil 59: 61-73.

Ezedinma, F.O.C. 1964. Effects of inoculation with local isolates of cowpea *Rhizobium* and application of nitrate-nitrogen on the development of cowpeas. Tropical Agriculture 41: 243-249.

De Freitas, J.R., Victoria, R.L., Ruschel, A.P. and Vose, P.B. 1984. Estimation of N_2-fixation by sugarcane, *Saccharum* sp., and soybean, *Glycine max*, grown in soil with ^{15}N-labelled organic matter. Plant and Soil 82: 257-261.

Fried, M. and Broeshart, H. 1975. An independent measurement of the amount of nitrogen fixed by a legume crop. Plant and Soil 43: 707-711.

Fried, M. and Broeshart, H. 1981. A further extension of the method for independently measuring the amount of nitrogen fixed by a legume crop. Plant and Soil 62: 331-336.

Fried, M., Danso, S.K.A. and Zapata, F. 1983. The methodology of measurement of N_2 fixation by non-legumes as inferred from field experiments with legumes. Canadian Journal of Microbiology 29: 1053-1062.

Fried, M. and Dean, L.A. 1952. A concept concerning the measurement of available soil nutrients. Soil Science 73: 263-271.

Fried, M. and Middelboe, V. 1977. Measurement of amount of nitrogen fixed by a legume crop. Plant and Soil 47: 713-715.

Goh, K.M., Edmeades, D.C. and Robinson, B.W. 1978. Field measurements of symbiotic nitrogen fixation in an established pasture using acetylene reduction and a ^{15}N method. Soil Biology and Biochemistry 10: 13-20.

Graham, P.H. 1981. Some problems of nodulation and symbiotic nitrogen fixation in *Phaseolus vulgaris*. A review. Field Crops Research 4: 93-112.

Habish, H.A. and Ishag, H.M. 1974. Nodulation of legumes in the Sudan. III. Response of haricot bean to inoculation. Experimental Agriculture 10: 45-50.

Ham, G.E. 1978. Use of ^{15}N in evaluating symbiotic N_2 fixation of field-grown soybeans. Pages 151-161 *in* Isotopes in Biological Dinitrogen Fixation. STI/PUB/478. International Atomic Energy Agency, Vienna, Austria.

Ham, G.E. and Caldwell, A.C. 1978. Fertilizer placement effects on soybean yield, N_2 fixation, and ^{32}P uptake. Agronomy Journal 70: 779-783.

Hardarson, G., Zapata, F. and Danso, S.K.A. 1984a. Field evaluation of symbiotic nitrogen fixation by rhizobial strains using ^{15}N methodology. Plant and Soil 82: 369-375.

Hardarson, G., Zapata, F. and Danso, S.K.A. 1984b. Effect of plant genotype and nitrogen fertilizer on symbiotic nitrogen fixation by soybean cultivars. Plant and Soil 82: 397-405.

Hardy, R.W.F., Holsten, R.D., Jackson, E.K. and Burns, R.C. 1968. The C_2H_2-C_2H_4 assay for N_2 fixation: laboratory and field evaluation. Plant Physiology 43: 1185-1207.

Harper, J.E. 1974. Soil and symbiotic nitrogen requirements for optimum soybean production. Crop Science 14: 255-260.

Haydock, K.P., Norris, D.O. and Mannetje, L. 1980. The relation between nitrogen percent and dry weight of inoculated legumes. Plant and Soil 57: 353-362.

Heichel, G.H, Barnes, D.K. and Vance, C.P. 1981. Nitrogen fixation of alfalfa in the seeding year. Crop Science 21: 330-335.

Henzell, E.F., Martin, A.E., Ross, P.J. and Haydock, K.P. 1968. Isotopic studies on the uptake of nitrogen by pasture plants. IV. Uptake of nitrogen from labelled plant material by Rhodes grass and siratro. Australian Journal of Agricultural Research 19: 65-77.

Hudd, G.A., Lloyd-Jones, C.P. and Hill-Cottingham, D.G. 1980. Comparison of acetylene-reduction and ^{15}N techniques for the determination of nitrogen fixation by field bean (*Vicia faba*) nodules. Physiologia Plantarum 48: 111-115.

Jakobson, I. 1985. The role of phosphorus in nitrogen fixation by young pea plants (*Pisum sativum*). Physiologia Plantarum 64: 190-196.

Jensen, E.S., Sorensen, L.H. and Engvild, K.C. 1986. Danish *Rhizobium leguminosarum* strains nodulating 'Afghanistan' pea (*Pisum sativum*). Physiologia Plantarum 66: 46-48.

Kohl, D.H. and Shearer, G.B. 1981. The use of soils lightly enriched in ^{15}N to screen for N_2 fixing activity in plants. Plant and Soil 60: 487-489.

Knowles, R. 1981. The measurement of nitrogen fixation. Pages 327-333 *in* Current Perspectives in Nitrogen Fixation (Gibson, A.H. and Newton, W.E., eds.). Australian Academy of Sciences, Canberra, Australia.

Ledgard, S.F., Morton, R., Freney, J.R., Bergersen, F.J. and Simpson, J.R. 1985a. Assessment of the relative uptake of added and indigenous soil nitrogen by nodulated legumes and reference plants in the ^{15}N dilution measurement of N_2 fixation: derivation of method. Soil Biology and Biochemistry 17: 317-321.

Ledgard, S.F., Simpson, J.R., Freney, J.R., Bergersen, F.J. and Morton, R. 1985b. Assessment of the relative uptake of added and indigenous soil nitrogen by nodulated legumes and reference plants in the ^{15}N dilution measurement of N_2 fixation: Glasshouse application of method. Soil Biology and Biochemistry 17:323-328.

Legg, J.O. and Allison, F.E. 1959. Recovery of ^{15}N-tagged nitrogen from ammonium-fixing soils. Soil Science Society of America Proceedings 23: 131-134.

Legg, J.O. and Sloger, C. 1975. A tracer method for determining nitrogen fixation in field studies. Pages 661-666 *in* Proceedings of the Second International Conference on Stable Isotopes (Klein, E.R. and Klein, P.D., eds.), US Energy Research and Development Administration, Washington, USA.

Legg, J.O. and Stanford, G. 1967. Utilization of soil and fertilizer N by oats in relation to the available N status of soils. Soil Science Society of America Proceedings 31: 215-219.

Olsen, F.J. and Moe, P.G. 1971. The effect of phosphate and lime on the establishment, productivity, nodulation and persistence of *Desmodium intortum*, *Medicago sativa* and *Stylosanthes gracilis*. East African Agricultural and Forestry Journal 37: 29-37.

Patterson, T.G. and La Rue, T.A. 1983. Nitrogen fixation by soybeans: seasonal and cultivar effects, and comparison of estimates. Crop Science 23: 488-492.

Phillips, D.A. and Bennett, J. 1978. Measuring symbiotic nitrogen fixation in rangeland plots of *Trifolium subterraneum* L. and *Bromus mollis* L. Agronomy Journal 70: 671-674.

Rennie, R.J. 1982. Quantifying dinitrogen (N_2) fixation by ^{15}N isotope dilution: the question of the non-fixing control plant. Canadian Journal of Botany 60: 856-861.

Rennie, R.J. 1984. Comparison of N balance and ^{15}N isotope dilution to quantify N_2 fixation in field-grown legumes. Agronomy Journal 76: 785-790.

Rennie, R.J. 1986. Comparison of methods of enriching a soil with ^{15}N to estimate dinitrogen fixation by isotope dilution. Agronomy Journal 78: 158-163.

Rennie, R.J., Dubetz, S., Bole, J.B. and Muendel, H.H. 1982. Dinitrogen fixation measured by ^{15}N isotope dilution in two Canadian soybean cultivars. Agronomy Journal 74: 725-730.

Rennie, R.J. and Rennie, D.A. 1983. Techniques for quantifying N_2 fixation in association with field legumes and greenhouse conditions. Canadian Journal of Microbiology 29: 1022-1035.

Rennie, R.J., Rennie, D.A. and Fried, M. 1978. Concepts of ^{15}N usage in dinitrogen fixation studies. Pages 107-130 *in* Isotopes in Biological Dinitrogen Fixation. STI/PUB/478. International Atomic Energy Agency, Vienna, Austria.

Richards, J.E. and Soper, R.J. 1979. Effect of N fertilizer on yield. protein content and symbiotic N fixation in faba beans. Agronomy Journal 71: 807-811.

Ruschel, A.P., Vose, P.B., Matsui, E., Victoria, R.L. and Tsai Saito, S.M. 1982. Field evaluation of N_2-fixation and N-utilization by *Phaseolus* bean varieties determined by ^{15}N isotope dilution. Plant and Soil 65: 397-407.

Ruschel, A.P., Vose, P.B., Victoria, R.L. and Salati, E. 1979. Comparison of isotope techniques on nodulating isolines to study the effect of ammonium fertilization on dinitrogen fixation in soybean, *Glycine max* (L.) Merrill. Plant and Soil 53: 513-525.

Sekhon, H.S., Kaul, J.N. and Dahiya, B.S. 1978. Response of lentil to *Rhizobium* inoculation and N fertilizer. Journal of Agricultural Science, Cambridge 90: 325-327.

Steele, K.W. and Littler, R.A. 1986. Field evaluation of some factors affecting estimation of nitrogen fixation in pastures by ^{15}N isotope dilution. FAO/IAEA Research Coordination Meeting on the Use of Nuclear Techniques in Pasture Management, 10-14 March 1986.

Talbott, H., Kenworthy, W.J. and Legg, J.O. 1982. Field comparison of the ^{15}N and difference methods of measuring nitrogen fixation. Agronomy Journal 74: 799-804.

Vallis, I., Haydock, K.P., Ross, P.J. and Henzell, E.F. 1967. Isotopic studies on the uptake of nitrogen by pasture plants. III. The uptake of small additions of ^{15}N-labelled fertilizer by Rhodes grass and Townsville lucerne. Australian Journal of Agricultural Research 18: 867-877.

Vallis, I., Henzell, E.F. and Evans, T.R. 1977. Uptake of soil nitrogen by legumes in mixed swards. Australian Journal of Agricultural Research 28: 413-425.

Vasilas, B.L. and Ham, G.E. 1984. Nitrogen fixation in soybeans: An evaluation of measurement techniques. Agronomy Journal 76: 759-764.

Wagner, G.H. and Zapata, F. 1982. Field evaluation of reference crops in the study of nitrogen fixation by legumes using isotope techniques. Agronomy Journal 74: 607-612.

Weber, C.R. 1966. Nodulating and non-nodulating soybean isolines: II. Response to applied nitrogen and modified soil conditions. Agronomy Journal 58: 46-49.

West, C.P. and Wedin, W.F. 1985. Dinitrogen fixation in alfalfa-orchardgrass pastures. Agronomy Journal 77: 89-94.

Williams, W.A., Jones, M.B. and Delwiche, C.C. 1977. Clover N-fixation measurement by total-N difference and ^{15}N A-values in lysimeters. Agronomy Journal 69: 1023-1024.
Witty, J.F. 1983. Estimating N_2-fixation in the field using ^{15}N-labelled fertilizer: Some problems and solutions. Soil Biology and Biochemistry 15: 631-639.
Witty, J.F. and Ritz, K. 1984. Slow-release ^{15}N fertilizer formulations to measure N_2-fixation by isotope dilution. Soil Biology and Biochemistry 16: 657-661.
Zaroug, M.G. and Munns, D.N. 1979. Nodulation and growth of *Lablab purpureus* (*Dolichos lablab*) in relation to *Rhizobium* strain, liming and phosphorus. Plant and Soil 53: 329-339.

Discussion

Cocks

Under what circumstances is it reasonable to use the difference method to measure nitrogen fixation?

Danso

Under conditions of very high N_2 fixation, low soil N, or when % fertilizer use efficiency for fixing and non-fixing crops are similar.

Materon

Please briefly state the basic differences in using C_2H_4 reduction, ^{15}N, and the difference method in estimating nitrogen intake.

Danso

The acetylene reduction method is instantaneous and an enzyme assay technique, while the ^{15}N technique is an integrated estimate of N_2 fixed. The total N difference method does not tell us how much N or the proportion of N that came from soil, fertilizer, and fixation, while we can get this from the ^{15}N method.

D.P. Beck and L.A. Materon (editors), Nitrogen Fixation by Legumes in Mediterranean Agriculture
 ISBN 90 247 3624 2

Isotope Dilution as a Method for Measuring Nitrogen Transfer from Forage Legumes to Grasses

R.W. WEAVER

Texas A & M University, Texas, USA

Summary. Isotope dilution provides a yield-independent method of measuring the contribution a legume makes to the nitrogen nutrition of a grass grown in mixture. The method has a major advantage over yield-dependent measurements because it is not necessary for the reference plant (grass growing in pure stand) to accumulate the same quantity of mineral nitrogen as the plant growing in mixture with the legume. The reference plant need only accumulate mineral and ^{15}N-enriched nitrogen in the same proportions as the plant growing with the legume. Using isotope dilution, transfer of nitrogen from legumes to grasses has been measured under both controlled conditions and in the field environment. Direct transfer of nitrogen does not always occur, but in some cases contributes a significant portion of the nitrogen contained in the grass.

Introduction

Legumes enrich the soil with nitrogen when most of their biomass is returned to the soil. However, a legume can deplete the soil of N if plant residues are removed and if the legume obtained much of its N from the soil as opposed to biological N_2-fixation. Legumes may also provide N directly to plants growing in association with them. The mechanism of transfer is poorly understood and may be from root exudates and decaying nodules and roots. Fortunately, measurement of the transfer is less complicated than understanding the process. Vallis *et al.* (1967) first applied the isotope dilution method, proposed by McAuliffe *et al.* (1958) for measurement of biological N_2-fixation, to measurement of biologically fixed N transferred from a legume to a grass. The purpose of this presentation is to describe the isotope dilution method for measurement of N transferred from legumes to grasses during active growth of the legume and to provide results of some experiments that have utilized the method. Soil enrichment with N due to grazing of a legume or incorporation of residues will not be discussed.

The method of isotope dilution has received considerable attention as a means for measuring N_2-fixation (Hauck and Weaver 1986; Rennie and Rennie 1983). The method uses ^{15}N to distinguish between sources of N for plant growth. These sources include mineral N from the soil, fertilizer N and biologically-fixed N. The method of isotope dilution allows the measurement of all three sources. To utilize the method, soil N must contain a different isotopic enrichment from biologically-fixed N_2. Although this often occurs naturally, the degree of enrichment is not adequate for quantitative measurements when the quantity of N transferred is relatively small. Therefore, the soil must be enriched with ^{15}N to enhance the sensitivity of the method for measuring transfer of N. Alternatively, the soil could be depleted of ^{15}N by adding ^{15}N- depleted fertilizer (Morris *et al.* 1986).

It is obvious that if the ^{15}N concentration of plant-available N in the soil is higher than the ^{15}N concentration in atmospheric N and the plant benefits from both sources, the plant will contain a concentration of ^{15}N lying between that of the two sources. This would occur whether the plant was fixing N itself or receiving biologically-fixed N from a legume. The equation that expresses this relatioship (first described by McAuliffe *et al.* (1958) is:

$$\%\text{N from fixation} = \left(1 - \frac{\%^{15}\text{N excess plant}}{\%^{15}\text{N excess plant available soil N}}\right) \times 100$$

$$\%^{15}\text{N excess} = \%^{15}\text{N sample} - \%^{15}\text{N reference}$$

The reference is often air (0.3663% ^{15}N) or the ^{15}N concentration in the legume when it is totally dependent on N_2-fixation. The reader is referred to Rennie and Rennie (1983) for a more complete description of the methodology.

The isotope dilution method is utilized by altering the soil ^{15}N content so that it is distinctly different from that of air. This is accomplished by adding ^{15}N enriched or depleted fertilizer to the soil. The nitrogen may be from any form of fertilizer N but there are advantages in using slow release materials (Witty 1983). Added N may be surface broadcast or incorporated into the soil. In the case of pot culture, the N may be provided in a nutrient solution. Nitrogen should not come in contact with plant foliage or uneven labelling may occur (Edmeades and Goh 1979). Generally, in field experimentation, microplots within larger plots are fertilized with the ^{15}N-enriched materials to reduce the expense of ^{15}N. The cost of the ^{15}N will vary according to plot size and rate of ^{15}N added, but generally does not exceed a few hundred dollars.

Isotope dilution methodology is a very good tool for measuring N transfer but some difficulties are inherent in the method (Vose and Victoria 1986; Chalk 1985). A major problem is obtaining the correct value for the ^{15}N concentration in plant-available soil N. Reliable soil analysis methods for measuring plant-available N in the soil are not available. Generally, this value is obtained by growing a non-nitrogen-fixing plant as a control to measure the ^{15}N concentration of the N available to the plant. It is critical that the control plant exploits the same isotopic concentration of mineral N as the plant receiving N from N_2-fixation. It is not necessary that the control plant utilizes the same quantity of soil N as the N_2-fixing plant. The control plant in experiments involving transfer of biologically-fixed N from a legume to a grass would normally be the same grass growing without the legume. Unless the legume influenced soil N availability in some manner, the grass would be a good reference. It has been demonstrated that a legume increased mineralization of native soil N when grown with a grass (Birch and Dougall 1967 and Vallis *et al.* 1977). This could result in apparent transfer of biologically-fixed N due to dilution of the ^{15}N in fertilizer N by increased mineralization of soil N.

Another problem is N_2-fixation by diazotrophic bacteria in the rhizosphere of the control plant. This would only be a problem if the plant associated with the legume was not benefitting in a similar manner. Generally, the magnitude of associative N_2-fixation would not be expected to pose problems.

Plants exhibit fractionation of N during distribution to various plant parts. This was reported by Moore (1974) and Moore and Graswell (1976) but was thought to be due to associative N_2-fixation by the grass and differential distribution of this N in the plant relative to fertilizer N uptake. Recently, Ismaili and Weaver (1986) also reported that fractionation occurs in Kleingrass but it is not due to associative N_2-fixation or changes in isotopic composition of plant-available N during growth of the plant (Ismaili and Weaver 1987). Fractionation should not be an important problem as long as the same genotype of the reference plant is the plant grown with the legume.

Results and Discussion

The purpose of this section is to review the results of some experiments in which isotopic dilution was used to measure N transfer from legumes to grasses. The experiments can be conveniently divided into those conducted under relatively controlled conditions of pot culture and those conducted under field conditions.

With respect to isotope dilution, the pot experiments have the distinct advantage of providing a uniform plant growth medium and restricting the roots to a defined volume. A well defined system was used by Ismaili and Weaver (1987). In their system, siratro and Kleingrass were grown in pure stand and mixtures in 20 liter buckets. The plant growth medium was vermiculite which was virtually free of combined N. Nitrogen enriched with ^{15}N was provided throughout the growth of the plant as ammonium sulfate in the nutrient solution. Grass plants grown in pure stand were exposed only to the ^{15}N-enriched N, while plants grown in mixed stands received ^{15}N-enriched and unenriched N ("atmospheric") transferred from siratro.

The results of the experiment clearly indicated that the grass growing with the legume contained a diluted ^{15}N concentration compared to the pure stand grass (Table 1). This was evident for plant shoots and roots. There were different concentrations of ^{15}N in plant parts, emphasizing the importance of utilizing similar plant parts for reference concentrations of ^{15}N.

In a similar experiment, Ismaili and Weaver (1986) utilized ground plant

Table 1. Percent ^{15}N excess in container-grown kleingrass (sown in pure stand and mixed with siratro) and the calculated percentage of total N in kleingrass shoots and roots that was transferred from siratro. Nitrogen enriched with ^{15}N was provided as $(NH_4)_2SO_4$ in nutrient solution (Ismaili and Weaver 1987).

Treatment	Plant part	% ^{15}N excess	%N from transfer +
Pure	Shoot	2.47	
Mixed	Shoot	2.17	12
Pure	Root	2.23	
Mixed	Root	1.51	32
LSD 0.05		0.10	

+ Calculated from the equation

$$\%\text{N transf.} = \left(\frac{1 - \%^{15}\text{N excess grass pure stand}}{\%^{15}\text{N excess grass mixed strand.}}\right) \times 100$$

material enriched with ^{15}N as the N source in a plant growth medium of vermiculite. The plant material provided slow release N as it was mineralized. Kleingrass grown in pure stand contained a significantly higher concentration of ^{15}N than when grown in mixture (Table 2). The extent of dilution was different for each clipping. Henzell *et al.* (1968) incorporated ^{15}N-labelled plant material into soil utilized in a pot experiment in which transfer of N from siratro to rhodesgrass was measured. The quantity of N transferred was approximately 0.1% of the total N in siratro and only amounted to approximately 1% of the N in the grass. Pot experiments by other investigators have also indicated little transfer of N to grasses grown in mixture with lucerne (Vallis *et al.* 1967), ladino clover (Broadbent *et al.* 1982), white clover (Haystead and Marriott 1979), subterranean clover (Ledgard *et al.* 1985), arrowleaf clover (Morris and Weaver 1987), and medic (Butler and Ladd 1985).

Measurement of N transfer from legumes to grasses in the field situation is particularly challenging. Isotope dilution provides a good methodology and was utilized in a field experiment described by Morris *et al.* (1986) to measure

Table 2. Percent ^{15}N excess in container-grown kleingrass sown in pure stand and in mixture with siratro and the calculated percentage of total N in kleingrass shoots that was transferred from siratro (Ismaili and Weaver 1986).

Clipping	Treatment (kleingrass)	%^{15}N excess	%N from transfer +
1	Pure	1.210	
	Mixed	1.181	2.4
2	Pure	1.168	
	Mixed	0.989	15.3
3	Pure	1.062	
	Mixed	0.823	22.5
LSD 0.05		0.042	

+ See Table 1 footnote.

Table 3. Percent ^{15}N excess in field-grown ryegrass sown in pure stand and in mixture with arrowleaf clover and the calculated percentage of total N in ryegrass transferred from arrowleaf (Morris and Weaver, unpublished).

Clipping	Treatment (ryegrass)	$\%^{15}N$ excess	%N from transfer [+]
1	Pure	0.242	
	Mixed	0.238	1.7
2	Pure	0.218	
	Mixed	0.191	12.4
3	Pure	0.172	
	Mixed	0.150	12.8
LSD 0.05		0.017	

[+] See Table 1 footnote.

competition between plant species for fertilizer N. Data on N transfer were collected but not reported. Arrowleaf clover and ryegrass were oversown on a bermudagrass sod and fertilized with ^{15}N depleted ammonium nitrate. Ryegrass was grown in pure stand and in mixture with arrowleaf clover. Nitrogen transfer was negligible at the first clipping but became biologically and statistically significant at the second and third clippings (Table 3). The method was relatively sensitive in detecting transfer because of low experimental error (CV = 6%). If arrowleaf clover had transferred as little as 7% of the N contained in the ryegrass it would have been statistically significant. The sensitivity of the method is enhanced because it is independent of yield.

Other research utilizing isotope dilution methodology on clovers grown in the field has not indicated N transfer during the year of establishment (Ledgard *et al.* 1985; Broadbent *et al.* 1982). It seems reasonable that for significant N transfer to occur, the legume must fix a considerable amount of its N. This occurred in a field experiment conducted by Haystead and Marriott (1978) in which companion grass obtained 6% of its N from N_2-fixation by the legume. When the clover fixed only 25 kg N/ha, transfer of N to the grass was not detected (Haystead and Lowe 1977). In another study, 1 year after establishment of a clover-ryegrass mixture, the ryegrass benefitted considerably from N_2-fixation by the clover (Broadbent *et al.* 1982). Approximately 60% of the N contained in the ryegrass was from N transferred from the clover. Vallis *et al.* (1977), also reported that a grass grown with various legumes for 2 years benefitted from N transfer.

Isotope dilution has proved to be a highly sensitive methodology for detecting "direct" transfer of N from legumes to associated grasses. Because the method is "yield-independent", the experimental error is greatly reduced compared to yield-dependent methods. Moreover, isotope dilution provides a much more realistic estimate of N transfer than subtraction methods because it cannot be assumed that a grass growing in pure stand will accumulate the same quantity of soil N as a grass grown with a legume.

References

Birch, H.F. and Dougall, H.W. 1967. Effect of a legume on soil nitrogen mineralization and percentage nitrogen in grasses. Plant and Soil 27: 292–296.

Broadbent, F.E., Nakashima, T. and Chang, Grace Y. 1982. Estimation of nitrogen fixation by isotope dilution in field and greenhouse experiments. Agronomy Journal 74: 625–628.

Butler, J.H.A. and Ladd, J.N. 1985. Symbiotically-fixed and soil-derived nitrogen in legumes grown in pots in soils with different amounts of available nitrate. Soil Biology and Biochemistry 17: 47–55.

Chalk, Phillip M. 1985. Estimation of N_2 fixation by isotope dilution: an appraisal of techniques involving ^{15}N enrichment and their application. Soil Biology and Biochemistry 17: 389–410.

Edmeades, D.C. and Goh, K.M. 1979. The use of the ^{15}N-dilution technique for field measurement of symbiotic nitrogen fixation. Communications in Soil Science and Plant Analysis 10: 513–520.

Hauck, Roland D. and Weaver, R.W. 1986. Field measurement of dinitrogen fixation and denitrification. Special Publication No. 18. Soil Science Society of America, Madison, Wisconsin, USA.

Haystead, A. and Lowe, A.G. 1977. Nitrogen fixation by white clover in hill pasture. Journal of the British Grassland Society 32: 57–63.

Haystead, A. and Marriott, Carol. 1978. Fixation and transfer of nitrogen in a white clover-grass sward under hill conditions. Annals of Applied Biology 88: 453–457.

Haystead, A. and Marriott, Carol. 1979. Transfer of legume nitrogen to associated grass. Soil Biology and Biochemistry 11: 99–104.

Henzell, E.F., Martin, A.E., Ross, P.J. and Haydock, K.P. 1968. Isotopic studies on the uptake of nitrogen by pasture plants. IV. Uptake of nitrogen from labelled plant material by Rhodesgrass and Siratro. Australian Journal of Agricultural Research 19: 65–77.

Ismaili, Mohammed and Weaver, R.W. 1986. Competition between siratro and kleingrass for ^{15}N-labelled mineralized nitrogen. Plant and Soil 96: 327–335.

Ismaili, Mohammed and Weaver, R.W. 1987. Nitrogen transfer from siratro to kleingrass, competition for mineral nitrogen and isotopic fractionation. Agronomy Journal 79. In press.

Ledgard, S.F., Simpson, J.R., Freney, J.R. and Bergersen, F.J. 1985. Field evaluation of ^{15}N techniques for estimating nitrogen fixation in legume-grass associations. Australian Journal of Agricultural Research 36: 247–258.

McAuliffe, Clayton, Chamblee, D.S., Chibe-Arango, H. and Woodhouse Jr. W.W. 1958. Influence of inorganic nitrogen on nitrogen fixation by legumes as revealed by ^{15}N. Agronomy Journal 50: 334–337.

Moore, A.W. 1974. Availability to rhodesgrasse (*Chloris gayana*) of nitrogen in tops and roots added to soil. Soil Biology and Biochemistry 6: 249–255.

Moore, A.W. and Craswell, E.T. 1976. Non-uniformity of ^{15}N labelling in plant material. Communications in Soil Science and Plant Analysis 7: 335–344.

Morris, D.R. and Weaver, R.W. 1987. Competition for ^{15}N depleted ammonium nitrate and N_2 fixation in arrowleaf clover-gulf ryegrass mixtures. Soil Science Society of America Journal 51: 115–119.

Morris, D.R., Weaver, R.W., Smith, G.R. and Rouquette, F.M. 1986. Competition for ^{15}N depleted ammonium nitrate between arrowleaf clover and annual ryegrass sown into bermudagrass sod. Agronomy Journal 78: 1023–1030.

Rennie, R.J. and Rennie, D.A. 1983. Techniques for quantifying N_2 fixation in association with nonlegumes under field and greenhouse conditions. Canadian Journal of Microbiology 29: 1022–1035.

Vallis, L., Haydock, K.P., Ross, P.J. and Henzell, E.F. 1967. Isotopic studies on the uptake of nitrogen by pasture plants. III. The uptake of small additions of ^{15}N labelled fertilizer by Rhodesgrass and Townsville lucerne. Australian Journal of Agricultural Research 18: 865–877.

Vallis, I., Henzell, E.F. and Evans, T.R. 1977. Uptake of soil nitrogen by legumes in mixed swards. Australian Journal of Agricultural Research 28: 413–425.

Vose, P.B. and Victoria, R.L. 1986. Re-examination of the limitations of nitrogen-15 isotope dilution technique for the field measurement of dinitrogen fixation. Pages 23-42 *in* Field Measurement of Dinitrogen Fixation and Denitrification (Hauck, Roland D. and Weaver, R.W., eds.). Special Publication 18. Soil Science Society of America, Madison, Wisconsin, USA.

Witty, J.F. 1983. Estimating N_2-fixation in the field using ^{15}N-labelled fertilizer: some problems and solutions. Soil Biology and Biochemistry 15: 631-639.

D.P. Beck and L.A. Materon (editors), Nitrogen Fixation by Legumes in Mediterranean Agriculture
© 1988 ICARDA – Printed in the Netherlands. ISBN 90 247 3624 2

Extension / Transfer of BNF Technology

DAVID H. HUBBELL
University of Florida, Florida, USA

Summary. In this presentation, I have interpreted the scope of my assigned topic very broadly. I have done this for two reasons. First, my field experience is limited strongly to the humid tropics, so I am poorly qualified to address the specific problems associated with BNF technology in semi-arid environments. Second, I have opted for a broad treatment of the subject because I believe, with genuine missionary zeal, that it needs broad treatment to achieve some new perspective. For those reasons, I will avoid presentation of detailed data. Instead, I intend to talk about concepts, principles, generalizations, approaches, ideas, etc. In essence, this is a position paper – a philosophical evaluation of the status of technology transfer based on past and present scientific endeavors as I perceive them.

Let us first ask about the origin of BNF technology transfer. It is essential that we know how it started if we are to understand where we are, how we got here, and where we are going. I will speculate that it began when one farmer showed another the benefits of transferring topsoil from a legume field to a non-legume field, prior to planting the latter with the legume. Such, perhaps, was the initial state of the art.

Where are we now? Who are the recipients of BNF technology transfer? Who is doing the transfer? What are we transferring? These are some of the questions I wish to dwell upon.

We are technologically developed to the extent that mere technology is no longer a limiting factor in most problem situations, provided that our desire for a solution (and funding) is sufficient.

Who are the recipients of this technology? In terms of both numbers of recipients and relative need, I would arbitrarily divide the recipients into subsistence, family, and large-scale farmers. Are these categories valid? I believe they are, but to a limited extent. The boundary between subsistence, family, and large-scale farmers is certainly ill-defined.

Large-scale farmers are perhaps most easily categorized; they generally possess sufficient land, capital, and education to make farming a significant profit-making venture. The family farmer might then be characterized as one who farms routinely at an economic level on the borderline of significance. He can feed his family adequately, with a small surplus left for sale. The subsistence category is most difficult to define but these farmers are "survivors". Characteristically, they have little education, no operating capital, and land which is insufficient in quality and quantity, thus keeping them chronically on the edge of collapse as farmers.

To which of these groups are we transferring technology? To which group(s) *should* we be transferring?

Clearly, according to individual situations, we are transferring to all three groups, as we should be. But what about relative emphasis or effort?

I regard the subsistence category as most critically in need of attention. They are numerically the greatest as well as being most in need. If they cannot be made self-sufficient, then society is morally bound to support them, which it does not do. This results in massive migration to the city, the consequences of which are all too familiar. The subsistence category is perhaps the most diverse, consisting of a complete spectrum of different situations. For that reason it cannot be dealt with as a category – it must be dealt with on an "individual farmer" basis. There is no simple, standard panacea applicable to the plight of the subsistence farmer. Each one has his own unique combination of limitations and these can only be identified and solved on an individual basis.

I believe most BNF technology transfer is directed at the large-scale farmer. Ironically, this is the category that, relatively speaking, needs it least. At the other end of the spectrum is the subsistence farmer, most in need and receiving least assistance. One reason for this anomaly may be that large-scale and family farm operations can be dealt with more easily and successfully as categories. Thcy are more amenable to standardized remedies or assistance programs which can be economical, short-term solutions.

This is not to say that the subsistence farmer is in all cases forgotten or neglected. There is substantial effort being made on their behalf, world-wide. I believe that provision of assistance to them, i.e., technology transfer, is largely ineffectual primarily because it is inappropriately conceived and executed.

Who is responsible for this perceived failure in technology transfer? Government administrators, who determine research priorities and provide funds, are the initial culprits and we, as the scientists who perform the work, become party to the crime. We obtain short-term research grants to fine-tune an already sophisticated technology. It is new, interesting, challenging, intellectually gratifying, and has real potential for transfer in package form to large-scale and family farms. The new technology transfers successfully because it is similar to what is already being used – the improvements come in small increments.

Can we transfer this technology, in its present form, to subsistence farmers? I think not, but we do persist in trying. The technology we have was developed for systems capable of using it. Complicated, mechanized, expensive technology is not transferable to a farmer familiar only with a planting stick operation.

This leads to the conclusion that neither technology per se, nor efforts to transfer it are limiting. What is limiting, perhaps, is the form in which we try to transfer it. How appropriate is it in terms of the capability of the target user (subsistence farmer) to understand, accept, and utilize it?

For subsistence farmers, BNF technology is generally not appropriate as it is commonly presented to them. Agricultural practice (technology) has been a slowly-evolving system, with changes or improvements coming as increments. Much of this evolutionary development came by trial-and-error on the part of the farmers and, subsequently, by agricultural scientists with a strong instinct for farmer-friendly research. But we seem to be losing sight of at least part of our objective. Our research in many cases seems to be obsessed with developing technology as an end rather than as a means to an end.

The subsistence farmer is practicing with a technology used 50, 100, or 1000 years ago. He cannot be expected to use a technology designed to convert his operation from minimum yield to maximum yield in one, two, or three growing seasons. We cannot telescope what is inherently a slow, step-wise evolutionary process.

One reason for this is that the agricultural practices of a culture are inextricably mingled with the social, political, economic, and religious traditions of that culture. Technological changes affecting any of these interrelated traditions arouse suspicion, antagonism, and mistrust – they are perceived to be not in the common weal and they self-destruct. This is most probable when changes are presented too rapidly or in a form too divergent from the familiar. Experience should tell us that we cannot force too many changes too fast. We must be patient and try to achieve small successes that build, one upon another. If this is done with subsistence farmers, on an individual basis by recognizing and solving the unique problems of each individual, then gradually this category of farmers is brought to a common denominator of capability. They progress to the level of family farmer and become more uniformly receptive to a more standardized and advanced technology.

Unfortunately, this approach is not popular with administrators or scientists. It is not easy, cheap, or quick. We seem to studiously ignore its most important or redeeming quality – it works!

Are there any tell-tale clues to indicate that our approaches and priorities with respect to technology transfer are in fact misguided or inappropriate? I think there are ample clues.

Let us consider a typical, government-funded grant program directed towards development and transfer of technology to a developing country. What are the criteria used ultimately to measure the success of the program? In general, we find such criteria as number of degrees awarded, number of journal articles generated, etc. Whatever happened to yield increase, number of farmers adopting a new practice, and increased income? Graduate training, for obvious reasons, is a critical component of the cooperative research grant program. How often are students selected from the host country and how often is the degree research done on-site in that country, on a problem relevant to that country's needs? The answer is rarely, again because it is not easy, cheap, or quick.

The training aspect of technology transfer deserves careful consideration because its potential long-term effects are critical. In the present context, I equate training with education. Individuals from developing countries must be trained according to the work they will do when their training is completed. Will they be extension agents, laboratory technicians, or research scientists? Virtually all of the training courses and symposia that I have been associated with have been conducted at the research scientist level, which is highly appropriate if that is what the participants need. But we currently need more soldiers in the trenches than generals on a distant hill. And the soldiers need different training. Designing a new tank will not qualify them to lead a bayonet charge. Joe Burton has made legume inoculation work for many years without using sophisticated serological methods and advanced microscopic techniques. His methods and materials were simple, direct, and inexpensive, and they worked in practice. I do not advocate desertion of basic research, but I do advocate that workers on *Rhizobium* at any level should be able to isolate and test strains, etc.

What are some suggested solutions to the plight of subsistence farmers around the world? I will try to tell you what I think should be done. I will not try to tell you how it should be done.

Existing technology in appropriate areas, such as biological nitrogen fixation, must be simplified and adapted to the needs of individual or highly localized situations. This can be accomplished best, perhaps only, by agricultural extension agents (or professionally qualified equivalents) who have no ancillary, interfering duties, such as tax collection, etc. Transferred technology must come in an appropriate form and from a familiar and accepted source if there is to be any hope of adoption of the technology.

The governments of developing countries must be educated in the critical importance of a teaching-research-extension program which is active and relevant to the needs of farmers at all levels, but with emphasis initially on subsistence farmers.

References

Feder, G., Just, R.E. and Zilberman, D. 1985. Adoption of agricultural inovations in developing countries: a survey. Economic Development and Cultural Change 33 (2): 255–298.

Randolph, R.H. and Koppel, B. 1982. A technology assessment of biological nitrogen fixation. *In* Biological Nitrogen Fixation Technology for Tropical Agriculture (Graham, P.H. and Harris, S.C., eds.). Papers presented at a workshop at CIAT, Cali, Colombia, 9–13 March 1981. Centro Internacional de Agricultura Tropical, Cali, Colombia.

Part V

Workshop synthesis and recommendations developed from group discussion

D.P. Beck and L.A. Materon (editors), Nitrogen Fixation by Legumes in Mediterranean Agriculture
 ISBN 90 247 3624 2

Workshop Synthesis and Recommendations Developed from Group Discussions

L.R. MYTTON

Welsh Plant Breeding Station, Aberystwyth, UK

Discussion Session Leaders: R. ROUGHLEY, R. BRADLEY, T.A. LIE, A. EAGLESHAM, S. DANSO, J. BROCKWELL

Introduction

This meeting has the specific task of considering those aspects of symbiotic nitrogen fixation which will improve economic yield and optimise inputs of nitrogen into agricultural systems. The synthesis is composed of two parts:

A. Identification of limitations to fixation and the need for further research.

B. Identification of appropriate existing technology for reducing these limitations and its transfer to national programmes.

To achieve these objectives ICARDA must continue its own biological nitrogen fixation research projects and be committed to a training programme. These two aspects form a central reference source for national programmes and where necessary can act as a catalyst in stimulating new work and co-ordinating existing efforts.

Nitrogen fixation research on pasture and grain legumes is well established at ICARDA and during the course of this workshop it has become evident that the centre has a strong commitment to these projects and to the development of training programmes. Further comments on the role of the centre are therefore not necessary and the task now is to define the needs for future research, to identify appropriate technology, and comment on training.

Research Objectives

Priorities for research can be established by identifying the size and extent of limiting factors. In general terms the problems are easy to define. When legumes are grown as a field crop the effects of plant cultivar (genotype), the environment, and *Rhizobium* are confounded. Experiments have to be devised which will separate these effects so as to reveal and quantify major limitations, i.e., is productivity limited by environmental variables, by deficiencies in the *Rhizobium* population, or by the inherent characteristics of the crop plant?

Details of these experiments will vary with location and with crop but they should include the following general comparisons:

(a) Compare treatments dependent upon native rhizobia with those receiving non-limiting applications of fertilizer nitrogen. Absence of root nodules has obvious implications. This will identify nitrogen limitations and act as a guide for the "need to inoculate". An inoculation treatment can often be a useful adjunct to the initial trial.

(b) Compare normal agricultural practices with treatments which remove water, and nutritional and management limitations (These treatments should not be confused with any economic requirements related to farm practice). This will give a general indication of the size of environmental limitations.

(c) Compare a range of genotypes. This will reveal genetic limitations as either specific to particular cultivars or general to the species. Local material should not be overlooked.

Because of the difficulties of specifying precise details for these experiments priority should be given to producing a manual which sets out suitable experimental procedures for the crops and environments appropriate to the region. Note should be taken of the approach developed by CIAT which is a sound example of a successful programme (see Sylvester-Bradley).

Each component of the comparisons listed above will now be considered in more detail.

Reducing Limitations to Fixation by Inoculation

These experiments will need suitable high quality inoculants. There are likely to be problems with commercial sources e.g. maintaining quality during transportation and storage within the region. ICARDA should therefore maintain a small collection of superior rhizobia and produce experimental inoculants for use by national programmes. It also needs to operate quality control procedures. Sterile packets of carrier should be produced and distributed for local production of inoculants.

Some national programmes may want to produce their own inoculants. Suitability of locally available inoculant carriers merits further research. Training and technology transfer are needed for national inoculant production and quality control.

Maintaining a large national *Rhizobium* collection is time consuming and therefore costly. Most field populations of rhizobia are genetically diverse and can be used as source material when needed. There is therefore little point in maintaining large germplasm collections of rhizobia. However, changes in land use and replacement of native legume communities could cause concomitant changes in *Rhizobium* populations. Where this occurs some form of genetic conservation is required. *In situ* conservation of threatened plant communities should be considered first before starting a *Rhizobium* collection. However, maintenance of a diverse regional collection of superior lyophilized *Rhizobium* cultures by ICARDA is essential.

Reducing Environmental Limitations

Environmental factors can limit nodulation and nitrogen fixation; identification and amelioration may produce substantial improvements. It is tempting to relieve the microbiologist of the burden of these investigations by suggesting that his primary responsibility is to ensure that, within the constraints of the environment and the crop, absence or unsuitability of *Rhizobium* is not limiting yield. In practice he has a major role to play in identifying environmental limitations but collaboration with plant physiologists and soil scientists is definitely desirable. Factors of particular relevance to the region are: moisture stress, salinity, high temperature, sitona damage, phosphorus nutrition, and effects of agrochemicals on the symbiosis. Studies on the interactions between nodulation, nitrogen fixation, and available soil nitrogen are also needed.

Choice of Plant Material

Individual plants differ in degree to which they are adapted to particular environments. This is the basis for crop improvement by plant breeding. Differences in compatibility with local rhizobia can be an important component of environmental adaptation (see Lie). Where particular cultivars or species are shown to have problems with fixation it should be recognised that inoculation and removing environmental limitations are not the only strategies for improvement. Identification of better adapted plants may be an alternative. It is therefore important that rhizobiologists work with plant breeders and agronomists in the initial assessment of plant genetic resources.

Plant germplasm collections, unlike *Rhizobium* collections, have been of great value to plant breeding and crop improvement. Extension of plant genetic resources is of continuing relevance to legumes. This approach has been emphasized by ICARDA scientists in this workshop. ICARDA has a vigorous plant introduction programme and the region can continue to rely on this resource. In addition, national programmes should be encouraged to screen and develop indigenous plants. This is probably most relevant to pasture legumes. Advantages may be better adaptation to local environments and local rhizobia without much need for further genetic improvement.

Measurement of Nitrogen Fixation

All of the above considerations assume that we can measure amounts of nitrogen fixed by legumes in the field. Acetylene reduction has been widely used for this purpose but the paper presented here by Witty provides convincing evidence that the "standard method" is not reliable for quantitative comparisons. Its continued use in this way cannot be recommended. The Danso review describes the strengths and weaknesses of using the stable heavy isotope of nitrogen (^{15}N) in estimating fixation. If used with care it offers reasonable accuracy. Its use in experiments is straight forward but analysis requires access to a mass spectrometer. This expensive item is not available to most programmes and analysis done by outside laboratories is expensive and time consuming. However, where accurate assessments of fixation are thought to be essential, the use of ^{15}N is recommended and national programmes should explore the possibilities of using the mass spectrometry facility of the IAEA at Vienna for analysis of samples.

The need for accurate measurement must be thoroughly assessed. In general, emphasis should be placed on simple indirect methods. If plants are grown in situations where soil nitrogen is relatively unavailable there is usually a good correlation between fixed nitrogen and total dry matter. Addition of a fertilizer nitrogen treatment may reveal nitrogen limitations. Available soil nitrogen can be reduced by previous cropping with plants with high nitrogen demand, or by incorporating organic matter low in nitrogen such as chopped straw. Where rhizobia are absent, or of reduced effectiveness, comparison of inoculated and uninoculated treatments is useful. Yield comparisons between legumes and comparable non-legumes are also valuable. Measurement of simple parameters such as dry weight, grain yield, and total nitrogen are highly recommended and may also be of agronomic value.

Nitrogen transfer from legume to companion crop (see Weaver) needs more research. Evidence for substantial and rapid movement is sparse. However, the information is important for rational development of inter-cropping legumes with non-legumes and is also relevant to pasture systems.

Identification of Existing Technology and Training

Technology should be appropriate to the needs of national programmes. Initial priorities will be for simple experiments which demonstrate that inoculation can be of benefit. The soil core method described here by Sylvester-Bradley is a good example. Strain selection for use as inoculants is the next step. This has also been considered by Sylvester-Bradley and by Thompson. Methods of inoculant production, quality control, and packaging are well established. Simple, inexpensive fermenters are available (Roughley). More attention needs to be paid to shelf- life of inoculants. They are probably less sensitive than early experiences suggested but long shelf-life under conditions of high temperature is of particular importance to countries of North Africa and West Asia. Both the inoculants and the methods of application need to be made simple and reliable to encourage farmers to use them. Where inoculation is beneficial it is important to establish convincing field demonstrations for farmers. Agriculturalists need to be alerted to the possibility that replacement of old, adapted, cultivars and landraces by new cultivars containing foreign germplasm may bring with it a requirement for inoculation.

Training

Practical achievements deriving from past training programmes conducted in developing countries have sometimes been disappointing. This has been a matter of much concern in the deliberations of this workshop's discussion groups. Two points emerged strongly. First, that future programmes could be strengthened by identifying and correcting past deficiencies. Second, that successful programmes should be analysed and their strengths incorporated into new schemes.

The following points were identified from participant's experiences as particular weaknesses:

(i) Failure to identify the precise training needed to achieve defined objectives in national programmes. There can also be a failure to recognise that training programmes have to be especially adapted to those needs.

(ii) Selection of inappropriate candidates for training. Factors to be considered are:

a) Unsuitable background training.
b) Inability to implement training or lack of facilities.
c) Lack of awareness by training staff of local conditions (i.e. agricultural, socio-economic).

Solutions to these problems are:

d) Obtain full background information on candidates and home institutions.
e) Ensure that training staff advise institutions on background qualifications needed.
f) Allow training staff to advise on selection of candidates.
g) Ensure that training staff are familiar with the particular local environmental and socio-economic factors.

Legume – *Rhizobium* programmes are most likely to succeed where the following trained personel work together as teams (see Vincent):

1. Microbiologists.
2. Agronomists (should also be involved in the national programme of legume selection).
3. Extension workers (identified and trained by the microbiologist and agronomist).

In some cases there would be advantages in locating training programs in particular countries with emphasis on local crops and conditions. Possibilities of conducting the course in the local language should be explored.

Factors of Importance in Successful Programmes

ICARDA was identified as providing notable successes arising from its training programmes. Critical points are:

(a) The first step has been to involve the national policy makers in identifying the need for the program. This provides the local incentive to achieve defined objectives. The main reason for the wider use of improved germplasm is probably that reliable improvements in yield can be demonstrated, which in turn motivates farmers to modify agricultural practices to make best use of the new material. But it should also be noted that ICARDA has ensured that the potential benefits of new cultivars have been demonstrated and published to make senior administrators of national programs aware of development.

(b) Training has held a strong practical content aimed specifically at national needs.

(c) Trainees have been supported by follow-up visits to inspect trials and assist with interpretation of data.

Technology Transfer to Farmers

The broad aim of development programmes must be to increase agricultural productivity throughout the farming community. In developing countries the majority of producers are often subsistence farmers who should therefore be a priority target for technology transfer. However, they are often unwilling to depart from systems which have sustained them for generations. Reluctance to do so is understandable. No change is without risk, and in a fragile environment it could be disastrous to relinquish proven systems for technologies which have only recently been introduced.

The method most commonly employed to solve this problem is to introduce the new technology to larger and more innovative farmers in the hope that the rest of the community will eventually follow their example. The strategy is perfectly reasonable and can be successful, but there are weaknesses in it and these often arise from the new materials and methods being introduced. The philosophy of improvement is often based on current technology in countries which have highly developed agriculture. It can easily be forgotten that both the materials and their methods of use are based on generations of fertility-building agricultural methods.

In these systems high inputs are vital and good management skills essential to sustain production. The ability of subsistence farmers to meet such demands is often limited and must be assessed. A complementary approach to improvement would be directed at a longer term strategy of fertility build-up and general improvement based on a staged progression through species and cultivars. These would have reasonable yields but be less demanding in nutritional requirements and need little change in management. The advantages of this approach are:

(i) It allows farmers to easily adopt such methods and will help to generate the capital to make further improvements.
(ii) The biological and economic risks are smaller.

In developing this system it is particularly important to make thorough evaluation of native germplasm before turning to introduced material. The ICARDA forage medic screening programme provides a particularly apt example (see Cocks).

Concluding Remarks

This synthesis is based on reports produced by different discussion groups formed by workshop participants. Incorporated into it are a good number of the views presented but it is not a catalogue of all the ideas expressed. In the final analysis, if the review does nothing more than provoke alternative suggestions it will have proved to be of value.

GPSR Compliance

The European Union's (EU) General Product Safety Regulation (GPSR) is a set of rules that requires consumer products to be safe and our obligations to ensure this.

If you have any concerns about our products, you can contact us on ProductSafety@springernature.com

In case Publisher is established outside the EU, the EU authorized representative is:

Springer Nature Customer Service Center GmbH
Europaplatz 3
69115 Heidelberg, Germany

Batch number: 10809439

Printed by Printforce, the Netherlands